PLATO'S COSMOLOGY

The *Timaeus* of Plato translated
with a running commentary

By

FRANCIS MACDONALD CORNFORD

LONDON
ROUTLEDGE & KEGAN PAUL LTD
BROADWAY HOUSE, 68–74 CARTER LANE, E.C.4

First published in Great Britain 1937
by Kegan Paul, Trench, Trubner & Co. Ltd

Reprinted 1948, 1952, 1956 and 1966
by Routledge & Kegan Paul Ltd
Broadway House, 68-74 Carter Lane
London, E.C.4

Printed in Great Britain by
Compton Printing Ltd
London & Aylesbury

ARMILLARY SPHERE

Drawing of the Heavens representing the Earth, from a book (Globe Terrestre).
À Paris, chez Dellamarche Géographe, du Feu Jacques au Collège de
Montaigu.

TO

ELEANOR MEREDITH COBHAM

ἣ ταῦτά τε σοφή ἐστι καὶ ἄλλα πολλά

PREFACE

THIS book is constructed on the same plan as an earlier volume in the series, *Plato's Theory of Knowledge*. It contains a translation of the *Timaeus* interspersed with a commentary discussing each problem of interpretation—and there are many hitherto unsolved —as it arises. My first aim has been to render Plato's words as closely as I can. Anyone who attempts to reproduce his exalted poetical style must face the certainty of failure, with the added risk of falsifying the sense, especially by misleading reminiscences of the English Bible. The commentary is designed to guide the reader through a long and intricate argument and to explain what must remain obscure in the most faithful translation; for the *Timaeus* covers an immense field at the cost of compressing the thought into the smallest space. Only with some such aid can students of theology and philosophy have access to a document which has deeply influenced mediaeval and modern speculation. I have tried not to confuse the interpretation of the text with the construction of theories of wider scope. The later Platonism is a subject on which agreement may never be reached; but there is some hope of persuading scholars that a Greek sentence means one thing rather than another.

The translation follows Burnet's text, except where I have given reasons for departing from it or proposed corrections of passages that are probably or certainly corrupt. For the interpretation I have consulted, in the first instance, the commentaries of Proclus and Chalcidius, the fragment of Galen's commentary lately re-edited by Schröder, the relevant treatises of Plutarch, and Theon of Smyrna, who preserves valuable extracts from Dercylides and Adrastus. The careful summary of the *Timaeus* in the *Didascalicus* of the Middle Platonist Albinus deserves more attention than it receives. Among the moderns I have drawn freely upon Martin's admirable *Études sur le Timée de Platon*, Archer-Hind's commentary, and the translations of Apelt, Fraccaroli, Rivaud, and Professor A. E. Taylor.[1]

More useful than any of these has been Professor Taylor's

[1] I regret that I did not learn that Mr. R. G. Bury's translation had appeared until it was too late to make use of it.

vii

Commentary. His wide learning and untiring industry have amassed a great quantity of illustrative material, and he has cleared up the meaning of many sentences hitherto misunderstood. These amendments will pass into the common stock of future editors and translators, and I have for the most part adopted them tacitly. It is unfortunate that I should so often have had to quote his views where it was necessary to give reasons for dissent. My notes, accordingly, do not indicate the extent of a debt which I here acknowledge with gratitude.

On many of the larger questions of interpretation, however, I differ widely from Professor Taylor. He has launched in this volume a new Taylorian heresy. After confounding the persons of Socrates and Plato in earlier books, he has now divided the substance of Plato and Timaeus. All the ancient Platonists from Aristotle to Simplicius and all mediaeval and modern scholars to our own day have assumed that this dialogue contains the mature doctrine of its author. Professor Taylor holds that they have been mistaken. He writes :

' It is in fact the main thesis of the present interpretation that the teaching of Timaeus can be shown to be in detail exactly what we should expect in a fifth-century Italian Pythagorean who was also a medical man, that it is, in fact, a deliberate attempt to amalgamate Pythagorean religion and mathematics with Empedoclean biology, and thus correctly represents the same tendency in fifth-century thought for which the name, e.g. of Philolaus stands in the history of philosophy. If this view is sound, it follows that it is a mistake to look in the *Timaeus* for any revelation of the distinctively Platonic doctrines, the ἴδια Πλάτωνος as Aristotle calls them (*Met.* A. 987a, 31), by which Platonism is discriminated from Pythagoreanism, or for a ' later Platonic theory ' which can be set in opposition to the type of doctrine expounded in the *Phaedo*. I shall set myself in commenting on the relevant passages to argue in detail that we do not, in fact, find any of the doctrines Aristotle thought distinctive of Plato taught in the *Timaeus* or in any other dialogue. But, on the other hand, what the *Timaeus* loses, if my view is a sound one, as an exposition of Platonism it gains as a source of light on fifth-century Pythagoreanism. If I am interpreting it on right lines, it is incomparably the most important document we possess for the history of early Greek scientific thought.'

Further on, Professor Taylor describes Plato's plan in more detail. ' The formula for the physics and physiology of the dialogue is that it is an attempt to graft Empedoclean biology on the stock of Pythagorean mathematics ' (p. 18). This fusion, he adds, could not be completely carried out. There were incongruities which lead Timaeus ' into a variety of real inconsistencies which culminate in an absolutely unqualified contradiction between a medical or physiological " determinism " (*Tim.* 86B–87B) and a religious and ethical doctrine of human " freedom " ', which is undoubtedly Pythagorean.

' Plato repeatedly warns us in this very dialogue that cosmology and physical science in general can never be more than " provisional ". It is at best made up of tales " like the truth ". Hence Plato was not likely to feel himself responsible for the details of any of his speaker's theories. All that is required by his own principles is that they shall be more or less " like " the truth, i.e. that they shall be the best approximations to it which could be expected from a geometer-biologist of the fifth century. In other words, we are entitled to say that Plato thought the view which arose from the fusion of Pythagoras with Empedocles the most promising line in fifth-century science and the one most directly connected with his own developments. It does not follow that *any* theory propounded by Timaeus would have been accepted by Plato as it stands. The way in which Timaeus is made at each chief new step in his narrative to insist on the highly provisional character of his speculations is a most significant feature of the dialogue, to which no one as yet seems to have done full justice. What Plato himself really thought about a good deal of Empedocles has to be learned not from our dialogue but from *Laws* x, where Empedocles more than anyone else is plainly aimed at in the exposure of the defects of " naturalism " ' (pp. 18–19).

According to this theory, then, Plato, having occasion to give an account of the nature of the visible world, concocted an amalgam of two philosophies belonging to the previous century, although he knew them to be incompatible and largely disapproved of one of them. All he wanted was something ' like the truth '. What he actually produced was not a picture that he himself could accept as more like the truth than any other, but the best that could be expected from an imaginary eclectic, of two or three generations earlier, attempting to combine irreconcilables.

I cannot think that this theory will be accepted. The improbability is so great that overwhelming proof must be required. The evidence, if it existed, could hardly have been overlooked by all those ancient authorities whose knowledge of Platonism and its antecedents was far greater than any we can ever hope to possess.

PREFACE

Professor Taylor rightly insists that the student should know what the men who had heard Plato's doctrines from his own lips or from his immediate disciples supposed him to mean ; and how he was understood by men of real learning like Posidonius, Plutarch, and Atticus, and even later by men versed in the earlier literature like Plotinus and Proclus. The chief value of his own commentary lies in the exhaustive summaries of these ancient opinions. But if his theory is sound, how is it that not one of them furnishes a single unambiguous statement to the effect that the doctrines of the *Timaeus* are not Plato's own? Aristotle was living and working with Plato when the dialogue was written. Why does he never use the *Timaeus* as ' a source of light on fifth-century Pythagoreanism ' or refer to it as ' a document for the history of early Greek scientific thought ', a subject in which he was much interested ? How is it that Theophrastus (as Professor Taylor remarks, p. 1) ' treats the whole account of the sensible qualities given in our dialogue as the views of Plato ', without a hint that they are really no more than the best that could be expected from a geometer-biologist of the previous century ? From all that we know of Theophrastus' History of Physical Opinions it is clear that he used the *Timaeus* as his main source for Plato's physical doctrine. Aristotle and Theophrastus must have known the true character of the work. Both wrote at length on the history of philosophy. Neither left on record so much as a suspicion that Plato was really fabricating a medley of obsolete theories for which he acknowledged no responsibility. Had such a suspicion been expressed in any of their works now lost to us, it could not have escaped the notice of the later ancient commentators, who studied the *Timaeus* line by line and sought for light upon its meaning in every available quarter. The discovery would then have robbed the dialogue of all authority. Not only would it have lost its value as an expression of Plato's mind, but to the ancients it would have been useless as a record of fifth-century speculation. Possessing the original documents on which it was based, they would have contemplated with more amazement than interest the ingenuity spent in conjuring out of them an incoherent system which nobody had ever held.

It is hard to understand how anyone acquainted with the literature and art of the classical period can imagine that the greatest philosopher of that period, at the height of his powers, could have wasted his time on so frivolous and futile an exercise in pastiche. What could have been his motive ? Nowhere, in all his seven hundred pages, has Professor Taylor really faced this question ; yet it surely calls for an answer. When an archaeologist unearths a temple in a sixth-century style of architecture, it never occurs to him to doubt whether the sculpture may not be the work of Praxiteles or Scopas, deliberately faking an archaic manner. He knows that such things were not done till the blaze of creative genius had died down; the foundations of Wardour Street were laid in Alexandria. Yet such a supposition would be every whit as probable as Professor Taylor's thesis.

The reader who does not accept that thesis will find himself somewhat bewildered by attempts to prove that Timaeus says one thing while Plato believes another. There are two other tendencies, running through the whole commentary, which seem to me to distort the picture. One is the suggestion that Plato (or Timaeus ?) is at heart a monotheist and not far from being a Christian.[1] The Demiurge is not fully recognised as a mythical figure, but credited with attributes belonging to the Creator of Genesis or even to the God of the New Testament. Another is the practice of translating Plato's words into the terms of Professor Whitehead's philosophy. That philosophy could not have existed before the Theory of Relativity; and its author, having very unfamiliar ideas to express, uses common words in senses so peculiar and esoteric that no one can follow him without a glossary. Consider the following definitions of an ' occasion ' and an ' event ':

> ' Each monadic creature is a mode of the process of " feeling " the world, of housing the world in one unit of complex feeling, in every way determinate. Such a unit is an " actual occasion ":
> it is the ultimate creature derivative from the creative process. The term " event " is used in a more general sense. An event is a nexus of actual occasions inter-related in some determinate fashion in some extensive quantum: it is either a nexus in its formal completeness, or it is an objectified nexus. One actual occasion is a limiting type of event. The most general sense of the meaning of change is " the differences between actual occasions in one event ". For example, a molecule is a historic route of actual occasions; and such a route is an " event ". Now the motion of the molecule is nothing else than the differences between the successive occasions of its life-history in respect to the extensive quanta from which they arise; and the changes in the molecule are the consequential differences in the actual occasions ' (*Process and Reality*, pp. 111–12).

It is true that Professor Whitehead has been profoundly influenced by Jowett's translation, and that his eternal objects have a definite affinity to Plato's eternal Forms. But there is more of Plato in the

[1] Examples will be found in the notes on 29D–30C and 69C, 3.

Adventures of Ideas than there is of Whitehead in the *Timaeus*. The modern reader is likely to be misled by the constant use of Whitehead's 'event' as equivalent to Plato's γιγνόμενον. Moreover, Plato expressly declares that his Forms 'never enter into anything else anywhere' (52A)—a cardinal point of difference between himself and Aristotle. Yet Professor Taylor writes: 'γένεσις . . . is, in fact, the "ingredience of objects into events", by which the "passage" of nature is constituted. . . . The famous Forms . . . are what Whitehead calls "objects", and the point of insistence upon their reality is that Nature is not made up of the mere succession of events, that the passage of nature is a process of "ingredience" of objects into events' (p. 131). According to Professor Taylor's main thesis, the philosophy of our dialogue belongs to a period which already seemed archaic to Aristotle : he regularly speaks of the fifth-century thinkers as 'the primitives' (οἱ ἀρχαῖοι). Even if we restore this philosophy to Plato, it cannot usefully be paraphrased in terms which have first acquired their technical meaning in our own life-time. It is puzzling to find the contents of Timaeus' discourse represented at one moment as more antique than Plato and at the next as more modern (and considerably more Christian) than Herbert Spencer. Accordingly, while every student must acknowledge a great debt to Professor Taylor's researches, there is still room for a commentary based on the traditional assumptions and attempting to illustrate Plato's thought in the historical setting of Plato's century.

Friends and colleagues have generously helped me with their advice on matters in which I needed a judgment more competent than my own. Sir Thomas Heath, whose masterly works on Greek mathematics I have constantly consulted and never in vain, has written long and careful answers to my inquiries. Professor Onians has allowed me to use freely the proofs of his valuable book, *The Origins of Greek and Roman Thought*. I am also specially indebted to Dr. W. H. S. Jones, Professor D. S. Robertson, Mr. R. P. Winnington-Ingram, and Mr. R. Hackforth. The late Professor H. S. Foxwell kindly gave me permission to reproduce the photograph of the Armillary Sphere in his possession. Dr. R. T. Gunther tells me its probable date is about 1780–1820. In 1790 C. F. Delamarche published *Les usages de la Sphère et des Globes céleste et terrestre, selon les hypothèses de Ptolémée et de Copernic, accompagnées de figures analogues.*

F. M. C.

LIST OF ABBREVIATIONS

A.-H. = Archer-Hind, R.D. The Timaeus of Plato, London, 1888.

Albinus = Ἀλκινόου (sic) διδασκαλικὸς τῶν Πλάτωνος δογμάτων, ed. Hermann, Platonis Dialogi, Lipsiae, 1892, vi, pp. 152 ff.

Apelt = Platon's Dialoge Timaios und Kritias übersetzt und erläutert von O. Apelt, Leipzig, 1922.

Chalcidius = Platonis Timaeus interprete Chalcidio cum eiusdem commentario, ed. J. Wrobel, Lipsiae, MDCCCLXXVI.

Fraccaroli = Il Timeo trad. da Giuseppe Fraccaroli, Torino, 1906.

Pr. = Procli Diodochi in Platonis Timaeum commentaria, ed. E. Diehl, Lipsiae, MCMVI.

Rivaud = Platon, Tome X, Timée, Critias, texte établi et traduit par Albert Rivaud, Paris, 1925.

Theon = Theon of Smyrna, τῶν κατὰ τὸ μαθηματικὸν χρησίμων εἰς τὴν Πλάτωνος ἀνάγνωσιν, ed. Dupuis, Paris, 1892.

Tr. = Taylor, A. E., A Commentary on Plato's Timaeus, Oxford, 1928.

TABLE OF CONTENTS

CONTENTS

CONTENTS

CONTENTS

INTRODUCTION

THE *Timaeus* belongs to the latest group of Plato's works : *Sophist* and *Statesman*, *Timaeus* and *Critias*, *Philebus*, *Laws*. The whole group must fall within the last twenty years of his life, which ended in 347 B.C. at the age of eighty or eighty-one. The *Laws* is the only dialogue that is certainly later than the *Timaeus* and *Critias*. It is probable, then, that Plato was nearer seventy than sixty when he projected the trilogy, *Timaeus*, *Critias*, *Hermocrates* —the most ambitious design he had ever conceived. Too ambitious, it would seem ; for he abandoned it when he was less than half-way through. The *Critias* breaks off in an unfinished sentence ; the *Hermocrates* was never written. Only the *Timaeus* is complete ; but its introductory part affords some ground for a conjectural reconstruction of the whole plan.

The conversation in this dialogue and its sequel is supposed to take place at Athens on the day of the Panathenaea. We are to imagine that, on the previous day, Socrates has been discoursing to Critias, his two guests from Italy and Sicily, Timaeus of Locri and Hermocrates of Syracuse, and a fourth unnamed person who is to-day absent through indisposition. The Panathenaic festival would provide an obvious occasion for the strangers' presence in Athens, as it does for the visit of Parmenides and Zeno in another of the late dialogues.[1]

The Athenian Critias is an old man, who finds it easier to remember the long-distant past than what happened yesterday, and speaks of his boyhood as ' very long ago ', when the poems of Solon could be described as a novelty. He cannot, therefore, be the Critias who was Plato's mother's cousin and one of the Thirty Tyrants. He must be the grandfather of that Critias and Plato's great-grandfather.[2] He tells us that he was eighty

---

[1] *Parm.* 127D. The comparison is made by Pr. i, 84. That ' the festival of the goddess ' (Athena) mentioned at 21A and 26E is the Panathenaea is clear from the context in both places and would never have been doubted but for the unfounded notion that Socrates is supposed to have narrated on the previous day the whole of the *Republic*, or a substantial part of it, as it stands in our texts. This will be considered below.

[2] See Burnet, *Gk. Phil.* i, 338, and Appendix. Tr., p. 23. Diehl, P.-W., *Real-Encycl.*, s.v. Kritias.

I

years younger than his own grandfather, the Critias who was Solon's friend.

Hermocrates, according to Proclus (on 20A) and modern scholars, is the Syracusan who defeated the Athenian expedition to Sicily in Plato's childhood (415–413 B.C.). Thucydides (vi, 72) describes him as a man of outstanding intelligence, conspicuous bravery, and great military experience. At his first appearance in the History (iv, 58) he delivers a wise speech at a conference of Sicilian states, advising them to make peace among themselves and warning them of the danger of Athenian aggression. Evidently at that date (424 B.C.) he was already a prominent figure in Sicilian politics. After the defeat of the Athenian expedition he was banished by the democratic party. He lost his life in an attempt to reinstate himself by force, probably in 407 B.C. In the present gathering of philosophers and statesmen he is pre-eminently the man of action. Since the dialogue that was to bear his name was never written, we can only guess why Plato chose him. It is curious to reflect that, while Critias is to recount how the prehistoric Athens of nine thousand years ago had repelled the invasion from Atlantis and saved the Mediterranean peoples from slavery, Hermocrates would be remembered by the Athenians as the man who had repulsed their own greatest effort at imperialist expansion. He had also attempted to reform from within his native city, Syracuse, the scene of Plato's own abortive essays towards the reconstruction of existing society.

There is no evidence for the historic existence of Timaeus of Locri. If he did exist, we know nothing whatever about him beyond Socrates' description of him as a man well-born and rich, who had held the highest offices at Locri and become eminent in philosophy (20A), and Critias' remark that Timaeus was the best astronomer in the party and had made a special study of the nature of the universe. This is consistent with his being a man in middle life, contemporary with Hermocrates.[1] The very fact that a man

---

[1] I cannot follow Tr.'s inference from Socrates' words that ' we cannot imagine him (Timaeus) to be less than seventy and he may be decidedly older ' (p. 17). Sir Arthur Eddington and Professor Dirac were both elected into chairs of mathematics at Cambridge in or about their thirtieth years. In the fifth century B.C. a man of that age might easily have read everything written in Greek on physics and mathematics. Nor did the Greeks wait till a man was nearing seventy before electing him to the highest offices. Tr. also says (p. 49) that ' the youth of Hermocrates explains why he remains silent throughout the dialogue. Proclus saw that his silence is significant, but did not interpret it correctly.' But Hermocrates does make a not unimportant contribution to the conversation on the only occasion offered him (20C), a fact on which Pr. comments. He also speaks in the introductory conversation of the *Critias* (108B) in terms which, with other passages, make it clear that he was to take the leading part in the third dialogue of the trilogy.

of such distinction has left not the faintest trace in political or philosophic history is against his claim to be a real person. The probability is that Plato invented him because he required a philosopher of the Western school, eminent both in science and statesmanship, and there was no one to fill the part at the imaginary time of the dialogue. Archytas was of the type required,[1] a brilliant mathematician and seven times *strategus* at Tarentum ; but he lived too late : Plato first met him about 388 B.C. In the first century A.D. a treatise *On the Soul of the World and Nature* was forged in the name of Timaeus of Locri. It was taken by the Neoplatonists for a genuine document, whereas it is now seen to be a mere summary of the *Timaeus*. In our dialogue, as Wilamowitz observes (*Platon* i, 591), Timaeus speaks dogmatically, but without any appeal to authority, and we may regard his doctrine simply as Plato's own. So in the *Sophist* Plato speaks through the mouth of an Eleatic, who is yet not a champion of Parmenides' system, but holds a theory of Forms unquestionably Platonic. Plato nowhere says that Timaeus is a Pythagorean. He sometimes follows Empedocles, sometimes Parmenides ; indeed he borrows something from every pre-Socratic philosopher of importance, not to mention Plato's contemporaries. Much of the doctrine is no doubt Pythagorean ; and this gave the satirist Timon a handle for his spiteful accusation of plagiarism against Plato. When the treatise ascribed to Timaeus had been forged, it was assumed that this was the book from which Plato had copied (Pr. i, 1 and 7).[2] As a consequence, all the doctrines which the forger had found in the *Timaeus* itself were supposed to be of Pythagorean origin. The testimony of later commentators is vitiated by this false assumption.

There is no ground for any conjecture as to the identity of the fourth person, who is absent. The only sensible remark recorded by Proclus is the observation of Atticus that he is presumably another visitor from Italy or Sicily, since Socrates asks Timaeus for news of him (Pr. i, 20). Plato may have wished to keep open the possibility of extending his trilogy to a fourth dialogue and held this unnamed person in reserve.[3] Socrates proposes that the three who are present (not Timaeus alone) shall undertake the whole task which the four were to have shared. He first recapitulates his own discourse of the previous day. Socrates, we are told, had been describing the institutions of a city on the lines of the *Republic*. He had ended by expressing his wish to see this city transferred from the plane of theory to temporal fact. He now

---

[1] As Frank observes, *Plato und d. sog. Pythagoreer*, 129.
[2] For the history of this document, see Tr., p. 39.
[3] So Ritter, *N. Unt.*, 181.

gives a summary of his own discourse, in response to Timaeus'
request to be reminded of the task to be performed by himself and
his friends. Later (20C) it appears that such a reminder was really
unnecessary, since the three have talked over the task required of
them and have come prepared with a plan for its fulfilment. The
summary is, in fact, entirely for the sake of informing the reader
of Plato's design to identify the citizens of the ideal state with the
prehistoric Athenians of Critias' romance.

From ancient times to the present day many false inferences
and theories have been founded on the situation imagined by Plato,
in spite of his own clear indication conveyed in the statement that
the summary actually given is complete : nothing of importance
has been omitted (19A, B). Plato could not have stated more plainly
that Socrates is not to be supposed to have narrated the whole
conversation in the *Republic* as we have it. It follows at once
that he did not intend the *Republic* to stand as the first dialogue
in his new series.[1] If he had, no recapitulation would have been
needed ; the stage should have been set in an introduction to the
Republic itself. But some scholars have seen evidence here for an
original edition of the *Republic*, containing only the parts summarised. Such speculations are baseless. The summary is confined to the external institutions of the state outlined in *Republic* ii,
369–v, 471. It is impossible to imagine an edition of the dialogue
omitting the whole of the analogy between the structure of the
soul and that of the state, the analysis of the individual soul into
three parts, and the discussion of the virtues of the individual and
of the state ; nor could the omission of these topics in the summary
be called a matter of no importance. The simple and natural
conclusion was drawn long ago by Hirzel.[2] No doubt Plato was
thinking of the contents of that part of the *Republic* and intending
his readers to recall them ; but he was not the slave of his own
fictions. There was nothing to prevent him from imagining
Socrates describing his ideal state on more than one occasion.
He tells us here that Socrates has outlined its institutions, and
nothing more, on the previous day. That day, moreover, was not
the day after the feast of Bendis (Thargelion 19 or 20), when the
conversation with Glaucon and Adeimantus at the house of Cephalus
took place, though nothing would have been easier than to mention
that date if Plato had meant to identify Socrates' discourse with

---

[1] As Pr., for example, imagined (i, 8). In consequence, he and other
critics were puzzled how to explain why the *Republic* was to precede the
Timaeus, and not follow it, as it obviously should (i, 200 ff.).

[2] *Der Dialog.* (1895), i, 257. So Ritter, *N. Unt.* 177, and Friedländer,
Plat. Schr. 600. Cf. also Rivaud, *Timée*, p. 19.

the narration of the *Republic*. The present occasion is ' the festival of Athena ',[1] and one to which the projected discourse of Critias is appropriate. As Proclus remarks (i, 172), the Panathenaic discourses regularly celebrated the Athenian victories by land and sea in the Persian Wars, while Critias celebrates Athens by recounting her victory over the invaders from Atlantis. Proclus himself had no doubt that the Lesser Panathenaea was meant ; he knew no more than that this festival ' came after ' the Bendidea and thought it took place ' about the same time ' (i, 84–5), whereas he knew that the Greater Panathenaea fell in Hecatombaeon (i, 26). Neither festival, in fact, came within two months of the Bendidea. Plato probably intended the Greater Panathenaea. There is no other indication of the dramatic date ; and it is unlikely that Plato had troubled himself about the question whether there was any such occasion on which Hermocrates could have visited Athens. The date is of no importance. In his earliest dialogues Plato was concerned to give the Athenians a true impression of Socrates' character and activity, and he was at great pains to recreate the atmosphere of the times. That interest was long past. In the latest group there was no motive to keep up the illusion that the conversations had really taken place. From all this it follows that the dramatic date and setting of the *Republic* have no bearing whatever on the dramatic date of the *Timaeus* trilogy. Also no ground remains for any inference that Plato meant the contents of the later books of the *Republic* to be superseded or corrected by the *Timaeus*.

The design of the present trilogy is thus completely independent of the *Republic*. What was that design ? The political question answered in the *Republic* had been : What is the least change in existing society necessary to cure the evils afflicting mankind ? Plato had imagined a reformed Greek city-state with institutions based, as he claimed, on the unalterable characteristics of human nature. It appeared to be just within the bounds of possible realisation. Referring to hopes founded on Dion or on the younger Dionysius, he had said that his state might see the light of day, if some prince could be found endowed with the philosophic nature, and if that nature could escape corruption. But towards the end of the *Republic* Plato seems less hopeful, and the state recedes as a pattern laid up in heaven, by which the merits and defects of all existing constitutions might be measured and appraised. Moreover, since that dialogue was written, Plato's Sicilian adventures

---

[1] 21A, ἐν τῇ πανηγύρει (the word implies an important festival) ; 26E, τῇ παρούσῃ τῆς θεοῦ θυσίᾳ. There was no such festival on Thargelion 21. The Plynteria came five days later.

had ended in disappointment. Accordingly, the discourse re-capitulated at the opening of the *Timaeus* covers only the outline of the state given in the earlier books of the *Republic*, ignoring all the later books, which had started from the question how it might be realised in the future and sketched its possible decline through lower forms of polity. The new trilogy is to transfer this state to the plane of actual existence, not in the future, but in the remote past, as the Athens of nine thousand years ago. This is the subject of the *Critias*, introduced at once as the central theme of the whole.

By way of preface, *Timaeus* is to recount his myth of creation, ending with the birth of mankind. The whole movement starts from the ideal world of the Demiurge and the eternal Forms, descending thence to the frame of the visible universe and the nature of man, whose further fortunes Critias will 'take over' for his story. Looking deeper, we see that the chief purpose of the cosmological introduction is to link the morality externalised in the ideal society to the whole organisation of the world.[1] The *Republic* had dwelt on the structural analogy between the state and the individual soul. Now Plato intends to base his conception of human life, both for the individual and for society, on the inex-pugnable foundation of the order of the universe. The parallel of macrocosm and microcosm runs through the whole discourse. True morality is not a product of human evolution, still less the arbitrary enactment of human wills. It is an order and harmony of the soul ; and the soul itself is a counterpart, in miniature, of the soul of the world, which has an everlasting order and harmony of its own, instituted by reason. This order was revealed to every soul before its birth (41E) ; and it is revealed now in the visible architecture of the heavens. That human morality is so based on the cosmic order had been implied, here or there, in earlier works ; but the *Timaeus* will add something more like a demonstration, although in mythical form.

In the next dialogue Critias will repeat the legend learnt by Solon from an Egyptian priest : how primitive Athens (now to be iden-tified with Socrates' ideal state) had defeated the invaders from Atlantis. In the very hour when freedom and civilisation were saved for the mediterranean world, the victorious Athenians had themselves been overwhelmed by flood and earthquake. Atlantis also sank beneath the sea and vanished. What was to follow ? The story was not to end with the cataclysm of the *Critias* ; and the Egyptian priest, discoursing at some length to Solon on these periodic catastrophes in which all but a small remnant of mankind perishes, has explained how the seeds of a new civilisation are

[1] Cf. Fraccaroli, p. 13.

preserved either on the mountains or in the river valleys, according as the destruction is by flood or fire. When it is by flood, as at the end of Critias' story, the cities on the plains are overwhelmed ; only the mountain shepherds survive, and all culture is lost. Taking up the story at this point, what could Hermocrates do, if not describe the re-emergence of culture in the Greece of prehistoric and historic times ? If so, the projected contents of the unwritten dialogue are to be found in the third and subsequent books of the *Laws*. There, after some preliminary ramblings about music and wine in Books i and ii, the Athenian settles down to business at the opening of Book iii with the question : What is the origin of society and government ? In the immensity of past time myriads of states have arisen and perished, reproducing again and again the same types of constitution. How do they arise ? Mankind has often been almost destroyed by flood, plagues, and many other causes ; only a small remnant is left. Imagine one such destruction—the Deluge. The herdsmen on the mountain-tops alone survived, while the cities on the plains or near the sea were overwhelmed. All arts and inventions perished ; all statecraft was forgotten. Here is exactly the situation with which the *Critias* was to end, described in language very like that of the Egyptian priest. The *Laws* continues the story. After the deluge came a very long and slow advance towards the present state of things. Before the metals were rediscovered there was an idyllic phase of society, resembling descriptions of the Golden Age, under the rule of patriarchal custom. Next came the beginnings of agriculture and the formation of more permanent settlements. The coalescence of various tribes led to the growth of aristocracies, or perhaps monarchies, with kings and magistrates. A third stage saw the blending of different types of constitution. Mankind, forgetting the dangers of flood, ventured down from the hills. Cities like Homer's Troy were built once more on the plains. (Here we reach what was for the Greeks the dawn of history.) Then followed the Trojan War ; and the troubles consequent upon the warriors' homecoming led to the migrations. Finally we reach the settlement of Crete and Lacedaemon. The Athenian recommends a study of this succession of social forms, to discover what laws preserve a city or tend to ruin it. The history of the Dorian states suggests that government should be a mixture of monarchy and democracy. It is then proposed to apply this principle by framing laws for a new colony. Book iv opens with the choice of a site, and the rest of the treatise outlines the institutions.

Since all this fits on exactly to the end planned for the *Critias*, it may well have been Plato's original purpose to use in the *Her-*

mocrates the material he had been collecting from a study of the laws of Greek states. The whole trilogy would then have covered the story of the world from creation, through prehistoric legend and all historic time, to a fresh project for future reform. But Plato was getting old. The composition of the *Critias* seems to have been interrupted; it stops in an unfinished sentence. After the interruption Plato might well feel that he could not complete all this elaborate romance about the invasion from Atlantis before starting upon the subject nearest his heart, which now fills ten books of the *Laws*.[1] There was, in fact, by this time far too much material for a continuation of the *Timaeus* trilogy, even with the assistance of the unnamed absentee. So he abandoned the *Critias*, and wrote the *Laws* in place of the *Hermocrates*.[2]

---

[1] In the same way (*si parva licet*) Mr. H. G. Wells has, with advancing years, grown impatient of the Utopian romance and taken to expressing his hopes and fears for the future through ever thinner disguises, ending with autobiography.

[2] For the conjecture here elaborated see Raeder, 379.

THE TIMAEUS

17A–27B. INTRODUCTORY CONVERSATION

AN account of the persons who take part in the conversation prefacing the discourse of Timaeus has already been given in the Introduction (pp. 1–3). We may proceed at once to the text.

SOCRATES. TIMAEUS. HERMOCRATES. CRITIAS

17A. SOCRATES. One, two, three—but where, my dear Timaeus, is the fourth of those guests of yesterday who were to entertain me to-day?

TIMAEUS. He suddenly felt unwell, Socrates; he would not have failed to join our company if he could have helped it.

SOCR. Then it will fall to you and your companions to supply the part of our absent friend as well as your own.

B. TIM. By all means; we will not fail to do the best we can. Yesterday you entertained us with the hospitality due to strangers, and it would not be fair if the rest of us were backward in offering you a feast in return.

SOCR. Well, then, do you remember the task I set you—all the matters you were to discourse upon?

TIM. We can remember some; and you are here to remind us of any that we may have forgotten. Or rather, if it is not too much trouble, will you recapitulate them briefly from the beginning, to fix them more firmly in our minds?

C. SOCR. I will. Yesterday the chief subject of my own discourse was what, as it seemed to me, would be the best form of society and the sort of men who would compose it.

TIM. Yes, Socrates, and we all found the society you described very much to our mind.

SOCR. We began, did we not? by separating off the farmers and all the other craftsmen from the class that was to fight in defence of the city?

TIM. Yes.

D. SOCR. And when we assigned only one occupation to each man, one craft for which he was naturally fitted, these, we said, who were to fight on behalf of all, must be nothing else

9

17D.   but guardians of the city against the assault of any that
would injure her, whether from within or from without,

18.   dealing justice to their subjects mildly, as to natural friends,
and showing a stern face to those enemies who meet them
in battle.

TIM.   Quite true.

SOCR.   There was, in fact, a certain temperament that we
said a guardian should have, at once spirited and philosophic
to. an exceptional degree, enabling them to show a right
measure of mildness or sternness to friend or foe.

TIM.   Yes.

SOCR.   And for their education, they were to be trained in
gymnastic and music and in all the studies suitable for them.

TIM.   Certainly.

B.   SOCR.   And the men so trained, we said, were never to regard
gold or silver or anything else as their private possessions.
Rather, as a garrison drawing from those whom they protect
so much pay for their services as would reasonably suffice
men of a temperate life, they were to share all expense and
lead a common life together, in the constant exercise of
manly qualities and relieved from all other occupations.

TIM.   So it was provided.

C.   SOCR.   And then we spoke of women. We remarked that
their natures should be formed to the same harmonious blend
of qualities as those of men ; [1] and they should all be given
a share in men's employments of every sort, in war as well
as in their general mode of life.

TIM.   That too was prescribed.

SOCR.   And then there was the procreation of children. Here,
perhaps, the novelty of our regulations makes them easy to
remember. We laid down that they should all have their
marriages and children in common. They were to contrive
that no one of them should ever recognise his own offspring,

D.   but each should look upon all as one family, treating as
brothers and sisters all who fell within appropriate limits of
age, and as parents and grandparents, or as children and
grandchildren, those who fell above or below those limits.

TIM.   Yes ; that, as you say, is easy to remember.

SOCR.   Then, in order that they might have the best possible
natural dispositions from birth, we said, you remember, that
the magistrates of both sexes must make secret arrangements

---

[1] συναρμοστέον refers to the proper blend of spirited and philosophic
elements mentioned above, which exist in women as in men (*Rep.* 456A).
For συναρμόττειν cf. *Rep.* 443D.

18E. for the contraction of marriages by a certain method of drawing lots, which would apportion both to the better men and to the worse partners like themselves and yet not lead to any ill-feeling, because they would imagine the allotment to be the result of chance.

TIM. I remember that.

19. SOCR. And further, the children of the better sort were to be educated, while those of the worse should be secretly dispersed through the rest of the community. The rulers were to keep the children under observation as they grew up, and from time to time take back again those who were found worthy, while the undeserving ones in their own ranks should take the places of the promoted.

TIM. Just so.

SOCR. Well, then, my dear Timaeus, have we now passed in review all the main points of yesterday's conversation ; or is there anything that we feel has been left out ?

B. TIM. No, Socrates ; you have exactly described what was said.

As I have argued in the Introduction, we are evidently not to imagine that Socrates has, on the previous day, narrated the whole conversation in the *Republic* or any part of it. There is, in fact, no part of the *Republic* of which it could be said that ' all the main points ' were covered by the above summary. Socrates now comes to the instructions he is supposed to have given on the previous day. He wishes the other three to draw a picture of his ideal State in actual existence. With his usual modesty, he represents this task as beyond his own powers. He had never been a man of action or taken part in politics.

19B. SOCR. I may now go on to tell you how I feel about the society we have described. I feel rather like a man who has been looking at some noble creatures in a painting, or perhaps at real animals, alive but motionless, and conceives

C. a desire to watch them in motion and actively exercising the powers promised by their form. That is just what I feel about the city we have described : I should like to hear an account of her putting forth her strength in such contests as a city will engage in against others, going to war in a manner worthy of her, and in that war achieving results befitting her training and education, both in feats of arms and in negotiation with various other states.

D. Now here, Critias and Hermocrates, my judgment upon

19D. myself is that to celebrate our city and its citizens as they deserve would be beyond my powers. My incapacity is not surprising ; but I have formed the same judgment about the poets of the past and of to-day. Not that I have a low opinion of poets in general ; but anyone can see that an imitator, of whatever sort, will reproduce best and most easily the surroundings in which he has been brought up ;

E. what lies outside that range is even harder to reproduce successfully in discourse than it is in action. The sophists, again, I have always thought, have had plenty of practice in making fine speeches on other subjects of all sorts ; but with their habit of wandering from city to city and having no settled home of their own, I am afraid they would hardly hit upon [1] what men who are both philosophers and statesmen would do and say in times of war, in the conduct of actual fighting or of negotiation. There remain only people of your condition, equipped by temperament and education for

20. both philosophy and statesmanship. Timaeus, for instance, belongs to an admirably governed State, the Italian Locri,[2] where he is second to none in birth and substance, and has not only enjoyed the highest offices and distinctions his country could offer, but has also, I believe, reached the highest eminence in philosophy. Critias, again, is well known to all of us at Athens as no novice in any of the subjects we are discussing ; and that Hermocrates is fully qualified in all such matters by natural gifts and education, we may trust

B. the assurance of many witnesses.[3] Accordingly this was in

---

[1] ἄστοχον. This unusual word recalls the description of rhetoric in the *Gorgias* 463A as a branch of Parasitism—' a profession which is not of the nature of an art, but demands a shrewd and virile spirit (ψυχῆς στοχαστικῆς καὶ ἀνδρείας) with a native cleverness in human relations '. Plato there seems to have echoed Isocrates' eulogy of rhetoric as demanding ' a virile and *imaginative* spirit ' (ψυχῆς ἀνδρικῆς καὶ δοξαστικῆς, κ. σοφ. 17), maliciously substituting στοχαστικῆς. In the *Euthydemus* (305C) Isocrates is evidently aimed at as one who is ' on the borderline ' between philosophy and statesmanship and fails to make the best of either.

[2] The constitution of Locri was attributed to Zaleucus (Ar., *Pol.* 1274A, 22). At *Laws* 638B the Athenian says that the Locrians are reputed to have the best laws of any western state. If Timaeus never existed, this would account for Plato's choice of Locri for his native place.

[3] At 20A, 8 read εἶναι ταῦτα ἱκανήν F Y, Pr., to avoid hiatus with ἱκανήν. So Blass (*Att. Bered.* ii, 458), who reckons hardly more than 50 cases of ' illegitimate ' hiatus in the *Timaeus*, some of which can be removed by adopting other MS. readings, as, for example, here and at 23A, 2 and 38A, 4. The rest, he thinks, should be regarded with suspicion, and some can be easily removed by conjecture, e.g. πάντα for ἅπαντα 78C, 1. According to Raeder's figures, the instances of illegitimate hiatus in *Lysis, Apol., Gorg.,*

20B. my mind yesterday when I was so ready to grant your request for a discourse on the constitution of society : I knew that, if you would consent to supply the sequel, no one could do it better ; you could describe this city engaged in a war worthy of her and acting up to our expectations, as no other living persons could. So, after fulfilling my part, I set you, in my turn, the task of which I am now reminding you. You agreed

c. to consult among yourselves and to requite my hospitality to-day. So here I am in full dress for the entertainment, which I am most eager to receive.

HERMOCRATES. Indeed, Socrates, as Timaeus said, we shall not fail to do our best, and we have no excuse for refusing. Yesterday, as soon as we had reached Critias' guest-chamber, where we are staying, and even while we were still on the

D. way there, we were considering this very matter. Critias then produced a story which he had heard long ago. Critias, will you repeat it now to Socrates, and he shall help us to judge whether or not it will answer the purpose of the task he is laying on us?

CRITIAS. It shall be done, if our remaining partner, Timaeus, approves.

TIM. Certainly I approve.

CRIT. Listen then, Socrates, to a story which, though strange, is entirely true, as Solon, wisest of the Seven, once

E. affirmed. He was a relative and close friend of Dropides, my great-grandfather, as he says himself several times in his poems ; and he told my grandfather Critias (according to the story the old man used to repeat to us) that there were great and admirable exploits performed by our own city long ago, which have been forgotten through lapse of time and the destruction of human life.[1] Greatest of all was one

21. which it will now suit our purpose to recall, and so at once pay our debt of gratitude to you and celebrate the goddess, on her festival, with a true and merited hymn of praise.

SOCR. Good. But what was this ancient exploit that your grandfather described on Solon's authority as unrecorded and yet really performed by our city?

Phaedo, Republic range between 35 and 45 per page of the Didot edition. In *Soph.* and *Polit.* the figures drop to 0·6 and 0·4, and the *Timaeus* shows only a slightly higher figure, 1·1. There is a slight further rise in *Philebus* (3·7) and *Laws* (5·8).

[1] i.e. the almost complete destructions of mankind outside Egypt by flood or fire, the φθοραὶ ἀνθρώπων of 22c and *Laws* 677A, one of which overwhelmed the actors in this exploit (φθορὰ τῶν ἐργασαμένων, 21D). Both Plato and Aristotle believed that such catastrophes occur.

21.   CRIT. I will tell you the story I heard as an old tale [1] from a man who was himself far from young. At that time, indeed, Critias, by his own account, was close upon ninety,

B.   and I was, perhaps, ten years old. We were keeping the Apaturia ; it was the Children's Day.[2] For us boys there were the usual ceremonies : our fathers offered us prizes for reciting. Many poems by different authors were repeated, and not a few of us children sang Solon's verses, which were a novelty in those days. One of the clansmen said—either because he really thought so or to please Critias—that he considered Solon to have shown himself not only extremely

C.   wise but, in his writings, the most free-spirited of poets. The old man—how well I remember it !—was much pleased and said with a smile :

'Yes, Amynander; if only he had taken his poetry seriously like others, instead of treating it as a pastime, and if he had finished the story he brought home from Egypt and had not been forced to lay it aside by the factions and other troubles he found here on his return, I believe no other

D.   poet—not Homer or Hesiod—would have been more famous than he.'

'And what was the story, Critias ? ' Amynander asked.

'It was about the greatest achievement ever performed by our city—one that deserved to be the most renowned of all, but through lapse of time and the destruction of the actors, the story has not lasted down to our time.'

'Tell it from the beginning ', said Amynander. 'How and from whom did Solon hear this tale which he reported as being true ? '

E.   'In Egypt,' said Critias, 'at the apex of the Delta, where the stream of the Nile divides, there is a province called the Saitic. The chief city of this province is Sais, from which came King Amasis. The goddess who presides over their city is called in Egyptian Neith, in Greek, by their account, Athena ; they are very friendly to Athens and claim a certain kinship with our countrymen. Solon said that, when he travelled thither, he was received with much honour ; and

22.   further that, when he inquired about ancient times from the priests who knew most of such matters, he discovered that neither he nor any other Greek had any knowledge of antiquity worth speaking of. Once, wishing to lead them on

---

[1] παλαιόν, i.e. the story was already old when Critias heard it from Solon ; and Critias himself was very old when he told it to his grandson.

[2] The day on which children were inscribed on the register of the clan.

22. to talk about ancient times, he set about telling them the most venerable of our legends, about Phoroneus the reputed first man and Niobe, and the story how Deucalion and Pyrrha

B. survived the deluge. He traced the pedigree of their descendants, and tried, by reckoning the generations, to compute how many years had passed since those events.

"Ah, Solon, Solon," said one of the priests, a very old man, "you Greeks are always children; in Greece there is no such thing as an old man."

"What do you mean?" Solon asked.

"You are all young in your minds," said the priest, "which hold no store of old belief based on long tradition, no knowledge hoary with age. The reason is this. There have

C. been, and will be hereafter, many and divers destructions of mankind, the greatest by fire and water, though other lesser ones are due to countless other causes. Thus the story current also in your part of the world, that Phaethon, child of the Sun, once harnessed his father's chariot but could not guide it on his father's course and so burnt up everything on the face of the earth and was himself consumed by the thunderbolt—this legend has the air of a fable; but the

D. truth behind it is a deviation of the bodies that revolve in heaven round the earth and a destruction, occurring at long intervals, of things on earth by a great conflagration. At such times all who live on mountains and in high regions where it is dry perish more completely than dwellers by the rivers or the sea. We have the Nile, who preserves us in so many ways and in particular saves us from this affliction when he is set free.[1] On the other hand, when the gods cleanse the earth with a flood of waters, the herdsmen and shepherds in the mountains are saved, while the inhabitants

E. of cities in your part of the world are swept by the rivers into the sea. But in this country the water does not fall from above upon the fields either then or at other times; its way is always to rise up over them from below. It is for these reasons that the traditions preserved here are the oldest on record;[2] though as a matter of fact in all regions where

23. inordinate cold or heat does not forbid it mankind exists at all

---

[1] The question from what, and by what, the Nile is 'set free' is discussed in the Appendix (p. 365).

[2] λέγεται, cf. λεγόμενον 21A, 5. Not 'are said to be': the Egyptian traditions *are* the oldest, because, although mankind is not completely destroyed anywhere, no records are kept elsewhere by the unlettered survivors of floods and conflagrations.

23. times in larger or smaller numbers. Any great or noble achievement or otherwise exceptional event that has come to pass, either in your parts or here or in any place of which we have tidings,[1] has been written down for ages past in records that are preserved in our temples ; whereas with you and other peoples again and again life has only lately been enriched with letters and all the other necessaries of civilisation when once more, after the usual period of years, the torrents from heaven sweep down like a pestilence leaving only the rude

B. and unlettered among you. And so you start again like children, knowing nothing of what existed in ancient times here or in your own country. For instance, these genealogies of your countrymen, Solon, that you were reciting just now, are little better than nursery tales. To begin with, your people remember only one deluge, though there were many earlier ; and moreover you do not know that the bravest and noblest race in the world once lived in your country.

C. From a small remnant of their seed you and all your fellow-citizens are derived ; but you know nothing of it because the survivors for many generations died leaving no word in writing. Once, Solon, before the greatest of all destructions by water, what is now the city of the Athenians was the most valiant in war and in all respects the best governed beyond comparison : her exploits and her government are said to have been the noblest under heaven of which report has

D. come to our ears."

On hearing this, Solon was astonished and eagerly begged the priests to tell him from beginning to end all about those ancient citizens.

"Willingly," answered the priest ; "I will tell you for your own sake and for your city's, and above all for honour of the goddess, patroness of our city and of yours, who has fostered both and instructed them in arts. Yours she

E. founded first by a thousand years, from the time when she took over the seed of your people from Earth and Hephaestus ; ours only in later time ; and the age of our institutions is given in the sacred records as eight thousand years. Accordingly those fellow-countrymen of yours lived nine thousand years ago ; and I will shortly describe their laws and the noblest exploit they performed ; we will go through the

24. whole story in detail another time at our leisure, with the records before us.

[1] Read ἀκοήν (AY, Pr.), with Blass and A.-H., to avoid hiatus. See note on 20A.

16

24. "Consider their laws in comparison with ours; you will find here to-day many parallels illustrating your own institutions in those days. First, there is the separation of the priesthood from the other classes; next the class of craftsmen—you will find that each kind keeps to its own craft without infringing on another; shepherds, hunters,

B. farmers.[1] The soldiers, moreover, as you have no doubt noticed, are here distinct from all other classes; they are forbidden by law to concern themselves with anything but war. Besides, the fashion of their equipment is with spear and shield, arms which we were the first people in Asia to bear, for the goddess taught us, as she had taught you first in your part of the world. Again, in the matter of wisdom, you see what great care the law has bestowed upon it here from the very beginning, both as concerns the order of the

C. world, deriving from those divine things the discovery of all arts applied to human affairs, down to the practice of divination and medicine with a view to health, and acquiring all the other branches of learning connected therewith.[2] All this order and system the goddess had bestowed upon you earlier when she founded your society, choosing the place in which you were born because she saw that the well-tempered climate would bear a crop of men of high intelligence. Being a lover of war and of wisdom, the goddess chose

D. out the region that would bear men most closely resembling herself and there made her first settlement. And so you dwelt there with institutions such as I have mentioned and even better, surpassing all mankind in every excellence, as might be looked for in men born of gods and nurtured by them.

"Many great exploits of your city are here recorded

---

[1] Isocrates' *Busiris* (certainly earlier in date than the *Timaeus*) mentions the Egyptian caste system, and is itself based on Herod. ii, 164–8. But it is not unlikely that Plato himself had visited Egypt.

[2] A.-H. suspects the soundness of the text here. The general sense seems to be that the Egyptians base all the arts applied to human life on the study of the heavens (for ἅπαντα ἀνευρών meaning the invention of arts, cf. Xenophanes frag. 18 οὔτοι ἀπ' ἀρχῆς πάντα θεοὶ θνητοῖσ' ὑπέδειξαν, ἀλλὰ χρόνῳ ζητοῦντες ἐφευρίσκουσιν ἄμεινον). Plato's language recalls Isocrates, *Busiris* 21: Busiris is τῆς περὶ τὴν φρόνησιν ἐπιμελείας αἴτιος. The leisure he provided for the priests enabled them to discover the art of *medicine* and to practise *philosophy*. The younger priests study *astronomy, calculation, and geometry* (perhaps the μαθήματα Plato mentions in the last clause). According to Diod. i, 82, 3 Egyptian physicians were bound to follow the treatment laid down by ancient physicians in sacred books, and condemned to death for departing from it. Aristotle (*Pol.* iii, 1286A, 13) says that they were allowed to alter the treatment after the fourth day.

24D. for the admiration of all; but one surpasses the rest in
E. greatness and valour. The records tell how great a power
your city once brought to an end when it insolently advanced
against all Europe and Asia, starting from the Atlantic ocean
outside. For in those days that ocean could be crossed,
since there was an island [1] in it in front of the strait which
your countrymen tell me you call the Pillars of Heracles.
The island was larger than Libya and Asia put together;
and from it the voyagers of those days could reach the other
islands, and from these islands the whole of the opposite
25. continent bounding that ocean which truly deserves the name.
For all these parts that lie within the strait I speak of, seem
to be a bay with a narrow entrance; that outer sea is the
real ocean, and the land which entirely surrounds it really
deserves the name of continent in the proper sense.[2] Now
on this Atlantic island there had grown up an extraordinary
power under kings who ruled not only the whole island but
many of the other islands and parts of the continent; and
besides that, within the straits, they were lords of Libya
B. so far as to Egypt, and of Europe to the borders of Tyrrhenia.
All this power, gathered into one, attempted at one swoop
to enslave your country and ours and all the region within
the strait. Then it was, Solon, that the power of your city
was made manifest to all mankind in its valour and strength.
She was foremost of all in courage and in the arts of war,
C. and first as the leader of Hellas, then forced by the defection
of the rest to stand alone, she faced the last extreme of
danger, vanquished the invaders, and set up her trophy;
the peoples not yet enslaved she preserved from slavery,
and all the rest of us who dwell within the bounds set by
Heracles she freed with ungrudging hand. Afterwards there
was a time of inordinate earthquakes and floods; there came
D. one terrible day and night, in which all your men of war
were swallowed bodily by the earth, and the island Atlantis
also sank beneath the sea and vanished. Hence to this day
that outer ocean cannot be crossed or explored, the way
being blocked by mud, just below the surface,[3] left by the
settling down of the island.'' '

---

[1] Serious scholars now agree that Atlantis probably owed its existence
entirely to Plato's imagination. See Frutiger, *Mythes de Platon*, 244 ff.

[2] The *Etym. Mag.* connects ἤπειρος with ἄπειρος: land not bounded by
sea as an island is. παντελῶς should be taken with περιέχουσα. The outer
continent is ' unbounded ' as forming a completely unbroken ring.

[3] Reading κατὰ βραχέος, ' at a slight depth '. See Appendix, p. 366.

25E.　　Now, Socrates, I have given you a brief account of the story told by the old Critias as he heard it from Solon. When you were speaking yesterday about your state and its citizens, I recalled this story and I was surprised to notice in how many points your account exactly agreed, by some miraculous

26.　　chance, with Solon's. But I would say nothing at the moment; after so long an interval, my memory was imperfect. So I resolved that I would not repeat the story until I had first gone over it thoroughly in my own mind. That is why I so readily agreed to the task you laid upon us yesterday; I thought that in any case like this the hardest part is to find some suitable theme as a foundation for one's design, and that that need would be fairly well supplied. Accordingly, as Hermocrates has told you, no sooner had I left yesterday than I set about repeating the story to our

B.　　friends as I recalled it, and when I got home I recovered pretty well the whole of it by thinking it over at night. How true is the saying that what we learn in childhood has a wonderful hold on the memory! I doubt if I could recall everything that I heard yesterday; but I should be surprised if I have lost any detail of this story told me so long ago. I listened at the time with much boyish delight, and

C.　　the old man was very ready to answer the questions I kept on asking; so it has stayed in my mind indelibly like an encaustic picture. Moreover, I told it all to our friends early this morning, so that they might be as well provided as myself with materials for their discourse.

　　To come to the point I have been leading up to: I am ready now, Socrates, to tell the story, not in summary, but in full detail as I heard it. We will transfer the state you described yesterday and its citizens from the region of

D.　　theory to concrete fact; we will take the city to be Athens and say that your imaginary citizens are those actual ancestors of ours, whom the priest spoke of. They will fit perfectly, and there will be no inconsistency in declaring them to be the real men of those ancient times. Dividing the work between us, we will all try to the best of our powers to carry out your injunctions properly. It is for you to consider, Socrates, whether this story will suit our purpose or we must

E.　　look for another in its stead.

　　Socr. How could we change it for the better, Critias? Its connection with the goddess makes it specially appropriate to her festival to-day; and it is surely a great point that it is no fiction, but genuine history. How and where shall we

19

26E. find other characters, if we abandon these? No, you shall speak and good luck [1] be with you; I have earned by my
27. discourse of yesterday the right to take a rest and listen.

CRIT. Then I will submit to you the plan we have arranged for your entertainment, Socrates. We decided that Timaeus shall speak first. He knows more of astronomy than the rest of us and has made knowledge of the nature of the universe his chief object; he will begin with the birth of the world and end with the nature of man. Then I am to follow, taking over from him mankind, whose origin he has described, and from you a portion of them who have received a supremely

B. good training. I shall then, in accordance with Solon's enactment as well as with his story, bring them before our tribunal and make them our fellow-citizens, on the plea that they are those old Athenians of whose disappearance we are informed by the report of the sacred writings. In the rest of our discourse we shall take their claim to the citizenship of Athens as established.

SOCR. I see that I am to receive a complete and splendid banquet of discourse in return for mine. So you, Timaeus, are to speak next, when you have invoked the gods as custom requires.

It has often been remarked that this introductory conversation, right down to Critias' last speech, might have been written for the *Critias* only, as if the task set by Socrates could have been completely fulfilled by the story of Atlantis. Plato's purpose may have been to indicate that, now as ever, his chief interest lies in the field of morals and politics, not in physical speculation. The whole cosmology of the *Timaeus* is only a preface to the legendary picture of the ideal state in action and to whatever were to have been the contents of the *Hermocrates*. Another motive for here anticipating the Atlantis story was suggested by Longinus (Pr. i, 83). The *Timaeus* is not easy reading; and the physiological and medical chapters towards the end would be repellent to many. The reader might be encouraged to persevere by the promise of an exciting romance to follow. It is, at any rate, well to remember that the unfinished state of the trilogy gives the *Timaeus* a prominence it would not have had in the completed design.

---

[1] Good luck is invoked here, the gods below (27c). Cf. *Laws* vi, 757E θεὸν καὶ ἀναθὴν τύχην καὶ τότε ἐν εὐχαῖς ἐπικαλουμένους. At *Epin.* 991D and 992A θεὸν καλεῖν and τύχην καλεῖν are treated as equivalent.

THE DISCOURSE OF TIMAEUS

27C–29D PRELUDE. *The nature and scope of Physics*

TIMAEUS' ' prelude ', marked off from what follows by Socrates'
expression of approval (29D), lays down the principles of the whole
discourse and defines the limitations of any treatment of physics.
It is constructed with great care. After the opening invocation
of the gods, the second paragraph states three general premises
concerning anything that is not eternal, but comes to be. These
premises are then applied successively to the visible universe. (1)
The eternal is the intelligible ; what comes to be is the sensible.
Since the world is sensible, it must be a thing that comes to be.
(2) Whatever comes to be must have a cause. Therefore the world
has a cause—a maker and father ; but he is hard to find. (3) The
work of any maker will be good only if he fashions it after an eternal
model. The world is good ; so its model must have been eternal.
Finally, the conclusion is drawn : any account that can be given
of the physical world can be no better than a ' likely story ', because
the world itself is only a ' likeness ' of unchanging reality.

27C. TIM. That, Socrates, is what all do, who have the least
portion of wisdom : always, at the outset of every under-
taking, small or great, they call upon a god. .We who are
now to discourse about the universe—how it came into being,
or perhaps had no beginning of existence—must, if our senses
be not altogether gone astray, invoke gods and goddesses
with a prayer that our discourse throughout may be above
all pleasing to them and in consequence satisfactory to us.[1]

D. Let this suffice, then, for our invocation of the gods ; but we
must also call upon our own powers,[2] so that you may follow
most readily and I may give the clearest expression to my
thought on the theme proposed.

[1] ἑπομένως ἡμῖν is usually taken to mean ' consistently with ourselves '
and translated ' consistent with itself '. But this should be ἑπομένως ἡμῖν
αὐτοῖς, and at 29C we are told not to expect αὐτοὺς ἑαυτοῖς ὁμολογουμένους
λόγους. Proclus rightly understood ἑπομένως as ' secondarily ' or ' conse-
quentially ' (as at Ar., *Met.* 1032A, 22 : the word ' being ' applies primarily
to substances, ἑπομένως to other categories) : he writes τοῦτο γάρ ἐστι τὸ
ἀκρότατον θεωρίας τέλος, τὸ εἰς τὸν θεῖον ἀναδραμεῖν νοῦν. . . . δεύτερον δὲ δὴ καὶ
ἑπόμενον τούτῳ τὸ κατὰ τὸν ἀνθρώπινον νοῦν καὶ τὸ τῆς ἐπιστήμης φῶς διαπεράνασθαι
τὴν ὅλην θεωρίαν (I, 221). ἡμῖν depends on κατὰ νοῦν, as at 17C καὶ μάλα γε ἡμῖν
. . . κατὰ νοῦν, 26D εἰ κατὰ νοῦν ὁ λόγος ἡμῖν οὗτος. ἑπομένως replaces the usual
ἔπειτα partly for euphony, partly perhaps to suggest that the discourse, if
pleasing to heaven, should *consequently* be satisfactory to us.

[2] τὸ ἡμέτερον, so A.-H. Cf. τὸ ἐμόν, ' my incapacity ' (19D, 3).

21

27D.   We must, then, in my judgment, first make this distinction : what is that which is always real and has no becoming,

28.   and what is that which is always becoming and is never
real ?   That which is apprehensible by thought with a
rational account is the thing that is always unchangeably
real ; whereas that which is the object of belief together with
unreasoning sensation is the thing that becomes and passes
away, but never has real being.[1]   Again, all that becomes
must needs become by the agency of some cause ; for without
a cause nothing can come to be.   Now whenever the maker
of anything looks to that which is always unchanging and
uses a model of that description in fashioning the form and
quality of his work, all that he thus accomplishes must be

B.   good.[2]   If he looks to something that has come to be and
uses a generated model, it will not be good.

So concerning the whole Heaven or World—let us call it
by whatsoever name may be most acceptable to it [3]—we
must ask the question which, it is agreed, must be asked
at the outset of inquiry concerning anything : Has it always
been, without any source of becoming ; or has it come to
be, starting from some beginning ?   It has come to be ; for
it can be seen and touched and it has body, and all such

C.   things are sensible ; and, as we saw, sensible things, that
are to be apprehended by belief together with sensation, are
things that become and can be generated.   But again, that
which becomes, we say, must necessarily become by the
agency of some cause.   The maker and father of this universe
it is a hard task to find, and having found him it would be
impossible to declare him to all mankind.   Be that as it
may, we must go back to this question about the world :

29.   After which of the two models did its builder frame it—after
that which is always in the same unchanging state, or after
that which has come to be ?   Now if this world is good and

---

[1] With Pr. (i. 240) I take ἀεὶ κατὰ ταὐτὰ ὄν (= τὸ ὄν ἀεί, γένεσιν δὲ οὐκ ἔχον
above) and γιγνόμενον καὶ ἀπολλύμενον, ὄντως δὲ οὐδέποτε ὄν (= τὸ γιγνόμενον μὲν ἀεί,
ὄν δὲ οὐδέποτε above) as the terms to be defined and τὸ νοήσει . . . περιληπτόν
and τὸ . . . δοξαστόν as the definitions demanded in the previous sentence.   Cf. the repetition of this statement below at 28B, 8 ' as we saw,
sensible things, apprehensible by belief together with sensation, are things
that come to be and can be generated '.

[2] καλόν, ' good ', ' satisfactory ', as at Gen. i. 8, ' God saw that it was good '
(εἶδεν ὁ θεὸς ὅτι καλόν, LXX).   The Greek word means also ' desirable ',
' beautiful ', and will be sometimes so translated.

[3] ' Heaven ' (οὐρανός) is used throughout the dialogue as a synonym of
cosmos, the entire world, not the sky.

29.   its maker is good, clearly he looked to the eternal; on the contrary supposition (which cannot be spoken without blasphemy), to that which has come to be. Everyone, then, must see that he looked to the eternal; for the world is the best of things that have become, and he is the best of causes. Having come to be, then, in this way, the world has been fashioned on the model of that which is comprehensible by rational discourse and understanding and is always in the same state.

B.   Again, these things being so,[1] our world must necessarily be a likeness of something. Now in every matter it is of great moment to start at the right point in accordance with the nature of the subject. Concerning a likeness, then, and its model we must make this distinction: an account is of the same order [2] as the things which it sets forth—an account of that which is abiding and stable and discoverable by the aid of reason will itself be abiding and unchangeable (so far as it is possible and it lies in the nature of an account to be incontrovertible and irrefutable, there must be no falling

C.   short of that);[3] while an account of what is made in the image of that other, but is only a likeness, will itself be but likely, standing to accounts of the former kind in a proportion: as reality is to becoming, so is truth to belief. If then, Socrates, in many respects concerning many things —the gods and the generation of the universe—we prove unable to render an account at all points entirely consistent with itself and exact, you must not be surprised. If we can furnish accounts no less likely than any other, we must be content, remembering that I who speak and you my judges

D.   are only human, and consequently it is fitting that we should, in these matters, accept the likely story and look for nothing further.

    SOCR. Excellent, Timaeus; we must certainly accept it as you say. Your prelude we have found exceedingly acceptable; so now go on to develope your main theme.

The chief point established in this prelude is that the visible world, of which an account is to be given, is a changing image or likeness (*eikon*) of an eternal model. It is a realm, not of being, but of becoming. The inference is that no account that we or

---

[1] 'These things' means the whole application to the world of the three foregoing premisses. There should be a full stop before τούτων δὲ ὑπαρχόντων αὖ as before τούτου δ' ὑπάρχοντος αὖ at 30C, 2.

[2] συγγενής in this sense, 31A, 1.

[3] Burnet's text. The uncertainty of the reading does not affect the sense.

anyone else can give of it will ever be more than ' likely '. There can never be a final statement of exact truth about this changing object.

(1) *Being and Becoming*. The first premiss lays down the Platonic classification of existence into two orders. The higher is the realm of unchanging and eternal being possessed by the Platonic Forms. This contains the objects of rational understanding accompanied by a rational account (μετὰ λόγου), namely, the discursive arguments of mathematics and dialectic which yield a securely grounded apprehension of truth and reality.[1] The lower realm contains ' that which is always becoming ', passing into existence, changing, and perishing, but never has real being. This is the world of things perceived by our senses. Sense-perception, as Proclus remarks (i, 249), is ' unreasoning ' in several ways. Sight tells us that an apple is red, smell, that it is fragrant, taste, that it is sweet ; judgment (not sense) tells us that it is an apple. If the sun looks to our eyes a foot in width, the reasoning which assures us that the sun is really larger than the earth will never make it look any bigger. Finally, sense can never apprehend what whiteness *is* ; sight is merely aware, by its own passive affection, that some object is white. The judgments we pass on objects of perception are also unreasoned. They can only state what is, at best, a fact when the judgment is made, though it may cease to be a fact when the object changes. The reason why can only be apprehended by the higher faculty of understanding.

The application of this premiss tells us that the visible world—the object of physics, as distinct from mathematics and dialectic—belongs to the lower order of existence. As having a visible and tangible body, it is an object of perception and of judgments based on perception. Accordingly, it belongs to the realm of ' things that become and can be generated '. It is not eternal, but has a beginning or source of becoming.

The ambiguity of the word ' becoming ' (γένεσις, γίγνεσθαι) gave rise to a controversy on the question whether Plato really meant, as he appears to mean, that the world had a beginning in time. (*a*) A thing *comes into existence* at some time, either suddenly or at the end of a process during which it has been developing (if it is a natural object that is born and grows) or has been fashioned (if it is a thing made by a craftsman). This sense of the word corresponds to the notion of a cause imaged as a father who begets his offspring, or as a maker who fashions his product out of his

---

[1] So at 51E rational understanding is ' always accompanied by a true account ' (ἀεὶ μετ' ἀληθοῦς λόγου), whereas ' true opinion ' can give no rational account of itself (is ἄλογον).

materials. The thing is not there at the beginning of the process; it is there at the end: we can say 'it has become'. (*b*) To 'become' can also mean *to be in process of change*. The word is used of events that 'are happening'; or changes that are 'going on'. It is true that in such 'becoming' something new is always appearing, something old passing away; but the process itself can be conceived as going on perpetually, without beginning or end. For this perpetual becoming the sort of cause needed is not a cause that will start the process at some moment and complete it at another, but a cause that can sustain the process and keep it going endlessly. For such a cause both the images, 'father' and 'maker', are inappropriate. We should need rather to think of some ideal or end, constantly exercising a force of attraction, and perhaps of some impulse in the thing itself, constantly aspiring towards the ideal.

Which kind of becoming did Plato mean to attribute to the physical world? On the surface, he speaks of becoming in the first sense, as if the ordered world came into existence at some time out of a previous state of disorder. It was made by a divine Craftsman, and completed once for all (ἀποτελεῖσθαι, 28B, 1). The question is immediately prejudged where he simply substitutes for the cause of becoming, mentioned in the second premiss, the maker, mentioned in the third. We may compare the division of production in the *Sophist* (265B) into the two kinds, divine and human. Is the coming into being of natural things out of not-being to be attributed to divine craftsmanship (θεοῦ δημιουργοῦντος), 'a causation which, working with reason and art, is divine and proceeds from divinity', or to 'Nature, giving birth to them as a result of some spontaneous cause that generates without intelligence'? Both speakers accept the alternative of divine craftsmanship. The suggestion in either case is that the world had a beginning of existence in time. The only question is, whether it was made upon a divine plan or grew by some blind spontaneous impulse. Similarly in the *Philebus* (26E) we hear that all things that become must have some cause (αἰτία), and this is immediately identified with 'the maker' (τὸ ποιοῦν); 'what becomes' and 'what is made' are two names for one thing. As in the *Timaeus*, the Craftsman (τὸ δημιουργοῦν) is substituted as the equivalent of 'the maker' and of 'the cause'·; and later (28D) this cause is said to be Intelligence, the King of Heaven and Earth.

On the other hand, the statement that the world 'has become' in this sense is formally contradicted by the language of the first premiss, which contrasts with the eternally real 'that which is *always* becoming, but never has real being'. This phrase can only mean what 'becomes' in the second sense, what is everlastingly in process of change. The application of the premiss to the visible world must mean that the world belongs to the lower order of existence so described. This is clear from the reason Plato gives for saying that the world 'has become': 'for it is visible and tangible and has a body and all such things are sensible,' and what is sensible belongs to the lower order, in contrast with the realm of eternal being. Modern authorities, accordingly, agree with Proclus, who contrasts the undivided and eternal being of the intelligible, which is not in time, with the everlasting existence in time of the world. The phrase 'it has become' he understands as meaning that the world possesses 'the existence that is measured by time', a derivative and dependent existence which is not self-sufficing. In this matter Proclus was following the main tradition of the Academy, from Xenocrates, Plato's second successor, onwards.[1] Speaking of contemporaries at the Academy, Aristotle writes: 'They say that in describing the generation of the world they are doing as a geometer does in constructing a figure, not implying that the universe ever really came into existence, but for purposes of exposition facilitating understanding by exhibiting the object, like the figure, in process of formation' (*decaelo*, 279*b*, 33). Professor Taylor finds that 'apparently this tradition was steadily maintained by almost all the Platonists down to the time of Plotinus (in the third century A.D.). Proclus mentions only two dissentients, Plutarch himself and Atticus, an acute and learned Platonist of the age of the Antonines.' Though Aristotle chose to criticise Plato's statement in its apparently literal meaning, his colleague Theophrastus recorded the Academic interpretation as at least possible.[2] This question is, of course, bound up with the question whether the Demiurge, as such, is mythical. If he was not really a 'maker', then there was no moment of creation. We shall presently argue in support of this position. For the present we may accept the Academic tradition.

(2) *The Cause of Becoming.* It follows that the 'cause' of this becoming must be a perpetually sustaining cause. The application of the second premiss merely states that the maker and father of the universe is hard to find and impossible to declare to all men. Plato, in fact, does not pretend to have solved the mystery of the universe; and had he done so, he would not (as the Seventh Letter declares) have set down the solution in writing for all men to read

---

[1] The evidence is collected by Tr., p. 67.

[2] See Tr., p. 69, note. Add the testimony of Albinus (' Alcinous ') : ' When Plato speaks of the world as " generated ", it is not to be understood that there ever was a time when the world did not exist' (*Didasc.*, ch. xiv).

and misunderstand. He was certain that the visible world exhibited the working of a divine intelligence aiming at what is good, and he held it to be of the utmost importance for the conduct of human life that this should be believed. The truth is best conveyed by the image of the divine maker, pictured as distinct (like the human craftsman) from his model, his materials, and his work. But he here warns us not to imagine that, in using this image, he has declared the true nature of the cause. It is to be taken, not literally, but as a poetical figure. The whole subsequent account of the world is cast in a mould which this figure dictates. What is really an analysis of the elements of rational order in the visible universe and of those other elements on which order is imposed, is presented in mythical form as the story of a creation in time. Plato had used a similar device in the *Republic*, where the analysis of the ideal State is cast into the form of a history, starting from the barest necessities of social life and adding storey upon storey to the fabric. He did not mean that any actual state ever came into existence by these stages. What the sustaining cause is, Plato does not tell us and could not tell us without stepping outside the framework of the very myth he is constructing.[1] This question, again, must be held in reserve till we have considered the status of the Demiurge.

(3) *Model and copy*. The third premiss and its application develope further the image of the craftsman and his model. If a craftsman copies an eternal model, his work will be good ; if the model is a generated thing, it will not be so. The reference is to *Republic* x, where the good type of craftsman is the carpenter who makes an actual bed, taking for his model ' the real bed '—a Form which he does not create or invent, but which exists in the nature of things. The bad type is the painter who takes a generated thing, the carpenter's bed, for his model, and produces only an appearance of a thing which itself is not wholly real, an image of an image. The same analogy is drawn in the *Sophist*, 265. The ' divine production of originals ' (the contents of the visible world, made by the Demiurge in the *Timaeus*) is parallel to the human craftsmanship which builds an actual house. In nature there are also dream-images, shadows, reflections, parallel to the painter's

---

[1] Tr. here outruns Plato's exposition : ' The physical world, then, has a maker. . . . This means, exactly as the dogma of creation does in Christian theology, that the physical world does not exist in its own right, but depends on a really self-existing being, the " best ψυχή ", God, for its existence.' I am not theologian enough to know what the orthodox interpretation of the dogma of creation is ; but myriads of Jews and Christians, from Moses to the present day, have believed that *in the beginning* God created the heavens and the earth, and have understood ' beginning ' in a temporal sense.

picture of a house, 'a man-made dream for waking eyes.' In the application here it is argued that, since the visible world is, in fact, good, its maker must have copied a model that is eternal. The world, then, is a copy, an image, of the real. It is not, indeed, like an artist's painting, at the third remove from reality; but on the other hand it is not wholly real. Plato will return to consider the nature of the model at 30C.

Physics only a 'likely story'. Hence follows the conclusion in the last paragraph: the visible world being only a likeness of the real, no account of it can be more than a likely story.

Here it is important to observe that the statement that the world is an image or likeness is independent of the symbolism of the Demiurge creating his work after a model. Not all images are made by artists. Among likenesses, Plato often instances reflections in water or in a mirror. For these all that is required is the thing reflected, the reflection, and the medium which holds it. If the world is an image of that sort, we can dispense with the maker in any literal sense. The realm of Forms will be the original, the visible world the reflection; and the medium will be that Receptacle of becoming which is later provided. We shall, in fact, find in the second part of the dialogue that the three factors needed are Being, Becoming, and Space (52D), and the symbol of the father is there transferred to Being, which serves as the model for Becoming (50D), as if the Forms themselves could be credited with the power to beget Becoming in the womb of Space, or to cast their reflections on that medium. It is true that this symbolism again cannot be taken literally: the Forms can possess no generating power. There must also be a rational soul to cause motion. But, however this moving cause may be mythically represented, the conclusion that the visible world is an image of the eternal remains. It is supported by many passages in other dialogues which are not mythical in form. It is, indeed, the cardinal doctrine of Platonism.

The doctrine carries with it the conclusion that since the world is only a likeness of the real, any account of it can be no more than a 'likely' story. This means that there can be no exact, or even self-consistent, science of Nature. The view is characteristically Platonic. There is no evidence that any of the earlier Pythagoreans doubted the possibility of physical science. On the contrary, Aristotle says that they did not distinguish sensible bodies from the solids of mathematics, as if they agreed with the physical philosophers in general that the visible world is the real.[1] In fact,

[1] *Met.* 989b, 29 ff. This is one of many grounds for rejecting the thesis that the *Timaeus* is merely reproducing fifth-century Pythagoreanism.

they ignored the distinction here drawn by Plato between the field of eternal truth, which includes mathematics, and the region of physics.

In Plato's view there can be no exact science or knowledge of natural things because they are always changing.[1] The objects of mathematical science are timeless and invariable; the things of sense are always in process of becoming. An 'account' must be of the same order as its objects. The objects of physics are of the lower order, apprehensible only by belief involving sense-perception. The substance of our account of them must be related to truth in the same way as Becoming to Being—the relation of a ' likeness ' to reality. This analogy was symbolised in *Republic* vi by the Divided Line, of which the lower part stands for belief (δόξα or πίστις) and its changing objects, the higher part for rational understanding and true reality. There is, accordingly, no such thing as a science of Nature, no exact truth to which our account of physical things can ever hope to approximate.

I here differ from Professor Taylor, who says that the cosmology of the *Timaeus* ' properly speaking is not " science " but " myth ", not in the sense that it is baseless fiction, but *in the sense that it is the nearest approximation which can " provisionally " be made to exact truth* ' (p. 59, my italics). Things which change or move or grow are always ' turning out to be more or less than we had supposed them to be ', and so, in all the natural sciences, we need ' to be perpetually revising and improving on the results ' we have reached about them. ' Physical " laws " are always being revised and " corrected " in the light of newly discovered " facts " or of more accurate measurements of " facts " which were already familiar.' This is a modernism. It implies that there is an exact truth in physics, to which we can constantly approximate. Plato denies this. The becoming which makes physical things unknowable cannot be reduced to their ' *turning out* to be more or less than we had supposed '. A similar confusion is suggested by Burnet's account of the *Timaeus* (*Greek Phil.* i, 340) : Our account of the world ' will be truth in the making, just as the sensible world is the intelligible world in the making '. The phrase ' in the making ' suggests that the sensible world is on the way to become, and might end by becoming, the intelligible world, and similarly that our accounts of it are on the way to become, and might end by becoming, truth. The one result is as impossible as the other.

[1] Aristotle, *Met.* A, 6 : ' Plato, having in his youth become familiar with Cratylus and with the Heraclitean doctrine that all sensible things are ever in a state of flux and there is no knowledge about them, continued to hold these views in later years.'

Plato's word ' likely ' (εἰκώς) has a history going back to Parmenides and Xenophanes, and even to Hesiod. It means ' probable ' or ' plausible '. In Parmenides' poem the goddess, after revealing the nature of the real, turns to the region of false appearance and mortal opinion ; this she calls a ' plausible ' world-order.[1] Xenophanes [2] had used the full phrase : ' Let these be taken as fancies, something like the truth.' Diels compares Parmenides' goddess to Hesiod's Muses, who ' know how to tell many fictions that are like the truth, or, when they will, to speak the truth itself '.[3] Poetry may be fiction that is like the truth, not wholly false. The cosmology of the *Timaeus* is poetry, an image that may come nearer to conveying truth than some other cosmologies. But the truth to which it can approximate is not an exact and literal statement of ' physical laws ', such as modern science dreams of ; it is the truth, firmly believed by Plato, that the world is not solely the outcome of blind chance or necessity, but shows the working of a divine intelligence. Plato would have claimed that, considered as an explanation of sensible appearances, his own theory of the simple primary bodies and their transformations was quite as plausible as the atomic theory of Democritus. He would also have claimed that it was a better explanation and nearer to the truth in that it attributes to intelligible design much that Democritus left to mere chance. This nearness to truth has nothing to do with the modern notion of ' approximation ' indicated, for example, in the following passage : ' The accuracy of the observations is dependent on the limits to the discriminative fineness of our senses, and on the delicacy of our " instruments of precision " . . . When all possible precautions have been taken, the measurements of physical magnitudes are necessarily approximate and would remain so even if we had not to allow for the possible modifications of every hypothesis in natural science by the discovery of new " appearances " '.[4]

---

[1] Parm. 8, 60, τόν σοι ἐγὼ διάκοσμον ἐοικότα πάντα φατίζω, ὡς οὐ μή ποτέ τίς σε βροτῶν γνώμη παρελάσσῃ. A possible interpretation of the second line would assimilate it to Plato's λόγους μηδενὸς ἧττον εἰκότας. Proclus (i, 345) rightly connects Parmenides' distinction between Truth and Belief with Plato's here.

[2] Xenoph. 35, ταῦτα δεδοξάσθω μὲν ἐοικότα τοῖς ἐτύμοισι.

[3] Hesiod, *Theog.*, 27

ἴδμεν ψεύδεα πολλὰ λέγειν ἐτύμοισιν ὁμοῖα,
ἴδμεν δ', εὖτ' ἐθέλωμεν, ἀληθέα γηρύσασθαι.

The phrase in *Odyssey*, 19, 203, means a false but plausible story.

[4] Tr., p. 73. I hope I am not misrepresenting Professor Taylor. These sentences come from a passage which professes to state Plato's conclusion, ' *as we should put it* '. If all that Plato meant by calling physics a ' likely story ' was that natural science must always be provisional and progressive, we should expect him to state what he believed to be the nearest approxima-

The *Timaeus* is a poem, no less than the *De rerum natura* of Lucretius, and indeed more so in certain respects. Both poets are concerned, in the first instance, with our practical attitude towards the world—what we should make of our life there and how face the prospect of death. Lucretius believed that atoms and void are the ultimately real things of which everything that exists is built. Plato denied reality to what is commonly called matter; his real things are the Forms, and the bodies we touch and see are not built of Forms, nor are the Forms in them (52B, C). Accordingly, for Lucretius reality is in the world of sensible things and he can offer statements about its nature which claim to be literally true; for Plato that whole world is an image, not the substance. You cannot, by taking visible things to pieces, ever arrive at any parts more real than the whole you started with. The perfection of microscopic vision can bring you no nearer to the truth, for the truth is not at the further end of your microscope. To find reality you would do better to shut your eyes and think.

There are two senses in which the *Timaeus* is a 'myth' or 'story' ($\mu\tilde{\upsilon}\theta o \varsigma$).[1] One we have already considered: no account of the material world can ever amount to an exact and self-consistent statement of unchangeable truth. In the second place, the cosmology is cast in the form of a cosmogony, a 'story' of events spread out in time. Plato chooses to describe the universe, not by taking it to pieces in an analysis, but by constructing it and making it grow under our eyes. Earlier cosmogonies had been of the evolutionary type, suggesting a birth and growth of the world, due to some spontaneous force of life in Nature, or, as in Atomism, to the blind and undesigned collision of lifeless atoms. Such a story was, to Plato, very far from being like the truth. So he introduced, for the first time in Greek philosophy, the alternative scheme of creation by a divine artificer, according to which the world is like a work of art designed with a purpose. The Demiurge is a necessary part of the machinery, if the rational ordering of the universe is to be pictured as a process of creation in time. But the important point is that, no matter whether you prefer to analyse the world or to construct it piece by piece, the account can never be more than 'likely', because of the changing nature of its object; it can never be revised and amended into exact truth.

We may here read a warning to the interpreter of the *Timaeus*.

tions to truth yet attained, not to be content with 'the best approximations to it which could be expected from a geometer-biologist of the fifth century'. Yet Tr. represents this as 'all that is required by his own principles that' his speaker's theories 'shall be more or less "like" the truth' (p. 19).

[1] Cf. Frutiger, *Mythes de Platon*, 173 ff.

31

Some have regarded the mythical character of the dialogue as a 'veil of allegory', which can be 'stripped off', and have imagined that they could state in literal terms the meaning which Plato has chosen to disguise. It is true that we can say, with a fair degree of certainty, that some features are not to be taken literally. We shall soon find reason to say this much of the Demiurge. But there remains an irreducible element of poetry, which refuses to be translated into the language of scientific prose. Plato declares that his account, so far from being exact, cannot even be consistent with itself. The inexactness and inconsistency are inherent in the nature of the subject; they cannot be removed by 'stripping off the veil of allegory'. An allegory, like a cypher, has a key; the *Pilgrim's Progress* can be retranslated into the terms of Bunyan's theology. But there is no key to poetry or myth.

Plan of the Discourse. The discourse on the nature of the universe and of man which now begins and continues without interruption to the end of the dialogue, is divided into three main sections.

(1) The first (29D–47E) is described as containing the works of Reason (τὰ διὰ Νοῦ δεδημιουργημένα, 47E), those elements in the visible world, and especially in the heavens, which most clearly manifest an intelligent and intelligible design. Here Plato approaches the world (so to say) from above, from the realm of the benevolent maker and the Forms which provide his model. The Demiurge himself is responsible for the main structure and ordered movements of the world's soul and body, and for the creation of the heavenly gods: stars, planets, and Earth. These created gods are then associated in the task of fashioning mankind and the other animals. A preliminary account of the human soul, disordered at its incarnation by the assaults of the material world, leads to the physical mechanism of sense-perception. This is contrasted with the rational purpose of sight and hearing, as revealing the order and harmony which our souls need to relearn and re-establish in themselves. The physical process whereby light acts upon the eyes or sound upon the hearing is a secondary and subordinate type of causation, the means by which the true purpose is attained. Such causation is connected with the notion of Necessity, as opposed to Reason.

(2) The second section (47E–69A) contains 'what comes about of Necessity' (τὰ δι' Ἀνάγκης γιγνόμενα, 47E). Making a fresh start, the discourse plunges into the obscure region of the bodily and of blind causation, approaching the world this time from below. A new factor, Space, is introduced, as the necessary condition or medium in which Becoming images reality. The unlimited and

unordered qualities and powers of the bodily are pictured as a chaos. The Demiurge imposes upon them a rational element of geometrical form in the shapes of the four primary bodies. The properties of these regular figures are then connected with certain qualities in the sensations we receive ; and so, from the opposite pole, we return to the point of contact between the human organism and the outer world, where the first part ended.

(3) In the third section (69A–end), the two strands of rational purpose and necessity are woven together in a more detailed account of the human frame, the working of its organs, and the disorders of body and soul.

I. THE WORKS OF REASON

29D–30C. *The motive of creation*

FORESHADOWING the contrast between rational purpose and the blind operation of Necessity, Plato opens with the creator's motive, the true reason (αἰτία) for the existence of an *ordered* world in the realm of Becoming.

29D. TIM. Let us, then, state for what reason becoming and
 E. this universe were framed by him who framed them. He was good ; and in the good no jealousy in any matter can ever arise. So, being without jealousy, he desired that all things should come as near as possible to being like himself. That this is the supremely valid principle of becoming and of the order of the world, we shall most surely be right to
30. accept from men of understanding. Desiring, then, that all things should be good and, so far as might be, nothing imperfect, the god took over all that is visible—not at rest, but in discordant and unordered motion—and brought it from disorder into order, since he judged that order was in every way the better.

Now it was not, nor can it ever be, permitted that the work of the supremely good should be anything but that which
 B. is best. Taking thought, therefore, he found that, among things that are by nature visible, no work that is without intelligence will ever be better than one that has intelligence, when each is taken as a whole, and moreover that intelligence cannot be present in anything apart from soul. In virtue of this reasoning, when he framed the universe, he fashioned reason within soul and soul within body, to the end that the work he accomplished might be by nature as excellent and

33

30B. perfect as possible.   This, then, is how we must say, according to the likely account, that this world came to be, by
the god's providence, in very truth [1] a living creature with
c. soul and reason.

The Demiurge. The dialogue yields no more information about
the Demiurge than is conveyed in this passage. Here, then, we
may take up the question, how far this figure is mythical and what
it really stands for. The temptation to read into Plato's words
modern ideas that are in fact foreign to his thought has proved too
much for some commentators.

Plato is introducing into philosophy for the first time the image
of a creator god. Recalling the punishment inflicted by jealous
Olympians upon Prometheus for his benefits to mankind, he denies,
as he had done before,[2] the current notion that the gods grudge
to man a perfection and felicity like their own. The kernel of
Plato's ethics is the doctrine that man's reason is divine and that
his business is to become like the divine by reproducing in his
own nature the beauty and harmony revealed in the cosmos,
which is itself a god, a living creature with soul in body and reason
in soul, as here described. Hence he repudiates the old maxim
warning man not to provoke nemesis by harbouring aspirations
too high for mortals. Near the end of the dialogue he explicitly
enjoins the duty of ' thinking thoughts immortal and divine ' and
endeavouring ' to possess immortality in the fullest measure that
human nature permits ' (90C). By calling the Demiurge ungrudging, he may also imply that the imperfection of the world is due
to Necessity, not to the deliberate withholding of any excellence
that it might possess.

This is all that is meant by the statement, in the first paragraph,
that the god is not jealous or grudging. The reader must be warned
against importations from later theology. Professor Taylor, for
instance, after pointing out that Timaeus is thinking of the common
Greek view that the divine ($\tau\grave{o}\ \theta\varepsilon\tilde{\iota}ov$) is grudging in its bestowal
of good things, proceeds : ' So just because God is good, He does
not keep His blessedness selfishly to Himself. He seeks to make
something else as much like Himself in goodness. It is of the very
nature of goodness and love to " overflow ". This is why there
is a world and why, with all its defects, it is " very good " ' (p. 78).
If this is intended as a paraphrase of Plato's words, it is misleading.
There is, in the first place, no justification for the suggestion,

---

[1] It is literally true (not merely ' probable ') that the world is an intelligent
living creature.

[2] *Phaedrus* 247A, $\phi\theta\acute{o}vo\varsigma\ \gamma\grave{a}\rho\ \check{\varepsilon}\xi\omega\ \theta\varepsilon\acute{\iota}ov\ \chi o\rho o\tilde{v}\ \H{\iota}\sigma\tau a\tau a\iota.$

conveyed by 'God' with a capital letter, that Plato was a monotheist. He believed in the divinity of the world as a whole and of the heavenly bodies. The *Epinomis* recommends the institution of a cult of these celestial gods. Neither in the *Timaeus* nor anywhere else is it suggested that the Demiurge should be an object of worship : he is not a religious figure.[1] He must, therefore, not be equated with the one God of the Bible, who created the world out of nothing and is also the supreme object of worship.[2] Still less is there the slightest warrant in Greek thought of the pre-Christian centuries for the notion of 'overflowing love', or love of any kind, prompting a god to make a world. It is not fair either to Plato or to the New Testament to ascribe the most characteristic revelations of the Founder of Christianity to a pagan polytheist.

The nature and position of the Demiurge cannot be finally determined without considering that central utterance of the whole dialogue which declares that the universe is produced by a combination of Reason and Necessity : 'Reason overruled Necessity by persuading her to guide the greatest part of the things that become towards what is best' (48A). When we come to that passage, we shall ask what Necessity stands for, how Necessity can be 'persuaded' by Reason, and why she should need to be persuaded. Further on still (52D), we shall find a more detailed picture of that chaos of disorderly motions and powers which the Demiurge has just been described as 'taking over' and reducing, so far as may be, to order. Necessity and chaos are represented as factors in the visible world which confront the divine intelligence, like the given materials which the human craftsman must use as best he can, though their properties may not be wholly suitable to his purpose. It will be argued that this second factor in the world

---

[1] The 'Maker' in some primitive mythologies has been similarly misinterpreted. Professor Nilsson writes : 'Just as man arranges matters as conveniently as he can to suit his simple needs, building a hut and making his few tools, and just as the advance of culture is brought about by culture-heroes, so, it is said, there was at the beginning of time some one, though much more powerful than man, who arranged the world as conveniently as possible to supply man with all that he needed. This creator, who is found among many primitive peoples, is called by the Australians characteristically enough "the Maker" (*Baiame*). He has also fixed the customs and institutions of the tribe. At first sight it would seem as though we had here a highly developed monotheistic type of divinity, but the idea is in reality due to the indolence of primitive habits of thought. The creator is a mythological, not a religious divinity ; and, therefore, he has no cult and no one troubles about him' (*A History of Greek Religion*, 1925, p. 72).

[2] The contrast between the Demiurge and the Christian Creator is developed in an interesting paper by Mr. M. B. Foster on *Christian Theology and Modern Science of Nature*, Mind XLIV, 439 ff. and XLV, 1 ff.

must not be explained away so as to give Plato's Demiurge the status of the omnipotent Creator of Jewish-Christian theology. We shall find that if Plato's language is to keep any substantial meaning, we must not ascribe to him either the belief in an omnipotent creator or the notion of natural law as a closed system of causes and effects. His Necessity is irregular and disorderly, and not inexorably determined, but open to the persuasion of Reason ; and Reason has need to persuade her, not having unlimited power to compel. This is not easy for us to understand ; but there is no need to explain it away. The omnipotent Creator and the modern notion of natural law were equally foreign to the minds of ancient Greece. Galen truly observed that, with respect to omnipotence, ' the doctrine of Moses differed from that of Plato and of all the Greeks who have correctly approached the study of Nature. For Moses, God has only to will to bring matter into order, and matter is ordered immediately. We do not think in that way ; we say that certain things are impossible by nature and these God does not even attempt ; he only chooses the best among the things that come about ' (*U.P.* xi, 14). To this I would add a quotation from Professor G. C. Field.[1] He points out that omnipotence is incompatible with the ordinary and familiar notion of purpose, which we never regard as a complete and sufficient explanation of anything : ' it is always purpose working in certain materials, or under certain conditions, which make it intelligible why this had to be done rather than that in order to fulfil the purpose '. He concludes that the appeal to purpose as a satisfying principle of explanation ' cannot claim to be decisively established, and if it points to anything, it points in the direction of a God or a Highest Purpose working in a universe which includes him as a part only of the whole, and a part which, however powerful and important, is at some point limited and restricted by other elements in the whole. I do not myself see any insuperable philosophic objection to such an idea. It appealed, if I interpret him aright, to Plato, in the final development of his doctrine.'

This conclusion is unquestionably consistent with what Plato actually says. Again and again, throughout the *Timaeus*, we are told that the benevolent Demiurge designed that such and such an arrangement should be ' as good as possible ', with the clear implication that his purpose was restricted by that other factor called Necessity. We must accept this, on pain of reducing much of his language to nonsense. There is nothing against it, except the desire to bring Plato into conformity with Christian doctrine or

[1] From an interesting essay on Modern Proofs of the Existence of God in *Studies in Philosophy* (1935), pp. 122 ff.

with some modern form of idealism. If this desire is brought into consciousness, it can be resisted ; for to yield to it is to do Plato no service. If we make his Demiurge omnipotent and at the same time attribute to him the modern conception of natural law, we shall involve him in the nineteenth-century ' conflict of religion and science ' ; for this arose largely out of the attempt to believe at once in the providence of an all-powerful God and in a completely determined chain of causes and effects which left no room for his intervention.

Here, then, we may conclude that Plato's Demiurge, like the human craftsman in whose image he is conceived, operates upon materials which he does not create, and whose inherent nature sets a limit to his desire for perfection in his work. He has been pictured as confronted with ' all that is visible ' in a chaos of disorderly motion. For this disorder he is not responsible, but only for those features of order and intelligible design which he proceeds to introduce, ' so far as he can '. These form the subject of the first part of the discourse. In the second part it will be made clear that the Demiurge is not the sole cause of Becoming. There are secondary causes, partly but not wholly amenable to the persuasion of Reason. Nor does the Demiurge create that Receptacle of Becoming in which the images of the Forms are mirrored. This is not mentioned among the works of Reason ; it is as independent of the Demiurge as the world of Forms. The Forms, again, he does not create ; they are not made or generated, but eternally real and self-subsisting. The function of the Demiurge is to contribute an element of order to Becoming, because an ordered world will be more ' like himself ', that is to say, better, than a disorderly one.

We shall be led to the conclusion that both the Demiurge and chaos are symbols : neither is to be taken quite literally, yet both stand for real elements in the world as it exists. If there was never a moment of creation, chaos cannot have existed before that moment ; and this part of the mythical imagery is not to be taken at its face value. But what was later called ' matter ' is the subject of the second part of the dialogue, not to be anticipated here. We can only remark that chaos, if it never existed before cosmos, must stand for some element that is now and always present in the working of the universe. Its nature will be disclosed in the analysis of ' what comes about of Necessity '.[1]

[1] Against Plutarch and Atticus, who took the pre-existing chaos literally, Proclus (i, 382) cites Porphyry and Iamblichus : ' They say that Plato, desiring to exhibit the Maker's providence descending into the universe, the government of reason and the presence of soul, and all the great benefits

It may equally be said of the Demiurge that, as a mythical symbol, he must stand for something that is seriously meant. He is mythical in that he is not really a creator god, distinct from the universe he is represented as making. He is never spoken of as a possible object of worship ; and in the third part of the dialogue the distinction between the Demiurge and the celestial gods, whom he makes and charges with the continuation of his work, is obliterated.[1] The evidences of design in the human frame are there attributed sometimes to ' the god ', sometimes to the celestial gods, who are the stars, planets, and Earth. On the other hand, there is no doubt that he stands for a divine Reason working for ends that are good. The whole purpose of the *Timaeus* is to teach men to regard the universe as revealing the operation of such a Reason, not as the fortuitous outcome of blind and aimless bodily motions. If this Reason is not a creator god, standing apart from his model and materials, where is it to be found ? Now this is precisely the question which Plato has refused to answer. It is a hard task, he says, to find the maker and father of this universe, and having found him it would be impossible to declare him to all mankind. This can only mean that the mythical imagery is not a ' veil of allegory ' that we can tear aside and be sure of discovering behind it a literal meaning which Plato himself would endorse. Commentators have not hesitated to essay this ' impossible ' task ; but the bewildering variety of their disclosures lends little encouragement for a further venture, and gives rise to a suspicion that each has found what he set out to look for.

We shall be on safer ground if we turn from the maker to consider what Plato says here about his work. The visible universe is a living creature, having soul (ψυχή) in body and reason (νοῦς) in soul. It is called a god (34B) in the same sense in which the term is applied to the stars, planets, and Earth—the ' heavenly gods '. All these gods are everlasting, coeval with time itself ; though theoretically dissoluble, because composite of reason, soul, and body, they will never actually be dissolved (41B). Man is also composed of reason, soul, and body ; but his body will be dissolved

---

these confer upon the cosmos, first contemplates the whole bodily frame by itself in its disharmony and disorder, so that you may see also by itself the order due to soul and to the disposition of the creator, and distinguish the nature of the bodily in itself from the nature of the created order. The cosmos itself exists everlastingly ; but the discourse distinguishes that which becomes from its maker and introduces in temporal order things that coexist simultaneously, because whatsoever is generated is composite.'

[1] On one such passage Tr. says : ' Passages like the present show how far he is from meaning his polytheistic phrases to be taken *au pied de la lettre* ' (p. 549). Substitute ' monotheistic ', and the remark will be equally true.

back into the elements, and the two lower parts of his soul are also mortal. Only the divine reason in him is imperishable. There is thus a contrast between macrocosm and microcosm, but also an analogy, which runs all through the discourse. The world itself, like the heavenly gods and man, is divine because it contains the divine element, reason. Reason, moreover, as Plato says here and elsewhere, ' cannot be present in anything apart from soul ' : if it is ' present ' in the body of the universe and in man's body, that body must be alive, endowed with soul, which is defined in the *Laws* and the *Phaedrus* as the self-moving source of all motion. The statement is consistent with the belief that the reason, as divine and immortal, can nevertheless exist in separation from the body and divested of the mortal parts of soul. There is, then, in the soul and body of the universe a divine Reason analogous to man's ; and we shall find that the unchanging movement of its thought is symbolised, or even visibly embodied, in the circular revolutions of the heavenly gods and of the universe as a whole.

We may ask how this divine Reason in the world is related to that divine Reason which is symbolised by the Demiurge. Can we simply identify the two ? In that case the Demiurge will no longer stand for anything distinct from the world he is represented as making. The desire for goodness will then reside in the World-Soul : the universe will aspire towards the perfection of its model in the realm of Forms, and the model will hold a position analogous to that of Aristotle's Unmoved Mover, who causes motion as the object of desire.[1] But this solution of the problem is no more warranted by Plato himself than others that can be supported by a suitable selection of texts. We shall do better to hold back from this or any other conclusion and confine our attention to the world with its body and soul and the reason they contain.

30C–31A. *The creator's model*

The visible world has been declared to be a living creature made after the likeness of an eternal original. This model is now further described. It can only be the ideal Living Creature in the world of Forms, not to be identified with any species of animate being, but embracing the ideal types of all such species, ' all the intelligible living creatures '.

30C. This being premised, we have now to state what follows next : What was the living creature in whose likeness he

---

[1] It has been observed that Aristotle's personified Nature, who aims at a purpose and does nothing in vain, may be regarded as equivalent to Plato's Demiurge.

30C. framed the world? We must not suppose that it was any creature that ranks only as a species [1]; for no copy of that which is incomplete can ever be good. Let us rather say that the world is like, above all things, to that Living Creature of which all other living creatures, severally and in their families, are parts. For that embraces and contains within

D. itself all the intelligible living creatures, just as this world contains ourselves and all other creatures that have been formed as things visible. For the god, wishing to make this world most nearly like that intelligible thing which is best and in every way complete, fashioned it as a single visible living creature, containing within itself all living

31. things whose nature is of the same order.

We have seen that, although the creator god, as such, is a mythical figure, the relation of likeness to model none the less subsists between the visible world and the intelligible. The model is not a piece of mythical machinery. The visible world, being ' in very truth ' a living creature with soul and body, has for its original a complex Form, or system of Forms, called ' the intelligible Living Creature '. This is a generic Form containing within itself the Forms of all the subordinate species, members of which inhabit the visible world. The four main families,[2] ' contained in the Living Creature that truly is ', are enumerated at 39E : the heavenly gods (stars, planets, and Earth), the birds of the air, the fishes of the sea, and the animals which move on the dry land. These main types, as well as the indivisible species of living creatures and their specific differences, are all, in Platonic terms, ' parts ' into which the generic Form of Living Creature can be divided by the dialectical procedure of Division (διαίρεσις). The generic Form must be conceived, not as a bare abstraction obtained by leaving out all the specific differences determining the subordinate species, but as a whole, richer in content than any of the parts it contains and embraces.[3] It is an eternal and unchanging object of thought, not itself *a* living creature, any more than the Form of Man is *a* man. It is not a soul, nor has it a body or any existence in space or time. Its eternal being is in the realm of Forms.

Plato does not say, here or elsewhere, that this generic Form of Living Creature contains anything more than all the subordinate generic and specific Forms and differences that would appear in

---

[1] μέρος or μόριον, ' part ', is Plato's normal term for ' species '.

[2] This is the probable meaning of γένη in καθ' ἕν καὶ κατὰ γένη (30A, 6) ; καθ' ἕν will mean the Forms of indivisible species, a class of Forms explicitly recognised at *Philebus*, 15A.

[3] Cf. F. M. Cornford, *Plato's Theory of Knowledge* (1935), pp. 268 ff.

the complete definitions of all the species of living creatures existing in our world, including the created gods. We have no warrant for identifying it with the entire system of Forms, or with the Form of the Good in the *Republic*, or for supposing that it includes the moral Forms of dialectic or the mathematical Forms, or even the Forms of the four primary bodies, whose existence is specially affirmed at 51B ff. Plato looks upon the whole visible universe as an animate being whose parts are also animate beings. The intelligible Living Creature corresponds to it, whole to whole, and part to part. It is the system of Forms that are, together with the Forms of the four primary bodies, relevant to a physical discourse, because they are the patterns of which the things we see and touch are sensible images, coming to be and passing away in time and space. We are not here concerned with the moral Forms, of which there are no sensible images (*Phaedrus* 250D).

The model, as strictly eternal, is independent of the Demiurge, whose function is to be the cause, not of eternal Being, but only of order in the realm of Becoming. However we may interpret the divine Reason symbolised by the Demiurge, this model is one among the objects of its thought. It is the ideal, whose perfection the visible universe, as a living being, is to reproduce in its own structure, so far as is permitted by the conditions of temporal existence in space. 'Intelligible' means that it is an object of . rational thought, divine or human. Plato gives no more ground for supposing that the divine Reason creates its objects by 'thinking' them than for supposing that our own reasons create these same objects when we think of them. The Forms are always spoken of as existing eternally in their own right.

31A–B. *One world, not many*

The concluding words of the last paragraph spoke of the world as a *single* living creature. This suggests the possibility that there should be more than one copy of the model—a plurality of visible worlds.

31A. Have we, then, been right to call it one Heaven, or would it have been true rather to speak of many and indeed of an indefinite number? One we must call it, if we are to hold that it was made according to its pattern. For that which embraces [1] all the intelligible living creatures that there are, cannot be one of a pair; for then there would have to be

---

[1] περιέχειν is used of the whole which 'includes' all its parts, e.g. *Soph.* 253D. This use has nothing to do with the Ionian use of περιέχον for the element which extends beyond and 'encompasses' the world, referred to in Tr.'s note.

31A.   yet another Living Creature embracing those two, and they
        would be parts of it; and thus our world would be more
        truly described as a likeness, not of them, but of that other
  B.   which would embrace them. Accordingly, to the end that
        this world may be like the complete Living Creature in
        respect of its uniqueness, for that reason its maker did not
        make two worlds nor yet an indefinite number; but this
        Heaven has come to be and is and shall be hereafter one and
        unique.[1]

There is no satisfactory evidence for the doctrine of a plurality
of coexisting worlds before the atomism of Leucippus in the second
half of the fifth century.[2] The Atomists' belief in innumerable
worlds, some always coming into existence, others passing away,
was an inference from their assertion of a strictly infinite void
partly occupied by an illimitable number of atoms in motion.
It was probable, they argued, that world-forming vortices would
arise at any number of different places. Granted that our world
is finite, that there is unlimited space outside its boundary, and
that there are materials left over, from which other worlds might
be formed, why should there not be any number of copies of the
same model? The world, according to Plato, is finite. On the
other hand, like Aristotle, he would have denied an unlimited void
outside; and he certainly denies that any materials are left over
(32c ff.). The point, however, is not argued on those grounds here.
He is not offering a proof that there cannot be more than one
world; he merely asserts that only one was made, because it
seemed better that the copy should be unique, like the model.
His argument is: (1) The model must be all-inclusive (παντελές),
containing all the species of animal that there are; otherwise our
world, being a copy of it, would not be as perfect as it might be.
(2) There cannot be a second all-inclusive model; for then the two
models would be duplicate instances of the same Form, and that
Form would become the true model. The model, therefore, is

[1] I cannot see in γεγονὼς ἐστιν καὶ ἐτ' ἐσται any more than 'has been
and is and shall be' or 'is at all times', though the word γεγονὼς preserves
the fiction of creation. Cf. 38c γεγονώς τε καὶ ὢν καὶ ἐσόμενος. Tr. discovers an allusion to a doctrine of γένεσις εἰς οὐσίαν in the *Philebus*, which
'Timaeus is not allowed to explain but only to imply', because 'the clear
conception of a γεγενημένη οὐσία is a result of Plato's own personal thought',
which a fifth-century Pythagorean has no business to know about. But
the doctrine of the *Philebus* should not be read into this simple phrase. All
the emphasis falls on 'one and unique', as in Tr.'s translation: 'sole and
single this our heaven came into being, sole it is, and sole it shall remain'.
[2] I have discussed this question in detail in *Classical Quarterly*, XXVIII
(1934), pp. 1 ff.

(like every other Form) unique. (3) The last sentence does not say that there cannot be more than one copy of a unique model (which is obviously untrue),[1] but that the creator made only one copy ' *in order that* ' the world should resemble its model ' in respect of its uniqueness '. Uniqueness is a perfection, and the world is the better for possessing it. One reason why it is better is given later : if the world were not unique, there would be body left outside it, whose ' strong powers ' might impair its life and even destroy it (33A). It is for this reason that this world ' having come into being one and unique, is and shall be so hereafter '. These final words deny both the innumerable coexisting worlds of the Atomists and the succession of single worlds which had figured in some Ionian systems and in Empedocles. Plato's single world is everlasting.

THE BODY OF THE WORLD

31B–32C. *Why this consists of four primary bodies*

THE next section (31B–34A) is concerned with the body of the Universe. Although soul is later declared to be prior to body, the making of the body is taken first for convenience. The present paragraph explains why not less than four primary bodies—fire, air, water, earth—were required, in order to give it the highest measure of unity. This attribute of internal unity follows naturally after the unity, in the sense of uniqueness, asserted in the previous paragraph. The primary bodies are here imagined as materials ready to be ' put together ' (συνιστάναι) by the builder's hand. The formation of them by the imposition of regular geometrical shape upon their unordered motions and powers belongs to the second part of the dialogue. There is no reference here to those geometrical shapes, of which nothing has yet been heard. All that the Demiurge does now is to fix their quantities in a certain definite proportion. This is an element of rational design in the structure of the world's body, and it belongs here among the works of Reason.

31B.   Now that which comes to be [2] must be bodily, and so visible and tangible ; and nothing can be visible without fire, or

---

[1] There is, accordingly, no ground for Tr.'s accusation that Plato has ' confused the principle of the " uniformity " of nature with the assertion that there is only one " stellar system " ' (p. 85).

[2] If τὸ γενόμενον means ' the world which came into being ' we should expect ἔδει, and perhaps τ' ἔδει should be read for τε δεῖ (cf. Chalcidius, *erat merito futurus* and 32B στερεοειδῆ γὰρ αὐτὸν προσῆκεν εἶναι). Pr. ii, 3[30] (lemma) has γιγνόμενον, which suits the present δεῖ. Contrast his paraphrase, ἐπειδὴ γὰρ ἔδει τὸν κόσμον ὄντα γενητὸν ὁρατὸν εἶναι καὶ ἁπτόν (ii, 17[7]).

31B.   tangible without something solid,[1] and nothing is solid without earth. Hence the god, when he began to put together the body of the universe, set about making it of fire and earth. But two things alone cannot be satisfactorily united

C.   without a third ; for there must be some bond between them drawing them together. And of all bonds the best is that which makes itself and the terms it connects a unity in the fullest sense ; and it is of the nature of a continued geometrical proportion [2] to effect this most perfectly. For whenever, of

32.   three numbers, the middle one between any two that are either solids (cubes ?) or squares [3] is such that, as the first is to it, so is it to the last, and conversely as the last is to the middle, so is the middle to the first, then since the middle becomes first and last, and again the last and first become middle, in that way all will necessarily come to play the same part towards one another, and by so doing they will all make a unity.

Now if it had been required that the body of the universe should be a plane surface with no depth, a single mean

B.   would have been enough to connect its companions and itself ; but in fact the world was to be solid in form, and solids are always conjoined, not by one mean, but by two. Accordingly the god set water and air between fire and earth, and made them, so far as was possible, proportional to one another, so that as fire is to air, so is air to water, and as air is to water, so is water to earth, and thus he bound together the frame of a world visible and tangible.

For these reasons and from such constituents, four in

C.   number, the body of the universe was brought into being, coming into concord by means of proportion, and from these it acquired Amity,[4] so that coming into unity with

[1] Solid, i.e. resistant to touch (Pr. ii, 12[11]).

[2] That ἀναλογία means this type of proportion *par excellence* will be explained below.

[3] The reason for taking the genitives εἴτε ὄγκων εἴτε δυναμέων ὡντινωνοῦν as depending on τὸ μέσον will be explained below (p. 47). Grammatically, the words can be construed : (1) ' Whenever of any three numbers, whether solids or squares, the middle one is such . . .' (So Heath, A.-H.), or (2) ' Whenever of any three numbers or solids or squares the middle one is such ' . . . , taking ' numbers ' to mean numbers that are neither squares nor solids.

[4] A reference to the *Philia* of Empedocles' system. But there is no contrary principle of *Neikos* in Plato's scheme, and hence no periodic destruction of the world. Cf. *Gorg.* 508A : the wise say that heaven and earth, gods and men, are held together by φιλία and κοσμιότης—a truth which has escaped Callicles because he has neglected geometry and not perceived the significance of *geometrical proportion* (ἡ ἰσότης ἡ γεωμετρική).

32C. itself it became indissoluble by any other save him who bound it together.

Empedocles had taken the four elements as given fact; Plato deduces the need of four primary and simple bodies by an argument. (1) There must be two (not one primary form of matter, as the Ionian monists had held), because fire is needed to make the world's body visible, earth to make it resistant to touch. Fire and earth had been commonly regarded as the two extreme elements, since fire belongs to the heavens, and air and water are between Heaven and Earth. (2) But two cannot hold together without a third to serve as bond. The three must be in proportion, and the most perfect bond is that proportion which makes the most perfect unity out of mean and extremes. (3) The most perfect type of proportion is the continued geometrical proportion (ἀναλογία), which Plato next proceeds to define. That geometrical proportion was the proportion *par excellence* and primary, all other types of proportion being derivable from it, was stated by Adrastus, the Peripatetic (early second century A.D.), who wrote a commentary on the *Timaeus*, parts of which are preserved by Theon of Smyrna.[1] If we ignore for the moment the words εἴτε ὄγκων εἴτε δυνάμεων, which specify certain classes of numbers,[2] the sentence simply gives a definition of a continued geometrical proportion with three terms. Take the progression 2, 4, 8 for purposes of illustration. The terms are related so that 'as the first is to the middle, so is the middle to the last (2 : 4 = 4 : 8), and conversely, as the last is to the middle, so is the middle to the first' (8 : 4 = 4 : 2). Then 'the middle becomes first and last, and again the last and the first both become middle' (4 : 8 = 2 : 4 or 4 : 2 = 8 : 4). Thus any of the three can stand as first or as last or as middle, and the unity they constitute is as perfect as possible. (4) Three terms, however, are not enough, because all the primary bodies are solids, and must accordingly be represented by solid numbers (a solid number

---

[1] The statement is repeated by Nicomachus (*Introd. Arith.* ii, 24, p. 126 Hoche), by Iamblichus (*in Nicom. Ar. Introd.*, p. 100 Pistelli, as 'an opinion of the ancients', and p. 104 citing our passage), and by Pr. ii, 20 (referring to Nicomachus). Cf. Heath, *Euclid*, ii, 292. Pr. records the (obviously correct) view that Plato here speaks of geometrical proportion only. Others, with whom Proclus himself agrees, made an unfortunate attempt to drag in arithmetical and harmonic proportion, connected with the false notion that δυνάμεις in our passage has a physical sense, and means the sensible qualities elsewhere called 'powers' (cf. *Chalcid*, p. 86, and *Occelus*, ii). Such qualities (pairs of opposites) form, in Plato's view, an ἄπειρον, and could not possibly stand as terms in a numerical proportion.

[2] These words are omitted by Tim. Locr/ 95, who has simply τριῶν ὡντινωνῶν ὅρων.

is the product of three numbers). To connect two plane numbers
a single mean is sufficient ; but if fire and earth, the extremes, are
to be connected, two means will be required.

As the ancients saw, this last statement is true only if the plane
and solid numbers in question are ' similar ' (i.e. having their sides
proportional)—a class which includes all squares and cubes. Some
held that Plato meant it to be taken for granted that the terms
in his proportion are all similar numbers [1] ; but he has not said so.
It has, accordingly, been inferred that the words εἴτε ὄγκων εἴτε
δυνάμεων, which serve no purpose in a mere description of a geometrical proportion with three terms, were inserted in order to
restrict the numbers in question to cubes and squares. Sir Thomas
Heath writes : [2]

> ' It is well-known that the mathematics of Plato's *Timaeus*
> is essentially Pythagorean. It is therefore *à priori* probable (if
> not perhaps quite certain) that Plato πυθαγορίζει even in the
> passage (32A, B) where he speaks of numbers " whether solid or
> square " in continued proportion, and proceeds to say that
> between *planes* one mean suffices, but to connect two *solids*
> two means are necessary. This passage has been much discussed,
> but I think that by " planes " and " solids " Plato certainly
> meant *square* and *solid numbers* respectively, so that the allusion
> must be to the theorems established in Eucl. viii, 11, 12, that
> between two square numbers there is one mean proportional
> number and between two cube numbers there are two mean
> proportional numbers.'

In a note Heath adds :

> ' It is true that *similar* plane and solid numbers have the
> same property (Eucl. viii. 18, 19) ; but, if Plato had meant
> similar plane and solid numbers generally, I think it would have
> been necessary to specify that they were " similar ", whereas,
> seeing that the *Timaeus* is as a whole concerned with regular
> figures, there is nothing unnatural in allowing *regular* or *equilateral*
> to be understood. Further, Plato speaks first of δυνάμεις and
> ὄγκοι and then of " planes " (ἐπίπεδα) and " solids " (στερεά) in
> such a way as to suggest that δυνάμεις correspond to ἐπίπεδα and
> ὄγκοι to στερεά. Now the regular meaning of δύναμις is *square*
> (or sometimes *square root*), and I think it is here used in the
> sense of *square*, notwithstanding that Plato seems to speak of
> *three* squares in continued proportion, whereas, in general, the

[1] See Pr. ii, 29[18] and 33[20] (quoting Democritus, the third-century Platonist).
[2] *Thirteen Books of Euclid*, ii, p. 294.

46

mean between two squares as extremes would not be square but oblong. And, if δυνάμεις are squares, it is reasonable to suppose that the ὄγκοι are also equilateral, i.e. the " solids " are cubes.'

Elsewhere [1] Heath writes :

' By *planes* and *solids* he [Plato in this passage] really means square and cube numbers, and his remark is equivalent to stating that, if p^2, q^2 are two square numbers,

$$p^2 : pq = pq : q^2,$$

while, if p^3, q^3 are two cube numbers,

$$p^3 : p^2q = p^2q : pq^2 = pq^2 : q^3,$$

the means being of course in continued geometric proportion. Euclid proves the properties for square and cube numbers in viii. 11, 12 and for similar plane and solid numbers in viii. 18, 19. Nicomachus (ii. 24, 6, 7) quotes the substance of Plato's remark as a " Platonic theorem ", adding in explanation the equivalent of Eucl. viii. 11, 12.'

This interpretation of the ambiguous words ὄγκοι and δυνάμεις as ' cubes ' and ' squares ' seems to be better supported than any other. It rules out the notion that ὄγκοι and δυνάμεις are alternatives to ἀριθμοί. They are subdivisions of ' numbers ', restricting the statement to cubes and squares, for the sake of the subsequent statement about one mean connecting squares, two means connecting cubes. The objection stated by Heath, that ' Plato seems to speak of *three* squares in continued proportion, whereas in general the mean between two squares as extremes would not be square but oblong ', can be obviated by construing the genitives εἴτε ὄγκων εἴτε δυνάμεων ὡντινωνοῦν not (as is commonly done) as in apposition to ἀριθμῶν, but as depending on τὸ μέσον. The effect is to make the limitation to cubes and squares apply only to the extremes. Here, as in many other places, Plato is compressing his statement of technical matters to such a point that only expert readers would fully appreciate his meaning.

The interpretation can be further supported by a consideration of Adrastus' treatment of geometrical proportion.[2] He says that geometrical proportion is the only proportion in the full and proper sense (κυρίως) and the primary one, because all the others require it, but it does not require them. The first ratio is equality $\left(\dfrac{1}{1}\right)$, the element of all other ratios and of the proportions they yield.

[1] *Greek Mathematics*, i. 89.
[2] Theon (p. 177, Dupuis) quotes the passage in full. It is presumably taken from Adrastus' commentary on our passage.

He then derives a whole series of geometrical proportions from ' the proportion with equal terms' (1, 1, 1) according to the following law :

Given three terms in continued proportion, if you take three other terms formed of these, one equal to the first, another composed of the first and the second, and another composed of the first and twice the second and the third, these new terms will be in continued proportion.

In this manner, from the proportion with equal terms arises the double proportion, and from that the triple, and so on, as follows. Take the equal proportion with the smallest possible terms, 1, 1, 1. Then take three terms according to the above rule :

$$1,\ 1 + 1 = 2,\ 1 + 2 + 1 = 4.$$

This is the double proportion, 1, 2, 4 . . . etc.   Now take 1, 2, 4 and proceed in the same way :

$$1,\ 2 + 1 = 3,\ 1 + 4 + 4 = 9.$$

This is the triple proportion 1, 3, 9 . . . etc.   By continuing the process we obtain :

$$
\begin{array}{ccc}
1, & 1, & 1 \\
1, & 2, & 4 \\
1, & 3, & 9 \\
1, & 4, & 16 \\
1, & 5, & 25 \\
1, & 6, & 36 \\
1, & 7, & 49 \\
1, & 8, & 64 \\
1, & 9, & 81 \\
1, & 10, & 100
\end{array}
$$

(Note that Adrastus stops at the perfect number 10.[1])   He then shows how the other, less perfect, kinds of proportion can be derived from these geometrical proportions.

The numbers in the third column are squares ($\delta\nu\nu\acute{\alpha}\mu\epsilon\iota\varsigma$), those in the second column are the roots of these squares.   Square roots also were sometimes called $\delta\nu\nu\acute{\alpha}\mu\epsilon\iota\varsigma$.   The underlying notion seems to be that any number (represented by a line) has, in itself and without the aid of any other factor, the *power* of multiplying itself or generating its own square by advancing as far as its own length into the second dimension.   Hence a line is said $\delta\acute{\nu}\nu\alpha\sigma\theta\alpha\iota$ the square

---

[1] Cf. Pr. i, 147, ἡ δὲ ἐσχάτη πρόοδος τῆς δεκάδος ὑπέστησε τὸν χίλια στερεὸν ἀριθμόν.

plane figure it thus generates.[1] So the root number is the *first* '*power*', δύναμις; the corresponding line is properly called δυναμένη. Δύναμις is more commonly applied to the square, in which this potency of the root is developed or deployed. Hence the square is the '*second power*'. The square contains the power that can be further deployed when the square advances into the third dimension and produces the cube, or *third power*.[2] If we now continue Adrastus' geometrical proportions, we shall next reach the cube. Taking the double and triple proportions, we have

$$1, 2, 4, 8$$
$$1, 3, 9, 27$$

These are the two series that Plato takes later (35B) as the basis for the harmony of the World-Soul. Both series emanate from unity, in which all the 'powers' concerned are conceived as gathered up. The series proceed through the first even, and the first odd, number to their squares and cubes. Plato's later use of these two progressions makes it probable that he had them in mind in our passage.[3] He would certainly choose a progression of what was held to be the most perfect type.[4]

Nicomachus, in his chapter on continuous geometrical proportion (ii, 24), repeats that this is the only proportion in the most proper sense (κυρίως καλουμένη) and gives the same examples: 'the numbers proceeding from unity according to the double proportion':

$$1, 2, 4, 8, 16, 32, 64 \ldots$$

and the triple proportion:

$$1, 3, 9, 27, 81, 243 \ldots$$

and so on with the quadruple proportion, etc. He points out that the terms in these proportions have the properties Plato mentions, and later speaks of 'the Platonic theorem, that the plane numbers

---

[1] Plato, *Theaet.* 148B, δυνάμεις, ὡς μήκει οὐ συμμέτρους ἐκείναις, τοῖς δ' ἐπιπέδοις ἃ δύνανται. Alex. in *Met.* 1019b, 32.

[2] The *Epinomis* 990D calls cube numbers τοὺς τρὶς ηὐξημένους καὶ τῇ στερεᾷ φύσει ὁμοίους. At *Rep.* 528B stereometry is described as concerned with '*cubic* increase (κύβων αὔξην) and that which has depth', as if the cube were the primary solid. See Stenzel, *Zahl u. Gestalt* 89 ff.

[3] Cf. also *Epinomis* 991A. 'The first progression of the double proceeds in the integer series (κατ' ἀριθμόν) in the ratio 1 : 2 ; double is the ratio of their second powers (ἡ κατὰ δύναμιν) ; the progression of the solid and tangible is again a double, the progression from one to eight' (trans. Harward). This progression 1, 2, 4, 8 is then used to construct the musical scale.

[4] It would not occur to the modern mathematician, who uses algebraic symbols, that one type of geometrical progression could be more perfect or better deserving of the name than another. For this reason algebraic symbols should not be employed in interpreting such a passage as ours.

are held together by one mean, the solids by two standing in proportion : for between two consecutive *squares* will be found only one mean preserving the geometrical proportion . . . and between two consecutive *cubes* only two '.

This is true of all proportions of the above pattern : e.g.

root	square	cube	square	solid	{square { cube	solid	square	cube
2	4	8	16	32	64	128	256	512...
	(2^2)		(4^2)		(8^2)		(16^2)	
		(2^3)			(4^3)			(8^3)

The special points of this pattern are : (1) All the plane numbers are squares ; there are no oblongs. Oblongs, such as 6 (2×3) appear only in geometrical progressions of a less perfect kind (e.g. $4 : 6 = 6 : 9$), which do not proceed by the self-multiplication of a single root number, but involve a second root. Also such progressions cannot be continued to four and more terms without introducing fractions. If Plato had the perfect pattern in mind, he could substitute ' plane ' for ' square ', as he does. Each two successive planes (squares) are connected by a single mean. (2) All the numbers which are not squares are solid ; and each two successive *cubes* are connected by two means. If ὄγκοι does mean ' cubes ', then the ' solids ' of the last sentence have been restricted to cubes by the insertion of εἴτε ὄγκων εἴτε δυνάμεων, and we must understand τὰ στερεά as meaning ' the solids above spoken of as ὄγκοι,' to the exclusion of the non-cube solids. The last sentence will then be true and all will be in order.[1]

[1] The only evidence I can find for ὄγκος as the older term for κύβος is in Simplicius, *Phys.* 1016, 23, commenting on Zeno's paradox of the Stadium, where Zeno appears to have used ὄγκοι for the bodies which pass one another on the race-course (Ar., *Phys.* 239b, 33). Simplicius records that Eudemus, in his account of Zeno's argument, substituted κύβοι for ὄγκοι. Eudemus may have understood ὄγκοι in Zeno as meaning ' cubes ' (the obviously appropriate figure). It may be added that some of the older terms in Greek mathematics have biological associations : χροία (skin) for *surface*, δύναμις (power) for *square*, αὔξη (growth) for *dimension*, σῶμα (body) for *solid*. These terms were applied to numbers as well as to figures. They were taken from living things and fit in with the Pythagorean conception of the unit as the ' seed (σπέρμα) or eternal root (ῥίζα) from which ratios grow or increase (αὔξονται) reciprocally on either side ' (*Iambl. in Nicom.*, p. 11 Pistelli). The unit contains potentially (δυνάμει) all the forms of even and odd number, ' as being a sort of fountain (πηγῇ) or root (ῥίζῃ) of both kinds ' (*ibid.*, p. 15). If the *seed* or *root* contains the latent *power* (δύναμις) of *growth*, its first *increase* is the line ; its second, the second *power* of the square, a *skin* (surface). The most natural term for the third increase would be ὄγκος, ' swelling ', ' bulk '. The square has the power of ' swelling itself out ' (ὀγκοῦσθαι) into the cube— the first *body* reached in the above progressions. When geometry became distinct from arithmetic, a fresh series of terms was borrowed from the

Plato has not indicated what are the quantities between which his geometrical proportion holds.[1] It cannot be connected with the construction of the four regular solids which are later assigned to the primary bodies; the proportion does not fit any of the sets of numbers there involved. It may be conjectured that the quantities in question are the total volumes of the four primary bodies. Empedocles had made his four elements equal in amount;[2] but since his time it had been realised that the world was much larger than had been supposed.[3] Since the heavenly bodies are composed mostly of fire, it is natural to suppose that the total volume of fire is much greater than that of earth. The largest number would then represent the volume of fire, the smallest that of earth. Plato would not imagine that anyone could know what the actual quantities were. He is only convinced that they must be linked in some definite proportion, evincing a rational design. This he asserts against the old Ionian belief in an indefinite quantity of matter, and the Atomists' belief in an infinite plurality of atoms. If body were thus indefinite and unlimited, there would be nothing to hold the world together; and in fact the Ionians and the Atomists had believed that their successive or coexistent worlds did fall to pieces and relapse into disorder. Plato's main point is emphasised in the concluding sentence: the world's body, consisting of neither

shapes of diagrams and of models in three dimensions: ἐπίπεδον (σχῆμα, plane figure) for *surface*; τετράγωνον (four-cornered figure) for *square*; διάστασις, διάστημα (extension, interval) for *dimension*; στερεόν (solid figure) for *body*; and perhaps we may add κύβος (die) for *cube* (ὄγκος). Theon (p. 159) gives, as sixth in his list of 11 *tetractyes*, ' the *tetractys* of things that are born and grow (τῶν φυομένων) : the *seed* is analogous to the unit or point, *growth* in length to the number 2 or the line, growth in breadth to the number 3 or the surface; growth in thickness to the number 4 or solid '.

[1] Theon (pp. 155 ff.), following Pythagorean sources, enumerates 11 *tetractyes*. (There should be only 10, the perfect number; Theon interpolates Plato's complex series composed of the two progressions 1, 2, 4, 8 and 1, 3, 9, 27, used for the harmony of the world-soul, 35B). The third is (1) point, (2) line, (3) surface, (4) solid. The fourth is ' the tetractys of the simple bodies : (1) fire, (2) air, (3) water, (4) earth '. ' For such is the nature of the elements *in respect of the fineness or coarseness of their parts* (κατὰ λεπτομέρειαν καὶ παχυμέρειαν), so that fire is to air as 1 to 2 ', and so on. But Plato gives no ground for this interpretation, which ignores the fact that 1, 2, 3, 4 is not a geometrical progression.

[2] Hirzel, *Themis* 309, observes : Gleichheit der elementaren Massen ahnte schon das älteste Denken in der Welt', and compares Hesiod, *Theog.* 126, Γαῖα . . . ἐγείνατο ἶσον ἑαυτῇ Οὐρανόν and Soph., *El.* 87, γῆς ἰσόμοιρ' ἀήρ. I owe this reference to Mr. J. S. Morrison.

[3] Anaxagoras supposed the Sun to be about the size of the Peloponnese, but Archytas estimated the distance of the Sun from the Earth as nine times the distance of the Moon. *Epinomis* 983A says that the Sun is larger than the Earth, and all the heavenly bodies are of stupendous size.

less nor more than four primary bodies, whose quantities are limited and linked in the most perfect proportion, is in unity and concord with itself and hence will not suffer dissolution from any internal disharmony of its parts. The bond is simply geometrical proportion. It is not a question of mechanical forces holding the world together. These belong to the second part of the dialogue and will be explained in due course at 58A.

32C–33B. *The world's body contains the whole of all the four primary bodies*

The next paragraph explicitly rejects the old Ionian conception of an indefinite circumambient mass of body, surrounding the cosmos and providing a reservoir of materials from which a series of successive worlds could be formed; and also the Atomists' conception of an unlimited quantity of matter scattered throughout an infinite void. In this respect the body of the world is once more all-inclusive, like its model. It must be (1) a whole and complete, consisting of parts each of which is whole and complete; (2) single or unique (not one of many coexistent worlds); (3) everlasting (not destroyed and superseded by another world), which it could hardly be, if it were exposed to assaults from outside.

32C. Now the frame of the world took up the whole of each of these four; he who put it together made it consist of all the fire and water and air and earth, leaving no part or power of any one of them outside. This was his intent:
D. first, that it might be in the fullest measure a living being
33. whole and complete, of complete parts; next, that it might be single, nothing being left over, out of which such another might come into being; and moreover that it might be free from age and sickness. For he perceived that, if a body be composite, when hot things and cold and all things that have strong powers beset that body and attack it from without, they bring it to untimely dissolution and cause it to waste away by bringing upon it sickness and age. For this reason and so considering, he fashioned it as a single whole consisting of all these wholes, complete and free from age and
B. sickness.

We are here given one of the reasons why the Demiurge thought it better that the visible world should resemble its model in respect of uniqueness (31B).[1] The primary bodies are described as 'hot

---

[1] Pr. i, 55[14]: 'The proportion does away with internal lack of symmetry, the uniqueness with external violence.'

and cold things and whatever has strong powers'. 'Powers' (δυνάμεις) means the qualities or properties of bodies considered as having the 'power to act and be acted upon' (δύναμις τοῦ ποιεῖν καὶ πάσχειν). Hotness is the property of fire that is manifest when fire makes something else hot or causes in sentient beings a sensation of heat. Coldness is the answering property of the thing which suffers the affection. The 'powers' of the primary bodies are these qualitative properties, as distinct from the quantitative element of form, the regular geometrical shapes later imposed upon these qualities by the Demiurge (53B). Outside the cosmos, fire and the rest, if they could exist at all, could only exist as unformed 'powers', as in the chaos described at 52D. They would then act upon the contents of the formed world and impair its health and stability.

The argument is Eleatic, or at least reminiscent of Melissus' proof (frag. 7) that the unchangeable Being cannot suffer pain : 'for if it did, it could not be completely real, since nothing that suffers pain could be for ever or have the same power as the healthy. Nor could it be alike, if it suffered pain ; since it would suffer pain when something was taken from it or added to it, and then it would no longer be alike.' Proclus (ii, 63) compares the description of the enfeeblement and wasting away of mortal living creatures when the particles of the body, instead of assimilating food from without, are broken down under its too powerful action (81C, D). Plato may also have in view the belief ascribed to Democritus that some of the innumerable worlds of his system are growing, others reaching their prime, others again in decay, and even that they destroy one another by collision.[1] Plato's world is saved from such calamities by its uniqueness. Aristotle appears to have repeated Plato's argument in his dialogue *On Philosophy* :[2] The cosmos must be ungenerated and indestructible, since the causes of destruction must be some power (δύναμις) either external or contained within it. There is nothing outside, since the cosmos contains everything. It is one, because if anything were left over, another like it might come into being ; whole, because all being is used up in forming it ; free from age and sickness, because bodies subject to sickness and age are upset by the strong assaults from outside of heat and cold and the other opposites, but no such power (δύναμις) is left outside the world. Nor can anything inside it cause its dissolution, since then the part would be stronger than the whole.

---

[1] Hippol. *Ref.* 1, 13 (*Vors.* A 40). Cf. Bailey, *Greek Atomists*, p. 146.
[2] Frag. 19 (Ps.-Philo, *de aetern. mundi*). Cf. Occelus Lucanus i.

33B–34A. *It is a sphere, without organs or limbs, rotating on its axis*

In the second part of the dialogue we shall be told how Necessity co-operates with Reason by the working of mechanical causes which keep the world's body in spherical shape (58A). Here we are concerned only with the rational desire of the Demiurge to give it the most perfect of forms and motions. The sphere is the most uniform of all solid figures, and the only one which, by rotating on its axis, can move within its own limits without change of place. This axial rotation symbolises the movement of Reason and is superior to all rectilinear motions.

33B. And for shape he gave it that which is fitting and akin to its nature. For the living creature that was to embrace all living creatures within itself, the fitting shape would be the figure that comprehends in itself all the figures there are ; accordingly, he turned its shape rounded and spherical, equidistant every way from centre to extremity—a figure the most perfect and uniform of all ; for he judged uniformity to be immeasurably better than its opposite.

Diels has quoted this description as the best commentary on Parmenides' comparison of his One Being, 'complete on every side', to 'the mass of a well-rounded sphere, equally poised from the centre in every direction '.[1] Proclus (ii, 71) suggests two explanations of the statement that the sphere embraces all other figures. Geometers have demonstrated that the sphere has a greater volume than any solid figure with plane sides, having the same perimeter. Also, the sphere is the only figure in which every equilateral polygon can be inscribed ; so the reference might be to the five regular solids mentioned later where the primary bodies are constructed. It is curious that Euclid xi, *def.* 14, defines the sphere, not in the usual terms, here quoted by Plato, as having its extremity everywhere equidistant from the centre, but by the mode of generating it : 'When, the diameter of a semicircle remaining fixed, the semicircle is carried round and restored again to the same position from which it began to be moved, the figure so comprehended is a sphere.' As Heath [2] points out, the last propositions of Book xiii show why Euclid put the definition in this form : ' it is this particular view of a sphere which he uses to prove that the vertices of the regular solids which he wishes to " comprehend " in certain spheres do lie on the surfaces of those spheres '.

[1] Parm., frag. 8, 42 (cited by Pr. ii, 69, on our passage).
[2] *Euclid* iii, 269.

33B. And all round on the outside he made it perfectly smooth,
 c. for several reasons. It had no need of eyes, for nothing
visible was left outside ; nor of hearing, for there was nothing
outside to be heard. There was no surrounding air to
require breathing, nor yet was it in need of any organ by
which to receive food into itself or to discharge it again when
drained of its juices. For nothing went out or came into it
from anywhere, since there was nothing : it was designed
 D. to feed itself on its own waste and to act and be acted upon
entirely by itself and within itself ; because its framer thought
that it would be better self-sufficient, rather than dependent
upon anything else.

It had no need of hands to grasp with or to defend itself,
nor yet of feet or anything that would serve to stand upon ;
so he saw no need to attach to it these limbs to no purpose.

34. For he assigned to it the motion proper to its bodily form,
namely that one of the seven which above all belongs to
reason and intelligence ; accordingly, he caused it to turn
about uniformly in the same place and within its own limits
and made it revolve round and round ; he took from it all
the other six motions and gave it no part in their wanderings.
And since for this revolution it needed no feet, he made it
without feet or legs.

Once more the argument is Eleatic, rather than Pythagorean.
Xenophanes had declared that his limited and spherical world had
no special organs of sense : ' it sees, thinks, and hears as a whole '
(frag. 24). The statement may possibly be directed against a
primitive doctrine which figures in some Orphic verses [1] frequently
quoted by the Neoplatonists : Zeus is first and last, one royal body,
containing fire water earth and air, night and day, Metis and Eros.
The sky is his head, the stars his hair, the sun and moon his eyes,
the air his intelligence (νοῦς), whereby he hears and marks all
things ; no sound nor voice escapes his ears, and so on. The
Pythagoreans certainly regarded the Heaven as a living creature
which breathed the circumambient air. Xenophanes [2] again had
denied this, like Plato here. Parmenides had said that the one
Being was not born and did not grow and Empedocles had echoed

---

[1] Kern, *Orph.* frag. 168. (Proclus ii, 82, quotes the fragment here, but as
evidence that the living world has sensation.) Epiphanius (*adv. haer.* i, 7)
attributes the doctrine to Pythagoras : ' he speaks of the god, i.e. the Heaven,
as a body and of the sun and moon and the other stars as his eyes and so
forth, as in a human being '.

[2] D.L. ix, 19 (*Vors.* 11, A1) μὴ μέντοι ἀναπνεῖν.

him.[1] All these statements must be taken as repudiating the primitive notion, traceable in the earliest Pythagorean cosmology, that the world starts from a seed and grows like a living thing by taking in, as nourishment, more and more of the body that environs it.[2]

A creature which requires no nourishment has no need to seek it by moving from place to place. So the sphere has no limbs, as Empedocles said : ' No two branches (arms or wings ?) spring from his back, no feet, no swift-moving knees, no parts of generation ; but he was a Sphere every way equal to itself ' (frag. 29). ' He always remains in the same place, altogether unmoved, nor does it beseem him to go from place to place ' (Xenophanes, 26).[3] There remains, as the only possible movement, the rotation proper to a sphere. That this is the only ' rational ' movement is here stated without any explanation. The point is argued for the first time in the Laws (897D ff.), where the Athenian asks : Of what nature is the motion of reason ? He replies that rotation in one place is most akin to the revolution of reason : both motions are ' regular and uniform, in the same place, round the same things and in relation to the same things, according to one rule and system '.[4] Motion that has not these characteristics, but involves change of place without order, system, or rule, is akin to all unreason (ἄνοια). So here the six rectilinear motions (up and down, forwards and backwards, to right and left) are associated with the irrational. They are ' wanderings ' in which the body of the universe, as a whole, has no share (ἀπλανές), though its constituents, the primary bodies, will be found to possess them.

It is clearly meant that this rational movement of rotation is not confined to the fixed stars ; it is a motion of the whole universe carrying with it all its contents, as the Laws explicitly declares.[5] Nothing has yet been said of the stars, the planets, and the Earth. We shall find that the planets are involved in this motion, though they have also independent motions of their own. The rotation

[1] Parm. 8, 6, τίνα γὰρ γένναν διζήσεαι αὐτοῦ | πῇ πόθεν αὐξηθέν ; Emped. 17, 32, τοῦτο δ' ἐπαυξήσειε τὸ πᾶν τί κε καὶ πόθεν ἐλθόν ;

[2] Cf. Aet. ii, 5, 1, ' Aristotle : If the world is nourished, it will perish ; but in fact it needs no nourishment ; hence it is everlasting '.

[3] Parmenides also (frag. 8, 26–33) seems to connect the immovableness of his Being with its perfection and its ' having no needs ' (οὐκ ἐπιδευές), a divine characteristic (Xenophanes, Vors. 11, A 32, ἐπιδεῖσθαι δὲ μηδενὸς αὐτῶν (τῶν θεῶν) μηδένα. Xen. Mem. 1, 6, 10 τὸ μηδενὸς δεῖσθαι θεῖον εἶναι. Eur. H.F. 1341. Cf. Ar. de caelo 1, 279a, 34.)

[4] Cf. below, 40A.

[5] 897C, ' If we are to assert that the whole course and motion of the Heaven and of all that it contains are of like nature to the motion and revolution and reflections of reason . . .'

of the whole must also affect the Earth, a point that will come up again when we have to consider whether the Earth has any proper movement (p. 130). Here the rotation of the world with all its contents, from axis to circumference, symbolises that reason penetrates and governs the entire universe. On the other hand, the six irrational motions do occur in nature. Since all physical motions are ultimately caused by the self-moving soul, this passage supports the view that the World-Soul has an element of unreason and, like our own souls, is not perfectly controlled by the divine reason it contains. Plato will deny that the so-called 'planets' really 'wander' from one course to another; but the primary bodies have rectilinear motions which are constantly changing their direction. These will be associated with 'what happens of Necessity' and the 'wandering cause' in the second part of the dialogue.

On the whole, this curiously archaic account of the world's body owes much more to the Eleatics and to Empedocles than to the early Pythagoreans. Where Xenophanes and Parmenides differed from the Pythagoreans Plato takes their side, except in Parmenides' denial of all motion. In particular, he rejects the primitive Pythagorean cosmogony, in which the living world expanded from a fiery seed by taking in the surrounding darkness, and, when formed, continued to breathe the vacant air from without. The sphere has always existed in its perfection and self-sufficiency, and outside it there is neither body nor void.[1] It everlastingly fills the whole of space.

THE WORLD-SOUL

THE next section, on the World-Soul, opens with a short summary enumerating the perfections which the world's body owes to divine forethought, and adding that its circular motion, already mentioned, is due to its soul, extending from centre to circumference. The soul is coeval with the body; both exist everlastingly. The composition of the soul is next described: it consists of certain intermediate kinds of Existence, Sameness, and Difference. When these constituents have been compounded, the mixture is divided in the proportions of a musical *harmonia*. Out of the stuff so compounded and divided the Demiurge then constructs a system of circles, representing the principal motions of the stars and planets. The

---

[1] Pr. repeatedly asserts that there is no void outside the cosmos for Plato any more than for Aristotle (ii, 73, 89, 91, etc.). In order to maintain his thesis, Tr. has to suppose that Plato is attributing to Timaeus a 'development within Pythagoreanism which repudiates prominent features of the original doctrine' (p. 100).

addition of these motions of soul to the bodily frame previously described starts the world upon its unceasing course of intelligent life. Finally, it is explained that, on the principle that like knows like, the composition of the World-Soul out of three elements, Existence, Sameness, and Difference, enables it both to know unchangeably real objects and to have true beliefs about changing things of the lower order of existence.

34A–B. *Summary. Transition to the World-Soul*

34A. All this, then, was the plan of the god who is for ever for the
B. god who was sometime to be. According to this plan he made it smooth and uniform, everywhere equidistant from its centre, a body whole and complete, with complete bodies for its parts. And in the centre he set a soul and caused it to extend throughout the whole and further wrapped its body round with soul on the outside; and so he established one world alone, round and revolving in a circle, solitary but able by reason of its excellence to bear itself company, needing no other acquaintance or friend but sufficient to itself. On all these accounts the world which he brought into being was a blessed god.

The statement (here and at 36E) that the soul is wrapped round the body of the world 'on the outside' does not mean that the soul extends beyond the body, but only that it reaches the extreme circumference. Similarly, the yellow colour of an orange might be said to cover it all over on the outside. At *Sophist* 253D the specific Forms are 'embraced on the outside' (ἔξωθεν περιεχομένας) by the generic Form, but the genus does not extend farther than the species it contains. Aristotle again speaks of 'the parts of animals on the outside' (τὰ ἔξωθεν μόρια τῶν ζῴων, *H.A.* 494a, 22), and Plotinus of 'the circumference on the outside' of a circle (ἡ ἔξωθεν περιφέρεια, *Enn.* ii, 2, 1). There may, however, be a suggestion that the presence of a rational soul is most clearly revealed at the circumference, where the diurnal revolution of the whole world is visibly manifested by the stars, unmodified by other motions.[1] This is the movement of the Same, which has the 'supremacy' over all the interior motions, as Albinus observes in explaining this phrase.[2]

34 B–C. *Soul is prior to body*

34B. Now this soul, though it comes later in the account we are
C. now attempting, was not made by the god younger than the body; for when he joined them together, he would not have

---

[1] Cf. *Tr.*, p. 105.   [2] *Didasc.*, ch. xiv. Cf. 36c.

> suffered the elder to be ruled by the younger. There is in us too much of the casual and random,[1] which shows itself in our speech ; but the god made soul prior to body and more venerable in birth and excellence, to be the body's mistress and governor.

The words ' elder ' and ' prior ' here obviously do not mean that the world's soul existed before its body. Plato's point is made at length in *Laws* X, where it is argued that all motion must have its source in a self-moving thing, which is precisely the definition of soul (896A). Accordingly, the characteristic motions of soul— wish, reflection, forethought, etc.—must be the motions whose operation is primary (πρωτουργοὶ κινήσεις, 897A) and which ' take over ' the secondary motions of bodies and control them. Soul itself may be associated with reason and guide all things aright, or with unreason. Plato is combating the atheistical view that the world order has arisen by chance and necessity from the blind working of lifeless powers in the bodily elements. That the world should have a body without a soul is as impossible as that it should have a soul without a body.

35A. *Composition of the World-Soul*

We now come to the composition and structure of the World-Soul. The next sentence states that it is compounded of three ingredients, which are described. The sentence (which, for convenience, I have divided into three numbered parts) is one of the most obscure in the whole dialogue, but not so obscure as it has been made by critics, who have altered the text and thereby dislocated the grammar and the sense. Proclus construed it in the only possible way, and his interpretation, once disengaged from the irrelevant intricacies of his own theology, is obviously correct.[2]

> 35A. The things of which he composed soul and the manner of its composition were as follows : (1) Between the indivisible Existence that is ever in the same state and the divisible Existence that becomes in bodies, he compounded a third form of Existence composed of both. (2) Again, in the case of Sameness and in that of Difference, he also on the same

---

[1] Because we are not wholly rational, but partly subject to those wandering causes which, ' being devoid of intelligence, produce their effects casually and without order ' (46E).

[2] This was pointed out by Professor G. M. A. Grube of Toronto in *Class. Philol.* xxvii (1932), p. 80. Other interpretations, ancient and modern, are reviewed by Tr. (pp. 106 ff.) ; but he has (very excusably) overlooked the valuable part of Proclus' discussion.

principle made a compound intermediate between that kind of them which is indivisible and the kind that is divisible in bodies. (3) Then, taking th᷾ three, he blended them all into a unity, forcing the nature of Difference, hard as it was to mingle, into union with Sameness, and mixing them together with Existence. [1]

The sentence falls into three clauses : (1) The first describes the compounding, out of indivisible, unchanging Existence and the divisible Existence which becomes in the region of the bodily, of a third kind of Existence intermediate between them. This intermediate sort of Existence is one of the three ingredients in the final mixture of the last clause. (2) The second clause states that the Demiurge proceeded on the same principle (κατὰ ταὐτά) also in the case of Sameness and in that of Difference. As there were two kinds of Existence, the indivisible and the divisible, so Sameness and Difference have each two corresponding kinds, described as ' that kind of them which is indivisible, and the kind that is divisible in bodies ' (τὸ ἀμερὲς αὐτῶν καὶ τὸ κατὰ τὰ σώματα μεριστόν). Accordingly, as before, the Demiurge made a third intermediate kind of Sameness (and again of Difference), composed of the indivisible and divisible kinds of Sameness (and of Difference). These intermediate kinds of Sameness and of Difference are the second and third ingredients in the final mixture.[2] (3) Finally, taking the

[1] The text is as follows : (1) τῆς ἀμερίστου καὶ ἀεὶ κατὰ ταὐτὰ ἐχούσης οὐσίας καὶ τῆς αὖ περὶ τὰ σώματα γιγνομένης μεριστῆς τρίτον ἐξ ἀμφοῖν ἐν μέσῳ συνεκεράσατο οὐσίας εἶδος· (2) τῆς τε ταὐτοῦ φύσεως αὖ πέρι καὶ τῆς τοῦ ἑτέρου καὶ κατὰ ταὐτὰ συνέστησεν ἐν μέσῳ τοῦ τε ἀμεροῦς αὐτῶν καὶ τοῦ κατὰ τὰ σώματα μεριστοῦ· (3) καὶ τρία λαβὼν αὐτὰ ὄντα συνεκεράσατο εἰς μίαν πάντα ἰδέαν, τὴν θατέρου φύσιν δύσμεικτον οὖσαν εἰς ταὐτὸν συναρμόττων βίᾳ, μειγνὺς δὲ μετὰ τῆς οὐσίας. Against all the MSS., editors have omitted αὖ πέρι after τῆς τε ταὐτοῦ φύσεως. But cf. τῆς δὲ Ἑρμοκράτους αὖ περὶ φύσεως (20A 7) ; τὸ δ' αὖ περὶ τῆς φρονήσεως (24B, 7). At the end, Jackson saw that μειγνὺς δὲ μετὰ τῆς οὐσίας goes with the other present participle συναρμόττων, not with the following aorist ποιησάμενος, and punctuated as above.

[2] Commenting on clause (2) Proclus (ii, 155) says that among the kinds, Existence ranks first, Sameness second, Difference third. As the intermediate sort of Existence is subordinate to intelligible Existence but superior to divisible Existence in the corporeal, so the Sameness of the soul is inferior to indivisible Sameness, but has a superior unity to divisible Sameness ; and this is true also of its Difference. He recognises what (in the terms of his own theology) he calls the ' demiurgic genus ' of Sameness (and of Difference), as having three species—the indivisible, the divisible, and the intermediate. He assigns to soul the intermediate species of both Sameness and Difference, and says they are combined (in the final mixture) with the intermediate species of Existence. ' For Plato says that, just as in the case of Existence, so in the case of Sameness and Difference the Demiurge compounded a third sort consisting of both, and " on the same principle " (reading κατὰ ταὐτά here

three ingredients, the Demiurge mixes them all into a unity. We may set out the full scheme of the Soul's composition as follows :

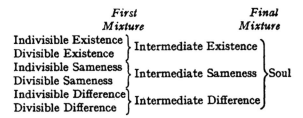

So much for the interpretation of the words ; it remains to consider what Plato's symbolism means. This passage is one of many in which he is writing for readers already versed in his own later thought, without regard for the uninstructed, who would be left wholly in the dark. The terms Existence, Sameness, Difference, would be simply unintelligible to anyone who had not read and understood the *Sophist*.[1] In that dialogue [2] these three ' kinds ' or Forms are singled out for the purpose of showing how Forms in general can be connected in true affirmative statements and disjoined in true negative statements. It was necessary to point out that the words ' is ' and ' is not ' are ambiguous : ' *is* ' can mean either ' *exists* ' or ' *is the same as* ' ; ' *is not* ' can mean either ' *does not exist* ' or ' *is different from* '. Non-existence has been ruled out of the discussion, because there are no true statements asserting that any Form does not exist. We are thus left with Existence, Sameness, Difference. It is carefully shown that these three Forms are wholly distinct. They are, indeed, ' all-pervading ', in that every one of them ' combines ' with every other and with every Form there is. You can say truly of any Form whatsoever (1) that it *exists*, (2) that it is the *same* as itself, and (3) that it is

and at 155[1] and 156[22] : so Tr.) : as in the former case the " compound of both " was a species of Existence, so in the case of these the intermediate is a species of Sameness or Difference.' This paraphrase clearly shows that he construed clause (2) in the only way consistent with the reading of the MSS. The confusions introduced by other commentators arise chiefly from omitting the words αὖ πέρι, and then imagining that τοῦ τε ἀμεροῦς αὐτῶν καὶ τοῦ κατὰ τὰ σώματα μεριστοῦ means the indivisible and divisible kinds (not ' of them ' (αὐτῶν), i.e. Sameness and Difference, but) of Existence. This reduces the second clause to a pointless repetition of the first, and leads to an identification of Sameness and Difference with Indivisible and Divisible Existence, which is flatly inconsistent with the *Sophist*.

[1] Tr.'s exposition of our passage is complicated by his not allowing Timaeus to know the contents of the *Sophist* (p. 128), though he does not hesitate to translate Timaeus' doctrine into the terminology of Whitehead (p. 131).

[2] For a fuller discussion see F. M. Cornford, *Plato's Theory of Knowledge* (1935), pp. 273 ff.

different from any other Form. But a main point of the argument is that no one of these three Forms can be identified with, or derived from, any other.[1] In this part of the *Sophist* ' Existence ' (τὸ ὄν) means, not ' that which exists ', but simply what is meant by the word ' exists ' in such a statement as ' Motion exists (partakes of Existence) '. Since the *Sophist* (as the ancient critics saw) provides the sole clue to the sense of our passage, the word οὐσία here must bear this meaning; it should not be rendered by ' essence ' or ' substance '. The upshot is that the soul has a sort of existence which is not simply identical with the real ' being ' of immutable and eternal things, nor yet with the ' becoming ' of the things of sense, but has some of the characteristics of both these sorts of Existence.

In the *Sophist* only Forms are in question, and the sort of Existence which Forms possess. This is evidently what Plato, in our passage, calls ' indivisible and always unchanging Existence '. When we say that a Form exists, we mean that it has the eternal and immutable being assigned to the higher order of existents at the opening of Timaeus' discourse (28A). With this Plato contrasts here, as before, the ' divisible Existence which becomes in bodies ' or in the region of the bodily. This belongs to that lower order of existents which is ' always becoming, but never has real being ', in the realm of the perceptible. The *Sophist* (240B) recognises images (*eidola*) as a class of entities which have ' some sort of existence ' (as ὄντα πως), but not the real being of the real things (ὄντως ὄντα) of which they are likenesses. These images of reality include all the contents of the visible world produced by the divine Demiurge, whose activity is compared in a later passage of the *Sophist*[2] to that of the human craftsman. They are those copies of the Forms which Timaeus (52A) describes as like the Forms whose names they bear, sensible, generated, perpetually in motion, coming to be in a certain place and vanishing out of it, apprehended by belief involving perception. As likenesses (εἰκόνες) they are contrasted with real things (τὸ ὄντως ὄν) and said to exist only as shifting appearances

---

[1] As Plutarch observes : αὐτοῦ Πλάτωνος ἐν τῷ Σοφιστῇ τὸ ὄν καὶ τὸ ταὐτὸν καὶ τὸ ἕτερον, πρὸς δὲ τούτοις στάσιν καὶ κίνησιν, ὡς ἕκαστον ἑκάστου διαφέρον καὶ πέντε ὄντα χωρὶς ἀλλήλων τιθεμένου καὶ διορίζοντος, *de anim. procr.* 1013D. *Soph.* 254D ff. It should be noted that in the whole account of the composition of the World-Soul, nothing is said about Motion and Rest. These two Forms are illegitimately imported into the interpretation of our passage by Proclus and other ancient and modern commentators, misled by the baseless notion that Motion and Rest together with Existence, Sameness, Difference are the five Platonic ' categories '. For this misinterpretation of the *Sophist*, see F. M. Cornford, *Plato's Theory of Knowledge* (1935), pp. 274 ff.

[2] 266A ff. See F. M. Cornford, *Plato's Theory of Knowledge*, p. 328 note.

in some medium (space), ' clinging to existence somehow or other, on pain of being nothing at all ' (52c).

Between these two orders he now inserts a third form of Existence, compounded of both, which is proper to the soul. All this is correctly pointed out by Proclus. Throughout his commentary, he speaks of soul as an intermediate entity, composed of the intermediate kinds of Existence, Sameness, and Difference.[1] He recognises three orders of Existence : ' intelligible and ungenerated things ; perceptible and generated things ; and intermediate things that are intelligible and generated. The first are altogether incomposite and indivisible and hence ungenerated ; the second composite and divisible and hence generated ; the intermediate kind are intelligible and generated, being by nature both indivisible and divisible, both simple and composite, though in different ways '.[2] ' That by indivisible Existence Plato means the intelligible Existence which, in its entirety, partakes of eternity, and by divisible Existence in bodies the Existence which is inseparable from corporeal bulk and has its being in the whole of time, he himself makes plain by speaking of the former as " unchanging ", of the latter as " becoming ", in order to call the soul not only at once indivisible and divisible, but also " intelligible " and " the first among things that become ".[3] There is a difference between the everlastingness which is eternal and the everlastingness which is spread out along the infinity of time ; and there is yet another, composed of both, such as belongs to the soul. For in its being the soul is unchangeable and eternal, but in respect of its thoughts it is in change and in time.' [4]

If this statement is substantially right, the World-Soul and all individual souls belong to both worlds and partake both of being and of becoming. As immortal and imperishable, the soul is ' most like the divine, immortal, intelligible, simple, and indissoluble (because incomposite) ; whereas the body is most like the mortal, multiform, unintelligible, dissoluble (because composite) and perpetually changing ' (Phaedo 78B). To that extent the soul is akin to the unchanging Forms in the eternal world. But the

---

[1] e.g. ii, 137, ἐπεὶ οὖν ἡ ψυχικὴ οὐσία μέση δέδεικται τῶν ὄντων, ἐκ τῶν μέσων εἰκότως ἐστὶ γενῶν τοῦ ὄντος, οὐσίας, ταὐτοῦ, θατέρου ; iii, 254[3], ψυχή ἐστιν οὐσία μέση τῆς ὄντως οὔσης οὐσίας καὶ γενέσεως, ἐκ τῶν μέσων συγκραθεῖσα γενῶν and in many other places.

[2] Pr. ii, 117[14].

[3] The reference is to 36E, 6, where soul is called ' invisible ' and ' the best of generated things '. On that passage Pr. remarks that soul belongs at once to both classes—things that eternally are and things that become, being the lowest in rank of the former class, since time has its place in soul (ii, 293[13]).

[4] Pr. ii, 147[22].

soul is unlike the Forms in that it is alive and intelligent, and life and intelligence cannot exist without change (*Soph.* 248E). All souls, therefore, must partake also of the lower order of existence in the realm of change and time.

The epithets 'indivisible' and 'divisible' call for some explanation.[1] The being of a Form is indivisible. A Form may, indeed, be complex and hence definable; but it is not 'composite' (σύνθετον), not 'put together' out of parts that can be actually separated or dissolved. Also every Form is unique; it cannot be multiplied. It is not extended in space, and never leaves its own intelligible region to pass into the multitude of things that become in the world of change (52A–C). There is a sense in which every soul is unique and everlastingly preserves its identity; the soul, too, or at least the immortal part of soul, is 'incomposite' and indissoluble. But souls do enter the world of time and change. They exist separately in different bodies, which exclude one another in space; and a soul may be conceived as permeating every part of the body it animates. To this extent it shares in the divided or dispersed (σκεδαστή, 37A) Existence of body; though it cannot be cut into pieces as the body can. The World-Soul is described as extended throughout the whole body from centre to circumference (34B, 36E). It is not clear that we have any right to explain this away. If we recognise such a thing as a soul, an animating principle of motion and consciousness somehow distinct from the bodily elements that continue to exist in a corpse, it is natural to think of it as extending to every part of the living creature. Such, then, is the intermediate form of Existence which, in the imagery of the myth, is produced by mixing the two original kinds of Existence, so as to form a third between them.[2]

It is less easy to see what is meant by the remaining ingredients, the intermediate kinds of Sameness and Difference. The question is best approached from the side of the cognitive functions of the soul, and the principle that like knows like.[3] Aristotle remarks

---

[1] Their meaning as applied to the soul is discussed by Plotinus from his own standpoint at *Enn.* IV, ii.

[2] There is a further question, too speculative to be here pursued, whether the intermediate existence of the soul is to be connected with the intermediate position of the objects of mathematics between the Intelligible and the Sensible in Plato's later '*Ableitungssystem*' as reconstructed by Robin and H. Gomperz. See Robin, *Place de la Physique dans la Philos. de Platon* (1919), pp. 51 ff., and P. Merlan in Philologus lxxxix, 197 ff.

[3] Cf. Crantor's explanation preserved by Plutarch *de anim. procr.* 1012F (summarised in Tr., p. 113). Plutarch's brief summary does not make it clear whether Crantor was really open to the objections Plutarch advances (1013B ff.); but Crantor appears to have misconstrued Plato's sentence like almost everyone else, except Proclus. Albinus in his *Didascalicus* starts his

that Plato in the *Timaeus* is among those who hold this principle and consequently teach that the soul is composed of the same ultimate elements as the things it knows. The doctrine is, in fact, stated below (37A), where Plato explains that the composition of the soul out of the three ingredients, Existence, Sameness, Difference, enables it both to know the objects of reason and to perceive the objects of sense, and to make judgments, involving the terms 'same' and 'different', about existents of both orders. As Proclus says, 'the soul, having an intermediate existence, also fills the gap between reason and irrationality. With the highest part of herself she consorts with reason; with the lowest she declines towards sensation' (i, 251).

In the *Sophist* 'Sameness' stands for the constant identity of a Form (Forms alone being there in question), or its positive content, in virtue of which it is always 'the same as itself'. A Form always is what it is; its sameness excludes any sort of change. This content, at the same time, makes it different from any other Form; for no two Forms are identical in content. A Form is defined by genus and 'differences'. These differences are both elements of positive content—part of what the Form is in itself—and what distinguish it from other Forms, constituting its 'otherness'. Any Form can be negatively described as what is not (is different from) any other Form.

What is meant by describing the Sameness which belongs to unchanging Forms as 'indivisible', we can only conjecture. Perhaps the meaning is that every Form is not only conceptually identical with itself, but numerically one and the same (unique). The Sameness that is 'divided' in the region of bodies must be the sort of Sameness that belongs to individual objects of sense. Such an object has, so long as it exists, some more or less constant identity which enables us to recognise it as 'the same thing' persisting, though in many respects it changes perpetually. But,

account of the soul (based on our passage) from the principle 'Like knows like': 'Since soul enables us to judge each kind of existents, the god naturally arranged the first principles of all things within the soul, in order that, since we always see each thing according to its affinity and likeness, we may posit the soul's reality in harmony with things. Plato, therefore, while declaring that there is an intelligible Existence which is indivisible, also posited another Existence which is divisible in the region of bodies, indicating that the soul can apprehend either by its thought. Perceiving, further, Sameness and Difference both in the realm of the intelligible and in that of the divisible, he made all these contribute to the composition of the soul. For either like is known by like, as the Pythagoreans hold, or, as Heraclitus thought, unlike by unlike' (ch. xiv). Albinus apparently did not confuse Sameness and Difference with indivisible and divisible Existence. Tim. Locr. 95E also avoided this confusion.

unlike Forms, any number of individual things may be conceptually identical, but numerically different. There are many men
or horses, all partaking of the same Form, Man or Horse. The
Sameness (conceptual identity) is dispersed or divided among all
the perceptible individuals. Both the indivisible and the divisible
kind must be represented in the composition of the soul, in order
that it may recognise both in their respective orders of Existence.
The two kinds of Difference could be explained on the same lines.

35B–36B. *Division of the World-Soul into harmonic intervals*

In the figurative language of the myth the compound of three
ingredients is spoken of as if it were a piece of malleable stuff—
say, an amalgam of three soft metals—forming a long strip, which
will presently be slit along its whole length and bent round into
circles. But first the strip is marked off into divisions, corresponding to the intervals of a musical scale (*harmonia*). The intention
is the same as in the previous paragraph. The soul must partake
of harmony as well as of reason (36E). Like knows like; and
just as the soul can recognise existence, sameness, and difference
because these are elements in its own composition, so the World-
Soul must contain the harmonious order which individual souls
ought to learn and reproduce in themselves.

The Demiurge begins by dividing the entire length into 'portions'
measured by the numbers forming two geometrical proportions of
four terms each: 1, 2, 4, 8 and 1, 3, 9, 27.

35B. And having made a unity of the three, again he divided this
whole into as many parts as was fitting, each part being a
blend of Sameness, Difference, and Existence.

And he began the division in this way. First he took one
portion (1) from the whole, and next a portion (2) double
of this; the third (3) half as much again as the second, and
three times the first; the fourth (4) double of the second;

c. the fifth (9) three times the third; the sixth (8) eight times
the first;[1] and the seventh (27) twenty-seven times the
first.

The numbers are evidently meant to be arranged in a single
series of seven terms starting from 1, because the unit had been
held by the Pythagoreans to contain within itself both the
'elements' of number, the even (or 'unlimited') and the odd
('limited' or 'limit'). 'The one consists of both these (since it

---

[1] 9 precedes 8, 'because 9 is a lower power, being the square of 3, while 8
is the cube of 2' (A.-H., *ad loc.*).

is both even and odd), and number proceeds from the one, and numbers are the whole Heaven.'[1] Accordingly, the two progressions advance, through the first even and the first odd number, to their squares and cubes. Theon reproduces Crantor's diagram, symbolising the procession from the one :

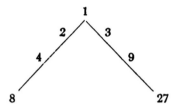

but in Plato's description the numbers are spoken of as measuring corresponding lengths of a single long strip of soul-stuff. We must imagine them as placed in one row at intervals answering to these lengths, in the order 1, 2, 3, 4, 8, 9, 27. The intervals are, of course, of very various lengths. They are presently to be filled in with additional numbers, until we finally obtain a series representing musical notes at intervals of a tone or a semitone. These notes can, for purposes of illustration, be taken as corresponding to the consecutive white notes on a piano covering a range of four octaves and a major sixth. This compass is determined solely by the decision to terminate the series with 27, the cube of 3.[2]

Modern commentators seem not to have taken sufficient notice of the fact that this decision has nothing whatever to do with the theory of musical harmony. Theon[3] remarks that Plato extends his diatonic system as far as to the fourth octave *plus* a fifth *plus* a tone, and quotes Adrastus as follows : ' If any one objects that it should not be extended so far, since Aristoxenus limits the extent of his diagram representing the different modes to two octaves and a fourth, while the moderns have their fifteen-mode diagram with maximum compass of three octaves and a tone,[4] the answer is that these latter

---

[1] Ar., *Met.* 986a, 17 (on the Pythagoreans). Cf. Theon (p. 155), discussing Plato's series, which he reckons as the second form of tetractys, formed by multiplication : ' One is taken as the first number because it is the principle of all numbers, even and odd and even-odd.'

[2] So Pr. ii, 170[18] : ' The advance to four octaves and a fifth (*sic*) is a necessary consequence of the 7 terms, the highest of which is 27.'

[3] p. 104. The same passage from Adrastus is quoted by Pr. ii, 170, with a few variants.

[4] The readings here vary. Mr. R. P. Winnington Ingram writes to me that he thinks the correct reading is : οἱ δὲ νεώτεροι τὸ πεντεκαιδεκάτροπον μέγιστον ἐπὶ τὸ τρὶς διὰ πασῶν καὶ τόνον, this being the total range of the notations (with the additions elsewhere ascribed to οἱ νεώτεροι) and therefore the most extended gamut known to Greek theory.

take only the practical point of view : they consider that performers cannot sing, nor could the hearers properly distinguish, notes beyond this compass.   Plato, on the other hand, is looking to the nature of things.   The soul must be composed according to a harmonia and advance *as far as solid numbers* and be harmonised by two means, *in order that, extending throughout the whole solid body of the world*, it may grasp all the things that exist.   For this reason Plato has extended its harmonia to that point, *though in a sense and in respect of its own nature, the harmonia might extend indefinitely.'* [1] Adrastus evidently saw that, from the musical standpoint, the extent of Plato's range of notes was really as accidental as the compass of the human voice or ear, which fixes a limit to the size of musical instruments.   The reason for stopping at the cube is that the cube symbolises body in three dimensions.[2]   We have already remarked that the two progressions 1, 2, 4, 8 and 1, 3, 9, 27 stand at the head of Adrastus' list of geometrical progressions of the primary and most perfect kind.   Continuous geometrical proportion was chosen as the most perfect bond to connect the four solid bodies forming the whole body of the world (31C).   It is obvious that these considerations are concerned with theories about the nature of number and with the functions of the soul as a bond holding the world's body together ;   they have nothing to do with music.   No one, setting out to construct a musical scale, would start by arranging the terms of two geometrical progressions in the series

$$1, 2, 3, 4, 8, 9, 27.$$

The single series 1, 2, 4, 8 would yield a compass of three octaves. Plato is not content with this because Pythagorean arithmetical theory demanded that the odd numbers should be represented, and also, perhaps, because he intends later to space the seven planets at distances corresponding to the terms, and so needs seven numbers.   The result is that his range of notes is extended to the compass of four octaves and a major sixth.   It is idle to look for any explanation of such a range in the science of harmonics.   This geometrical framework of the whole *harmonia* is determined by arithmetical and physical preoccupations, as Adrastus seems to have clearly perceived.

[1] In Proclus the last sentences appear in a shorter form : ' Looking to the nature of things, Plato composed the soul of all these (numbers), in order that it may advance so far as the solid numbers, since it is to be the patron of bodies.'

[2] *Epinomis* 991A, ' The first progression of the double proceeds in the integer series in the ratio 1 : 2 ; double is the ratio of their second powers ; the progression to the *solid and tangible* is again a double, the progression from one to eight ' (trans. Harward).

It follows that Plato's series of notes does not form a closed system. If a pianist plays the white notes on a piano from C to C he is playing the diatonic scale in the major mode ; if from A to A, he is playing the diatonic scale in the minor mode. Either octave forms a closed system whose structure is repeated in any other octave in the same mode. But Plato's series of notes is simply a section of the diatonic scale, which might be indefinitely prolonged in either direction. Its limits are determined by considerations which, from the musical point of view, are as arbitrary as the decision of a pianist, playing the white notes on his instrument, to stop at the end of four octaves and a major sixth, or the decision of the piano-maker to extend the compass to seven octaves. The seven notes which the Demiurge starts with can be represented, nearly enough for purposes of illustration, by the following passage in C major : [1]

It should be immediately obvious that, in starting with these notes, Plato is not laying down the framework of a scale on musical principles. The notes are chosen because they correspond to the terms of two geometrical proportions ending with cube numbers.

If Plato had intended merely to construct a musical scale, he would have started, as the Pythagoreans did, with the traditional *tetractys*—the arithmetical progression, 1, 2, 3, 4.[2] This series (which adds up to the perfect number, 10) contains the numbers forming the ratios of the perfect consonances : 2 : 1 (octave), 4 : 3 (fourth), 3 : 2 (fifth). These ratios, together with 9 : 8 (the interval of the tone, which occurs between the fourth and the fifth) and the ratio of the semitone, are in fact the ratios he will presently use to fill in the intermediate notes. Theon, in his chapter *On the*

---

[1] Following A.-H., I have represented the original notes by minims. The double bars separate octaves. The fact that the ancient intervals differed slightly from ours is no objection to the use of a notation which is anyhow, in practice, differently interpreted by a violinist and a pianist. Nor does it matter that, strictly, the notes should be written in descending order.

[2] Cf. Burnet, *Gk. Philos.* I, 47, for a simple account of the Pythagorean use of this *tetractys* in constructing the *harmonia*. The *Epinomis* 991A–B actually constructs it from the progression 1, 2, 4, 8 (only), by inserting the harmonic and arithmetical means. The progression is prolonged to the cube number to represent ' the solid and tangible '.

Tetractys and the Decad, enumerates ten *tetractyes* (sets of four things) which these four numbers were supposed to symbolise :

Numbers : 1, 2, 3, 4.
Magnitudes : point, line, surface (i.e. triangle), solid (i.e. pyramid).
Simple Bodies : fire, air, water, earth.
Figures of Simple Bodies : pyramid, octahedron, icosahedron, cube.
Living Things : seed, growth in length, in breadth, in thickness.
Societies : man, village, city, nation.
Faculties : reason, knowledge, opinion, sensation.
Parts of the Living Creature : body, and the three parts of soul.
Seasons of the Year : spring, summer, autumn, winter.
Ages : infancy, youth, manhood, old age.

Some of these are obviously primitive ; others show Platonic influence. They are all interpretations of the primitive *tetractys*, 1, 2, 3, 4, and there are ten of them, 10 being the perfect number. But Theon interpolates, after the first, an eleventh so-called *tetractys*, composed of ' the numbers with which Plato constructs the soul in the *Timaeus* '. The first *tetractys* of numbers at the head of the above list was formed by addition : 1, 2, 3, 4. The second (here added) is, Theon observes, formed by multiplication ; and in order to accommodate both even and odd numbers, it consists of two *tetractyes* : 1, 2, 4, 8 and 1, 3, 9, 27, which have the number 1 in common. Theon remarks that the numbers furnish the ratios of the perfect consonances and of the tone. Further, he says, the terms represent point, line (linear number), surface (square), solid (cube). The geometrical progression thus duplicates the original (arithmetical) *tetractys* of magnitudes : point, line, surface (triangle), solid (pyramid), in which line, surface, and solid are represented by points or dots. The substitution of two geometrical *tetractyes* (1, 2, 4, 8 and 1, 3, 9, 27) for one is obviously an artificial expedient to fit Plato's series of seven numbers into the scheme. Plato himself arranges the seven in a single row.[1] The point which concerns us is that Plato's set of seven numbers has no primary concern with the musical scale, which had been completely and more satisfactorily constructed on the basis of the primitive arithmetical *tetractys*, 1, 2, 3, 4.

Starting, then, with these seven notes, it remains for the Demiurge to fill in the intervening notes. This is effected by inserting, between the numbers forming the two sets of ' double and triple intervals ', the harmonic and arithmetical means. The effect is to combine the two remaining types of proportion with the perfect

---

[1] Plut., *de anim. procr.* 1027D, asks whether the numbers are to form one row, as Theodorus of Soli said, or be arranged as in Crantor's diagram. Pr. ii, 237¹⁵, νενοήσθωσαν οὖν οἱ ἀριθμοὶ πάντες ἐφ' ἑνὸς γεγραμμένοι κανόνος.

and primary geometrical type. At this point, for the first time, terms associated with music begin to be used.

35C. Next, he went on to fill up both the double and the triple
36. intervals, cutting off yet more parts from the original mixture and placing them between the terms, so that within each interval there were two means, the one (harmonic) exceeding the one extreme and being exceeded by the other by the same fraction of the extremes, the other (arithmetic) exceeding the one extreme by the same number whereby it was exceeded by the other.[1]

These links gave rise to intervals of $\frac{3}{2}$ and $\frac{4}{3}$ and $\frac{9}{8}$ within the original intervals.

When we insert the harmonic and arithmetical means between each two successive terms of the original series, we obtain:

harm.   arith.

$$1 \quad \frac{4}{3} \quad \frac{3}{2} \quad 2 \quad \frac{8}{3} \quad \left[3\right] \quad 4 \quad \frac{16}{3} \quad 6 \quad 8$$

$$1 \quad \left[\frac{3}{2}\right] \quad \left[2\right] \quad 3 \quad \frac{9}{2} \quad \left[6\right] \quad 9 \quad \frac{27}{2} \quad 18 \quad 27$$

Omitting the numbers in brackets, which occur in both series, we obtain the single series:

$$1 \quad \frac{4}{3} \quad \frac{3}{2} \quad 2 \quad \frac{8}{3} \quad 3 \quad 4 \quad \frac{9}{2} \quad \frac{16}{3} \quad 6 \quad 8 \quad 9 \quad \frac{27}{2} \quad 18 \quad 27$$

If we now fill in the corresponding notes, the result is as follows:

$$1 \quad \frac{4}{3}\frac{3}{2} \quad 2 \quad \frac{8}{3}3 \quad 4 \quad \frac{9}{2} \quad \frac{16}{3} \quad 6 \quad 8 \quad 9 \quad \frac{27}{2} \quad 18 \quad 27$$

As the last sentence remarks, this 'gives rise to intervals of a fifth ($\frac{3}{2}$) or a fourth ($\frac{4}{3}$) or a tone ($\frac{9}{8}$) within the original intervals'. The final step, taken in the next sentence, is to fill up every tetrachord with two intervals of a tone ($\frac{9}{8}$) and a remainder ($\frac{256}{243}$) nearly equivalent to our semitone.

B. And he went on to fill up all the intervals of $\frac{4}{3}$ (i.e. fourths) with the interval $\frac{9}{8}$ (the tone), leaving over in each a fraction.

[1] If we take for illustration the extremes 6 and 12, the harmonic mean is 8, exceeding the one extreme (6) by one-third of 6 and exceeded by the other extreme (12) by one-third of 12. The arithmetic mean is 9, exceeding 6 and falling short of 12 by the same number, 3.

36B.   This remaining interval of the fraction had its terms in the numerical proportion of 256 to 243 (semitone).

By this time the mixture from which he was cutting off these portions was all used up.

If we take the first octave (two disjunct tetrachords), the result can be illustrated (approximately) as follows, though Plato would have thought of the tetrachord in the shape A G F E, rather than C D E F:

The process, continued throughout the remaining tetrachords, completes the whole range of notes from 1 to 27. The upshot is that Plato has constructed a section of the diatonic scale, whose range is fixed by considerations extraneous to music. The harmonic and arithmetic means have their place in musical theory as determining the intervals of the fourth and the fifth. The two geometrical progressions merely impose an arbitrary limit to the compass. They are introduced in order that the type of proportion which was regarded as primary and most perfect may be represented, and for other non-musical purposes.

It should be noted that nothing is said, here or elsewhere in the *Timaeus*, of any music of the heavens that might be audible to human ears. Plato, no doubt, had in mind this old Pythagorean fancy; for it figures in the vision of Er in *Republic* x. But in the *Timaeus* the harmony resides in the structure of the soul; it is not connected with audible tones whose pitch had been imagined as depending on the relative speeds of the planetary motions.[1]

36B–D. *Construction of the Circles of the Same and the Different and the planetary circles*

Timaeus now speaks as if the Demiurge had made a long band of soul-stuff, marked off by the intervals of his scale. This he proceeds to slit lengthwise into two strips, which he puts together by their middles and bends round into two circles or rings, corresponding to the sidereal equator and the Zodiac.

---

[1] Tr. (p. 164) imports the music of the heavens into the *Timaeus*, and then attributes to Timaeus a form of the doctrine which is in ' absolute contradiction ' with his astronomy.

36B.   This whole fabric, then, he split lengthwise into two halves; and making the two cross one another at their centres in the

c.   form of the letter X, he bent each round into a circle and joined it up, making each meet itself and the other at a point opposite to that where they had been brought into contact.

He then comprehended them in the motion that is carried round uniformly in the same place, and made the one the outer, the other the inner circle. The outer movement he named the movement of the Same; the inner, the movement of the Different. The movement of the Same he caused to revolve to the right by way of the side; the movement of the Different to the left by way of the diagonal.

Plutarch (*de audiendo* 43A) mentions young men who show off their knowledge of mathematics by propounding problems such as the meaning of 'by way of the side', or 'by way of the diagonal'. The terms were, no doubt, unfamiliar to the layman. The plane of the Zodiac is inclined to the plane of the equator as the diagonal of a rectangle to its side. The rectangle in question is to be 'inserted between the summer and winter Tropics' (Pr. ii, 261[22]).

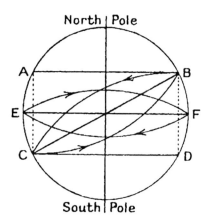

In the diagram, *AB* is a diameter of the summer Tropic, *CD* a diameter of the winter Tropic, *CB* the diagonal of the rectangle obtained by joining *AC*, *BD*. The movement of the Same is a movement of the whole Sphere from East (Left) to West (Right) in the plane of the Equator (*EF*), which is parallel to the planes of the Tropics and so is 'by way of the sides' *AB*, *CD*. The movement of the Different is in the reverse sense and in the plane of the diagonal *CB*, which is a diameter of the Ecliptic, a great circle touching the summer Tropic at a point (*B*) in Cancer, and the winter Tropic at a point (*C*) in Capricorn. The Zodiac is a broad band, containing the twelve signs, along the centre of which runs the Ecliptic. Adrastus (Theon, p. 245) similarly describes the Zodiac as ' inclined to the three parallel circles, the equinoctial, and the winter and summer tropics '.

As Proclus remarks (ii, 258[30]), in the traditional ' Tables of Opposites ', ' Right ' stood in the column of superior things, ' Left ' in the column of the inferior. This is probably Plato's reason for making the circle of the Same revolve ' to the right ', the other circle ' to the left '. The Same must have the superior motion. (Cf. Heath, *Aristarchus*, 163.)

36C.   And he gave the supremacy to the revolution of the Same
  D.   and uniform ; for he left that single and undivided ; but the inner revolution he split in six places into seven unequal circles, severally corresponding with the double and triple intervals, of each of which there were three. And he appointed that the circles should move in opposite senses to one another ; while in speed three should be similar, but the other four should differ in speed from one another and from the three, though moving according to ratio.

The language of the myth has here described the construction of a material model of the revolutions of the heavenly bodies, an armillary sphere.[1] The Demiurge takes a band of some pliable stuff, cuts it lengthwise into two strips, makes them touch at their middles and bends them round to form two rings, inclined to one another. He then takes one of the rings and cuts it up into seven smaller rings of unequal size, which he fits inside about the common centre. One expression, in particular, is appropriate only to a material model : the second ring or ' circle ' is said to be ' inside ' the first. Plato is not imagining strictly geometrical circles, such as would appear on the surface of a celestial globe, for these would have the same diameter. But in a material model, made (say) of copper bands, one band would naturally be fastened ' inside ' the other. That the Academy possessed an armillary sphere may be inferred from Timaeus' later remark (40C) that the intricate movements of the planets cannot be explained without a visible model.[2] Plato probably had it before him as he wrote. Theon[3] tells us that he had himself made a ' sphere ' (σφαιροποιία) to illustrate the Spindle

[1] Pr. ii, 281[19] : Plato all but speaks of the divine Craftsman as using the tools of Hephaestus, forging the whole heaven, giving it a pattern of figures, turning the bodies on a lathe, and shaping each to its proper form.

[2] So Wilamowitz, *Platon.* ii, 390. *Ep.* ii, 312D, mentions such a sphere (σφαιρίον). Cf. Apelt, note 89 (p. 163).

[3] Theon, p. 238, quoting *Timaeus* 40C.

of Necessity in the Myth of Er. This Spindle is not the cosmos, but a model of a primitive kind,[1] with a shaft of adamant for axis, and a 'whorl' composed of a blend of adamant and other substances. The whorl consists of eight concentric hemispheres, fitted into one another like a nest of bowls, and capable of moving separately. The upper half of each sphere is cut away so that the internal 'works' may be seen. The rims of the hemispheres correspond to the eight circles of the *Timaeus*. The outermost represents the equator of the sphere of the fixed stars, or more strictly the motion of that sphere, which carries round with it the whole of its contents, including the seven inner circles, from east to west. The inner circles revolve at different speeds in the opposite sense. All this is in agreement with our passage. A point of difference is that the Spindle does not provide for the seven inner circles being inclined at an angle to the outermost. But it must be remembered that the Spindle is, as Stewart remarks, a vision within a vision, and Plato could hardly be expected to distort its shape to provide for the obliquity of the planetary orbits. It is naïve to infer that he was ignorant of features which a mythical image could not accommodate.

The model made by the Demiurge is of a less primitive pattern, forming what the ancients called an 'armillary sphere' (κρικωτή σφαῖρα), in which the motions of the outermost sphere and of the planets are represented by rings (κρίκοι).[2] No doubt the 'sphere' at the Academy was of this kind, a simpler construction than the 'mechanical sphere' of Archimedes, which is said to have reproduced simultaneously all the celestial motions. The outermost ring corresponds to the equator of the sphere of the fixed stars. It is

---

[1] This was pointed out by J. A. Stewart, *Myths of Plato*, 165. Cf. Heath, *Aristarchus*, 155.

[2] Pr. ii, 249[21], mentions a dispute whether the two original circles are without breadth (in which case how can one of them be slit up ?) or are rings (κρίκοι), 'situated on the surface of the sphere as in armillary spheres'. At iii, 145[26], he mentions the armillary sphere with the ἀβάκιον and the astrolabe (also formed of rings) as instances of the 'visible models' required to illustrate the planetary motions. Daremberg and Saglio, s.v. *Astronomia*, give pictures and descriptions of astronomical instruments. Among the titles of Democritus' mathematical works is Ἐκπετάσματα (projections of the armillary sphere on a plane, Diels-Kranz, *Vors.*[5], ii, 141, 25 note). The eighteenth-century armillary sphere represented in the frontispiece to this book has the Earth in the centre fixed to the stand. The sphere, which revolves round the Earth, consists of the arctic and antarctic circles, the two tropics, and the equator, supported by meridian circles, to which the band of the zodiac is attached on the outside. There are no planetary rings, such as can be seen in more complicated patterns, figured and described by Dr. R. T. Gunther, *Early Science in Oxford*.

a ring, not a sphere, simply because a complete metal globe would hide from view the inner rings. So both hemispheres are cut away, leaving only the equatorial band. This symbolises the revolution of the sphere as a whole, which involves every star in the heavens and all the contents of the universe. As Aristotle, summarising our passage, says, ' the revolutions (φοραί) of the heaven are regarded as the motions (κινήσεις) of the soul ' (de an. 407A, 1). The ' outer revolution ' (ἡ ἔξω φορά, 36C) is the same as the movement of the whole body of the universe described earlier (34A), not a movement of the fixed stars only. It has the ' supremacy ' over the other circles in the sense that (as in the Spindle of Necessity) it carries round with it all the contents of the sphere, including the planets, though these have also motions of their own in the opposite sense. It may be added that this motion of the whole body of the world [1] must affect also the Earth at the centre, which would accordingly rotate with the heavens unless the motion were somehow counteracted. We shall return to this point in discussing the rotation of the Earth (p. 130).

When the motion of the Same is considered as a motion of the World-Soul, apart from the physical motions of the world's body, its ' supremacy ' may be understood as the supremacy of Reason in the World-Soul, regulating its other motions, its judgments and desires. For the Soul has other motions, symbolised by the circle of the Different ; and since the Different is associated with the planets and the Wandering Cause (πλανωμένη αἰτία), the possibility remains that even the World-Soul is not wholly rational. The sphere of the fixed stars, where the motion of the Same is conspicuously manifested, is actually called ' the intelligence of the supreme ' at 40A. But we are here concerned to explain the astronomical meaning of our passage.

The inner ring, the circle of the Different, before it is subdivided, must be identified with the Zodiac, rather than with the ecliptic, the great circle bisecting the signs of the Zodiac longitudinally and traced by the Sun's annual journey through the signs. The Sun is one of the seven planets, and its motion, parallel to the ecliptic, corresponds to one of the seven rings subsequently formed. ' Whereas each of the other circles has for its circumference a single line, the Zodiac has a certain breadth, like the circular frame of a timbrel, and on it are displayed the signs. The name " circle through the middle of the signs " is given to the great circle (ecliptic) which touches the two tropics at a single point in each and bisects the equinoctial. The two circles which limit the breadth of the Zodiac

---

[1] Pr. ii, 259³³, ἐν τῷ παντὶ τὸ μὲν ἀπλανὲς πάντων ἐστὶ κρατητικόν, καθ' ἕνα κύκλον τὰ πάντα περιάγον.

are smaller.' Within these limits the seven planets move in their several orbits.[1]

In an armillary sphere the two rings would have to be attached to a vertical (meridian) ring supporting them and itself revolving on the axis perpendicular to the plane of the equator. This feature of the model is so obviously necessary that Chalcidius saw a reference to it in the text. When the Demiurge had brought the two rings into contact with one another, ' he comprehended them in the motion which is carried round uniformly in the same place'.[2] Chalcidius understood that he ' bound the two circles round with another outside circle, whose revolution is always uniform '.[3] This is ' a meridian circle on the surface of the sphere of the fixed stars, touching both poles '. Its revolution (the movement of the Same) would describe the figure of that sphere, as Chalcidius remarks. The equatorial circle will still symbolise the plane of this revolution. Plato's phrase suits this view remarkably well,[4] though on the surface it may mean no more than ' he set the two circles revolving '. I am inclined to think that Chalcidius rightly divined what Plato was imagining—a feature of his model which it would not suit his purpose to mention as a third ring. It is rather the trace left by the ' carrying round ' of a meridian circle, namely the surface of the sphere considered as symbolising a motion. This image would help to explain the later statement that the fixed stars, which are scattered all over the sphere, were ' set in the intelligence of the supreme (i.e. the rational revolution of the Same) to keep company with it ' (40A). The stars are not set in the equator, but in the motion symbolised by the sphere's surface.

At this point there is some obscurity about the procedure of the Demiurge. He first sets the Zodiac in contact with the equator and gives it a movement in the opposite sense. But he then divides the broad band of the Zodiac into seven smaller rings, and sets these at intervals between the centre and the circumference of the sphere. In an armillary sphere the Zodiac would naturally be a permanent feature attached to the equator and moving with

---

[1] Theon, pp. 218, 214 (after Adrastus).

[2] At 36c, 2, καὶ τῇ κατὰ ταὐτὰ ἐν ταὐτῷ περιαγομένῃ κινήσει πέριξ αὐτὰς ἔλαβεν. A.-H. understood that the two circles are ' encompassed by a moving spherical envelope, being the circumference of the entire sphere of soul revolving κατὰ ταὐτὰ καὶ ἐν ταὐτῷ'. He does not refer to Chalcidius.

[3] Chalcid. *Comment*, p. 163: *Ut si quis . . . hos . . . ipsos (circulos) exteriore alio circulo, cuius motus conversioque idem semper et uniformis sit, circumliget, id est aplani.* The diagram printed by Wrobel is absurd. Chalcidius must have intended a diagram like that on p. 73 above.

[4] Cf. Euclid's definition of the Sphere (quoted above, p. 54) as the figure ' comprehended ' (περιληφθέν) by a (meridian) semi-circle, which is ' carried round ' (περιενεχθέν) to its starting-point.

it at the circumference of the sphere ; for the signs of the Zodiac, of course, move with the other fixed stars and the ecliptic is not the orbit of any individual body. But Plato's rings symbolise motions and nothing else. The bodies which have the motions are not mentioned at all at this stage ; they are fashioned later and set in the motions here provided, as the maker of an armillary sphere might first construct the planetary rings and then attach to them balls representing the planets. We are not to suppose that in the actual heavens there are material rings, like the star-rings in Anaximander, carrying round bodies set in them. What appears in the model as a material ring corresponds simply to a motion in the World-Soul. Now, since there is no physical motion corresponding to the Zodiac, the Demiurge does not require it as a permanent feature of his celestial mechanism for astronomical purposes. Accordingly he takes the Zodiacal band and subdivides it into the seven rings which do correspond to individual motions of the planets. The meaning can only be that a single motion— the motion of the Different—is, from the physical point of view, distributed among all the seven orbits where it actually takes place (with additional modifications). The result is that all the seven planets possess in common a motion contrary in sense to that of the fixed stars, as well as possessing the motion of the Same, from East to West, which they share with the fixed stars, thanks to the ' supremacy ' of the Same. Every planet, accordingly, has a composite or double motion.[1]

On the other hand, the significance of this passage is not confined to the construction of a celestial mechanism. The two original motions are motions of the World-Soul, associated with its cognitive faculty of making judgments involving Sameness and Difference. From this point of view Plato continues to speak of the motion of the Different as a single motion proper to the World-Soul as a whole.[2] It remains as a permanent feature in the con-

---

[1] This is clearly stated by Dercylides (Theon, p. 325), among others. He says, ' the planets move slowly with a motion contrary to that of the fixed stars, the interior motion being carried round by the exterior motion '. So too Adrastus (*ibid.*, p. 221) : ' Sun, Moon, and all the other planets are carried round with it by the universe in its daily motion from E. to W. ; but they appear, day by day, to have several other movements. They have a *proper* movement in the reverse order of the signs, which carries them in the opposite sense to the whole and is called their movement in longitude.'

[2] At 38c the Demiurge sets the bodies of the planets ' in the circuits *in which the revolution of the Different was moving*, in seven circuits seven bodies '. Here, as Proclus remarks, the motion of the Different is still regarded as single, although it has been distributed. Considered, not as a set of physical motions, but as a motion of the World-Soul, the Different is not subdivided into seven motions.

stitution of the World-Soul, though not in the structure of the physical heavens, except in so far as it is symbolised by the Zodiac. For the planets, it becomes, by subdivision, an *imparted* motion, not due to their individual souls, which have self-motions of their own, as will appear later. Similarly the motion of the Same is both a proper self-motion of the World-Soul, manifested physically as the axial rotation of the whole body of the world, and also an imparted motion. It is imparted to the individual fixed stars as a ' forward ' motion of translation (40B), since each star moves from place to place while the whole sphere is rotating in the same place ; and further to the planets and (as we shall argue) to the Earth, constituting one element in such individual motions as they may have.

The seven planetary rings are described as 'unequal', that is to say, of different diameters, so that they can fit one inside another round a common centre.[1] The distances between them correspond in some unspecified way to the six intervals between the seven terms of the series, 1, 2, 3, 4, 8, 9, 27 (' the double and triple intervals '). The simplest view is that these figures measure the radii of the successive orbits : the radius of the Moon's orbit = 1, that of the Sun's = 2, and so on.[2] Probably Plato intentionally left the meaning vague. He would not commit himself to any estimates that had actually been made on very insufficient data. He would be sure only that the distances were not casual and undesigned, but approximated to some simple numerical proportions, though these would not be exactly reproduced in the sensible copy of the ideal. Some more elaborate interpretations of later Platonists may have been inspired by the wish to accommodate Plato's intervals to the results of later more accurate observations.[3] Since the *Timaeus*, in contrast with the Myth of Er, says nothing about any music of the heavens, it is unnecessary to speculate about the connection between these distances and the harmony of the World-Soul.

There remains a well-known difficulty in the last sentence. So far, we have learnt that all the planets have a double motion,

[1] Tr. (p. 152) dismisses this meaning of ' unequal ' because ' we hardly need to be told that seven concentric circles forming a " nest " are unequal in radius or circumference ; that is obvious '. But the word informs us precisely of the fact that the seven circles do form a nest, which is not otherwise stated.

[2] Heath (*Gk. Math.* i, 313 ; *Aristarchus*, p. 163) observes that the meaning is uncertain and that in any case the figures have no basis in observation. Pr. ii, 212[12], mentions various ancient views, most of which are certainly wrong.

[3] For instance, the theories of Chalcidius (p. 167) and of the Platonists in Macrobius (*Somn. Scip.* ii, 3, 14), mentioned by Heath, *Gk. Math.* i, 313.

compounded of the motion of the Same, which they share with the fixed stars, and the opposite motion of the Different, distributed among their seven circles. But we are now told that some of the seven circles have a motion contrary to that of others :

'He appointed that the circles should move in opposite senses to one another ; while in speed three should be similar, but the other four should differ in speed from one another and from the three, though moving according to ratio.'

The natural sense of this statement is as follows : (1) The circles are the seven planetary circles mentioned just before. (2) Some of them have a motion contrary to that of the rest. (3) Three have a similar [1] speed. (It appears later (38D) that these three are the Sun, Venus, and Mercury.) The other four (Moon, Mars, Jupiter, Saturn) have different speeds from one another and from the three. (4) It is *not* stated or implied that the three with similar speed are the set which move in one sense, the four with different speeds the set which move in the opposite sense. The two clauses are distinct : one (κατὰ τἀναντία μὲν . . .) refers to the sense of the movements intended ; the other (τάχει δὲ . . .) to relative speeds.

Commentators have been led to depart from this natural interpretation partly by another set of difficulties connected with the statement at 38D that Venus and Mercury 'possess the tendency contrary to that of the Sun '.[2] As will appear, the contrary tendency there invoked is to account for the fact that Venus and Mercury, although (as we are here told) they keep near the Sun and finish their annual course in the same period, sometimes drop behind the Sun and then get in front of him again.[3] The tendency, in fact, is invoked to explain retrogradation. There is, as we shall see, some connection between the contrary power (or tendency) ascribed to Venus and Mercury as against the Sun and the contrary tendency in our passage of some of the circles as against others. But it is impossible to interpret our passage as meaning that Venus and Mercury have a movement contrary to that of *all the other five*

---

[1] ' Similar ' or ' corresponding ' (ὁμοίως) means that their actual velocities in their orbits are such that all three complete their orbits in the same period (the solar year). They have the same *angular* velocity.

[2] See the views discussed in Heath, *Aristarchus*, pp. 165 ff.

[3] The Sun, Venus, and Mercury keep together in a group. The true reason is, of course, that the orbits of Venus and Mercury are embraced by the Earth's orbit, so that an observer looking from the Earth towards the Sun will never see them at a greater distance from the Sun than the radii of their respective orbits, a distance which the ancients estimated at 50° for Venus and 20° for Mercury. Mars, Jupiter, and Saturn are outside the Earth's orbit, and the Moon goes round the Earth. Consequently these four may be seen at any angular distance from the Sun.

planets, without a flagrant contradiction of easily observed phenomena. Venus and Mercury, accordingly, cannot be simply identified with either of the sets of circles here said to move in contrary senses. We must, for the present, ignore that later statement and consider, independently of it, the question how some of the circles can go contrary to the rest, and which circles are meant. In the whole of this discussion we shall not be concerned with retrogradation, which can be left entirely out of account.

The temptation to construe the sentence unnaturally is chiefly due to its supposed inconsistency with the earlier statement that the motion of the Different, contrary to the motion of the Same, is distributed among all the seven circles. This difficulty leads some to the desperate expedient of supposing that ' the circles ' means, not the seven circles mentioned in the first part of the sentence, but the two original circles of the Same and the Different.[1] Others see that this construction is really impossible and give up the problem as insoluble.[2]

There is one possible meaning consistent with the text, which, however obscurely it may be expressed, must be preferred to meanings which the Greek words cannot bear and to sheer nonsense. One element of obscurity we can eliminate at once by substituting the moving bodies, the planets themselves, for the moving circles of which Plato speaks. Plato does not mean that there really are revolving material rings, to which the planets are fastened. The planets move freely ; the circles only mark their orbits and symbolise their motions. He speaks of circles because his plan demands that the creation of the planetary bodies shall not be described till later. It must also be premised that the science of mechanics was still unborn. Plato had not the notions of force or of mass. In *Republic* vii he regards the science of the motion of a body in three dimensions ($\phi o \varrho \grave{a} \ \beta \acute{a} \theta o v \varsigma$, 528E) as a sort of pure astronomy, for which the observed behaviour of stars and planets will provide illustrations and problems. The bodies dealt with in this science are simply geometrical solids with no physical properties except extension and position in space, and the object is to study the relative speed and slowness of their motions. So also in the *Gorgias* Socrates speaks of astronomy as concerned with the relative speeds of stars, sun, and moon (451C). As a consequence of this

---

[1] So Pr. ii, 264[14], after mentioning other views ; Apelt ; Tr.

[2] Cicero rightly understood that the seven circles revolve contrariis *inter se* cursibus. Fraccaroli (pp. 193 ff.) agrees, and Heath (*Aristarchus*, p. 163) recognises that the words ' can only mean that a certain number of the seven revolve in one direction, and the rest in the other '. But neither offers any solution.

point of view, where we should think of the composition of forces, Plato thinks of the composition of motions. It is natural to him to regard the actual composite motion of a body as the resultant, not of two forces, but of two motions, a faster and a slower, taking place in contrary directions. This conception is the key to the present problem.

The solution can only be that the actual motion of some planets is the resultant, not only of the two motions previously mentioned (the motions of Same and of the Different), but also of a third here added. (1) The motion of the Same carries round the entire universe with all its contents, relatively to absolute space.[1] If that motion operated alone, there would be no change in the relative positions of any parts of the universe. It can accordingly be ignored in the present discussion. (2) The motion of the Different, as we saw, was a single motion, shared out among the seven planetary circles. As single, it will affect the bodies afterwards placed in those circles as if all the seven circles moved together, like a solid disc, with ' similar speed ', i.e. with the same angular velocity. This distribution of the single revolution of a disc to larger and smaller circles within its circumference is described at *Laws* 893c : ' We observe in the case of this revolution that such a motion carries round the greatest and the smallest circle together, dividing itself proportionately to lesser and greater, and being itself proportionately less and greater. This, in fact, is what makes it a source of all sorts of marvels, since it supplies greater and smaller circles at once with velocities high or low answering to their sizes— an effect one might have imagined impossible ' (trans. Taylor). The revolution of the Different may be illustrated by the motion of a moving staircase, on which seven passengers are standing.[2] Suppose that the staircase is moving downwards. If this were all, the seven planets, though shifting (eastwards) against the background of the fixed stars (represented by the stationary walls

---

[1] The expression ' absolute space ' is justified by the fact that Plato certainly regards the rotation of the whole universe as a real motion, with a period of 24 hours, although there is nothing outside—not even empty space—to which the motion can be relative. The world rotates in its own *place* ; the place does not rotate with it. For this distinction between a body and its ' place ', see below, p. 195.

[2] The ancient commentators used a similar (but less convenient) comparison. The Same was represented by the movement of a ship (westwards), the motion of the planets by passengers walking along the deck towards the stern (eastwards). Chalcidius, p. 166 : *ut in navigando, cum ad destinata uenti pulsu naui uolante e regione prorae quidam ex nauigantibus ad puppim recurrunt.* Hyginus, *Poet. Astron.* iv, 13, *necesse est eum (solem) contra mundi inclinationem currere. Quare autem euenit, ut ante diximus, quod uidetur cum mundo sol uerti, eius similis haec est causa, ut si quis in nauiculae rostro sedens*

enclosing the staircase), would keep their relative positions, all being equally subject to the motion of the staircase. The present passage explains why they do not.

Let us take first the differences of speed, which are, in fact, sufficient to account for the changes of relative position. I suggest that we may take the Sun with his two companions, Venus and Mercury, as proceeding at the standard speed, against which the speeds of the remaining four will be measured. The Sun is obviously pre-eminent among the planets [1] and his period, the year, is the most important. The year is the cycle of life on Earth, which moves in that period through its round of birth, maturity, death, and rebirth. This movement of life was connected by Aristotle with the ' inclined circle ' (the ecliptic) marking the Sun's apparent annual track through the signs. The ancients thus attribute to the motion of the Sun all those seasonal changes which we, on the heliocentric theory, attribute to the annual revolution of the Earth.[2] Already, in the *Republic* (509B), the Sun has been called the cause of the *becoming* (birth, γένεσις), growth, and nourishment of all visible things, ' though not himself γένεσις' ; just as the Good is the cause of the *being* (οὐσία) of intelligible things, though itself ' beyond being '. This association of the Sun and its inclined circle with becoming and mutability and so with ' the Different ' suggests that the movement of the Sun (shared by his two companions) is *the actual movement of the Different*, with a speed unmodified by any individual variations. Obviously, if any planet exhibits the actual motion of the Different, it must be either the Sun or the Moon. Not to mention their superior conspicuousness, these are the only two planets which go steadily forward, without stations or retro-

---

inquirat (inde quaerat, Scheff.) *ad puppim transire, et nihilominus ipsa nauis iter suum conficiat : ille quidem uidebitur contra nauiculae cursum ire, sed tamen eodem perueniet quo nauis.*

[1] *Epin.* 986E, ' Of these three (Sun, Venus, Mercury) it must needs be that the one with an intelligence equal to the task (the Sun) leads the way '. Albinus, *Didasc.* xiv, ἥλιος μὲν ἡγεμονεύει πάντων (τῶν πλανητῶν), δεικνύς τε καὶ φαίνων τὰ σύμπαντα.

[2] Ar., *de gen. et corr.* ii, 10, 336a, 32 ff. ' It is not the primary motion (of the First Heaven) that causes coming-to-be and passing-away, but the motion along the inclined circle ; for this motion not only possesses the necessary continuity, but includes a duality of movements as well.' The lifetime of every living thing has a period, which in some cases is a year, in others shorter or longer. Coming-to-be occurs as the Sun approaches, decay as it retreats. With the revolution of the Sun the seasons come to be in a cycle, and so the becoming of living things, initiated by the seasons, is also cyclical. Cf. Adrastus (Theon, p. 242) : In the sublunary region there is becoming and perishing, growth and diminution, every sort of qualitative change and variety of locomotion. Of all these things the planets are the cause, and chiefly the Sun and Moon, by virtue of their composite movements.

gradations. It seems likely that the motion of one or the other will be compounded solely of the Same (which is common to all) and the Different. Both were associated with the mutability of earthly things, and the Moon, with her phases, had strong claims, which were duly recognised. But the Sun's claim is stronger because his period embraces the whole round of seasonal life. Every year is a repetition of the last one, whereas the months are very different in character : June is not a repetition of December. That is why the ecliptic is the trace of the Sun's apparent annual path, not of the Moon's apparent monthly path, through the signs. The solar year, then, will be the period of a revolution of the Different, just as twenty-four hours is the period of a revolution of the Same. We may thus compare the Sun, Venus, and Mercury (the ' three with similar speed ') to a group of passengers who *stand still* on one step of the moving staircase, which carries them slowly downwards. The staircase is bent round in a continuous band. Imagine this to be circular, and that the passengers can travel round and round. This group of three will then, at the end of a year, be back again at their initial position.

There are four more passengers on the staircase. The remaining planets are the Moon, who is between the Earth and the Sun group, and the three outer planets, Mars, Jupiter, Saturn. All these differ in speed from the Sun group and from one another. The Moon revolves rapidly in her orbit—the smallest of all—round the Earth. She moves much faster [1] than the Sun, completing over twelve monthly rounds to one of his yearly revolutions. The three outer planets are slower than the Sun. Mars was estimated in antiquity to take a little less than 2 years, Jupiter about 12 years, Saturn a little less than 30. [2]

There is thus a contrast between the behaviour of the Moon and that of the outer three, causing a phenomenon which Theon describes as follows :

> ' The conjunctions of the planets with the Sun and their appearances and disappearances, which we call their risings and settings, are not the same for all the planets. The Moon, after her conjunction with the Sun, since she has a swifter movement than his towards the antecedent signs (eastwards), always makes her first appearance or ' rising ' in the evening and disappears or ' sets ' in the morning. Saturn, Jupiter, and Mars, on the contrary, since they reach the antecedent signs more slowly than

[1] Boeckh pointed out that ' faster ' and ' slower ' as applied to the planets here does not mean absolute velocity. The faster planet is the one which completes its circuit in the shorter time, i.e. has the higher angular velocity.

[2] Theon, p. 222.

the Sun, as if overtaken and passed by him, always set in the evening and rise in the morning (after their conjunction).'[1]

To return to our illustration : three passengers (Sun, Venus, Mercury), *as a group*, stand still on the staircase and move with it. The other four, being alive, can walk either up or down the staircase and so get farther and farther from the stationary group. If the staircase is bent round in a circle, they will pass through all angles of divergence till they rejoin the group (conjunction with the Sun). But they do not all walk the same way. One (the Moon) runs down the staircase, so fast that he overtakes and passes the group nearly thirteen times while the group is making one circuit. The other three move *the opposite way*, mounting the staircase, at different rates of speed. They are, of course, all the time being carried downwards by the staircase ; but by walking upwards at lesser rates of speed they slow down this movement and get away from the stationary group. In respect of their individual voluntary motion, the three who are mounting can be said to be moving in the contrary direction to all the other four, for they alone are moving against the motion of the staircase. These three also will pass through all angles of divergence before they rejoin the group (conjunction with the Sun). But their behaviour will contrast with that of the Moon in the manner described by Theon.

Here a diagram may be useful.

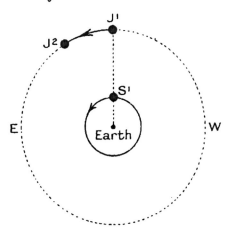

The outer circle is the orbit of Jupiter, the inner circle the orbit of the Sun. Suppose that on 1 January 1934 the Sun at S^1 and

---

[1] Pr. ii, 264[4], reproduces this : ' Saturn, Jupiter, and Mars make their first appearance after conjunction with the Sun as morning stars because the Sun moves in the direction of the antecedent signs more quickly than they ; the Moon, on the contrary, first appears in the West because, moving more quickly than the Sun, she is seen to the East of the Sun.

Jupiter at J^1 were in conjunction. If the two rings were both subject only to the motion of the Different from West to East, they would move ' with similar speed ', i.e. the same angular velocity. Then at the end of a year both planets would have completed one revolution, remaining in conjunction all the time, and returned to their original positions. But this is not what happens. By 1 January 1935 the Sun will have completed one revolution and be back at S^1; but Jupiter will have moved only a twelfth part of his course, from J^1 to J^2. Jupiter must therefore have counteracted the common motion of the Different. Instead of allowing this motion to swing him round in perpetual conjunction with the Sun, he slows it down by an additional motion in the opposite sense (westwards) rapid enough to let the Different carry him only as far as J^2. If we imagine his orbit as a moving circular platform on which he is walking, the platform will complete its revolution *eastwards* in one solar year, but Jupiter will have walked along it *westwards* $\frac{11}{12}$ths of its length. This individual motion is contrary to that of the Sun (with his companions Venus and Mercury) and to that of the Moon. It is symbolised by Jupiter's individual circle. The planet, while subject to the *westward* motion of the Same in the plane of the *equator* and also to the *eastward* motion of the Different in the plane of the *ecliptic*, has its own motion *westwards* in the plane of the *ecliptic*, counteracting the Different.

To sum up : if we leave out of account the motion of the Same, which affects all the seven planets equally, the proper movements of the planets, *relatively to one another*, are as follows : (1) The Sun, Venus, and Mercury, taken as a group with ' similar speed ', complete their course together in a solar year. Their proper motion is identical with that of the Different. (2) The Moon has an additional motion which carries her faster in the same sense. (3) The three outer planets move in the same sense inasmuch as they share in the motion of the Different. But they have, individually, the power of counteracting that movement in various degrees, and so slowing it down. These three planets are the set which have additional, individual motions in the opposite sense to the others. (It should be noted that these additional motions are strictly contrary to the Different, to which the Same, being in another plane, is not strictly contrary.) So, and only so, can it be true that two sets of circles (or bodies moving in those circles), though all moving in one sense with the common motion of the Different, have individual motions ' in opposite senses *relatively to one another* ' (κατὰ τἀναντία ἀλλήλοις).

We can now see why the changes in the relative positions of the planets are not ascribed merely to differences of speed, though that would be a possible way of representing the facts. The additional motion of the three outer planets is contrary to the motion of the Different, which is exhibited without modifications by the Sun group ; whereas the Moon's motion is in the same sense as the Different, which it merely accelerates. The result will be that, in returning to conjunction with the Sun, the Moon will overtake the Sun as it were from behind, whereas the Sun himself will overtake and pass the three outer planets. This is the pheno-menon noted by Theon : ' The Moon after her conjunction with the Sun, since she has a swifter movement than his towards the antecedent signs (eastwards), always makes her first appearance or " rising " in the evening and disappears or " sets " in the morning. Saturn, Jupiter, and Mars, on the contrary, since they reach the antecedent signs more slowly than the Sun, as if overtaken and passed by him, always set in the evening and rise in the morning (after their conjunction).'

The third force which modifies the motion of some of the planets is left unexplained. The reason is that the planets themselves have not yet been mentioned at all.[1] Later we shall learn that, like the fixed stars, they are divine living creatures with souls ; and these souls must have the power of self-motion, since that is the very definition of soul. It is, presumably, the self-motion of the planets that enables them either to counteract the motion of the Different to some extent or to reinforce it. If this is the explanation, it could not be given here in a passage which describes only a system of motions without reference to the bodies that have them. It is consistent with the statement of Dercylides, who maintained that, according to Plato, all the planets had a ' voluntary and unforced motion ' and blamed Aristotle, Menaechmus, and Callippus for introducing spheres to which they attached the heavenly bodies, as though these were inanimate and needed material spheres to carry them round.[2]

The interpretation offered above is confirmed by the description

---

[1] Pr. ii, 265⁵, μόνους γοῦν τοὺς κύκλους ἐν τῇ ψυχῇ θεὶς ἄνευ τῶν ἀστέρων—οὔπω γὰρ ὑπέστησαν—τούτους ἔφατο κινεῖσθαι.

[2] Theon, p. 327, πᾶσι δὲ τὴν κίνησιν προαιρετικὴν καὶ ἀβίαστον εἶναι. Aristotle is accused by Ritter (*Platon* ii, 372) of a ' depravation ' of Eudoxus' system of geometrical spheres. But Eudoxus was a mathematician concerned only with making a map of the celestial motions on the assumption that they must all be reducible to circular movements, as Plato taught. Aristotle was a physicist, concerned with making these motions work mechanically. Since he believed action at a distance to be impossible, the only way by which the movement of the Same (or any other revolution) could be communicated to an inferior body was by means of material spheres in actual contact with one another.

of the Spindle of Necessity in the Myth of Er (*Rep.* 617A), where the counter-movement of the three outer planets is explicitly mentioned, though its significance has not been understood.

' The Spindle turns round as a whole with one motion ; and within the whole, as it revolves, the seven circles revolve slowly in the opposite sense.'

Here, as in the *Timaeus*, the two main motions—of the Same, affecting the whole, and of the Different, shared by all seven circles —are first mentioned. Next come the different speeds of the seven circles and the changes in the relative positions of the planets :

' And of these circles themselves, the eighth (Moon) moves the most swiftly ; second in speed and all moving together, the seventh, sixth, and fifth (Sun, Venus, Mercury) ; third in speed moves the fourth (Mars), *as it appeared to them, with a counter-revolution*, fourth, the third (Jupiter), and fifth, the second (Saturn).' [1]

Adam and Heath rightly recognise that ἐπανακυκλεῖσθαι (as distinct from ἀνακυκλεῖσθαι) means ' counter-revolution '. But counter to what ? The movement of all the seven circles contrary to the fixed stars was mentioned in the previous sentence ; it is shared by all the planets. Why should Plato, in an exceedingly compressed account, mention it again, precisely at the point where the three outer planets are introduced, after the group of three which keep together ? I can only understand it as a reference to the doctrine of our passage, that the three outer planets (to all of which, I take it, the phrase applies) appear to have a movement contrary to the Moon and to the Sun, Venus, and Mercury, modifying the movement shared by all. The word ἐπανακύκλησις occurs,

---

[1] 617B, τρίτον δὲ φορᾷ ἰέναι, ὡς σφίσι φαίνεσθαι, ἐπανακυκλούμενον τὸν τέταρτον, τέταρτον δὲ τὸν τρίτον, καὶ πέμπτον τὸν δεύτερον. Adam (*ad loc.*) : ' The revolution relatively to that of the whole is retrograde ; hence ἐ π α ν α κυκλούμενον.' Heath (*Aristarchus*, p. viii) : ' what is meant is a simple circular revolution in a sense contrary to that of the fixed stars, and there is no suggestion of retrogradations '. Heath (*Gk. Astron.*, p. 48) translates accordingly : ' third in the speed of its counter-revolution the fourth appears to move '. Theon (p. 236) quoting *Rep.* 617B (not very accurately) has τρίτον δὲ φορᾷ ἰέναι, ὅν φασι (for ὡς σφίσι) φαίνεσθαι ἐπανακυκλούμενον ⟨τὸν τέταρτον⟩ μάλιστα τῶν ἄλλων. Burnet (E.G.P.³, p. 304 n.) thought that μάλιστα τῶν ἄλλων might be a line that had dropped out of the text of Plato. If so, I should understand it as meaning that, while οἱ ἄλλοι, the three outer planets, all have the counter-revolution, it is most apparent in the case of Mars, who takes only two years to complete his orbit. Burnet took ἐπανακυκλούμενον to mean retrogradation. But retrogradation is not confined to Mars or to the three outer planets—a fact which Plato recognises later (38D).

so far as I know, only once elsewhere in Plato, in a passage which bears out my interpretation. After describing the individual motions of the heavenly bodies and of Earth, due to their living souls, Plato says that all the effects resulting from so complicated a system of motions cannot be understood in detail without a visible model. These effects include the ways in which they gain upon and pass one another, their conjunctions and oppositions, and '*the counter-revolutions of the circles relatively to one another*'.[1] In the Myth of Er, the outer planets moved '*as it appeared to them* (the souls), with a counter-revolution'. Plato is not wasting words : there is a sense in which the counter-revolution is only apparent. The souls, watching the turning circles in their vision, see the Moon speeding ahead of the Sun group, while the outer three drop behind and get farther and farther away. They would 'appear' to be moving in the contrary direction, like our three passengers who walked the opposite way to the rest ; but their actual motion is (as we have already been told) governed by the movement of the Different. The bodies stationed in the circles are really moving the same way as the others, though more slowly as against the standard speed set by the Sun.

On the other hand, as I shall try to show later (p. 108), this power of the planets' individual souls to counteract the motion of the Different is invoked by Plato for another purpose. In our passage and in *Republic* x it explains a peculiarity of the three outer planets in contrast with all the rest. The effect is a slowing down of the planet's main motion, without real change of sense. But there is also the very striking phenomenon of retrogradation. As we watch the planets against the background of the fixed stars, all, except the Sun and Moon, appear at times to stand still, move backwards a certain distance, and then go forward again. This topic, however, had better be reserved till we reach the point where Plato introduces it (38D).

Here, it remains to point out that in this description of the composite motion of the planets there is nothing inconsistent with the *Laws* or the *Epinomis*. At *Laws* 821B the Athenian, addressing men supposed to be totally ignorant of astronomy, remarks that nearly all Greeks falsely say that Sun and Moon and certain other stars are never travelling along the same path (ὁδόν), and so call

---

[1] 40C, τὰς τῶν κύκλων πρὸς ἑαυτοὺς ἐπανακυκλήσεις. See below, p. 135. The phrase has been understood as 'the returning of the circles upon themselves'; but a model would not be needed to show that a circle returns upon itself. ἑαυτούς is a frequent substitute for ἀλλήλους, a word which Plato might well avoid, since he has to use it three times in the same sentence. It is unfortunate that there is a lacuna in the sentence where Pr. (*in remp.* ii, 226¹⁹) commented on the ἐπανακύκλησις in the Myth of Er.

them 'wanderers' (planets). 'The truth is precisely the opposite : each is always travelling in a circle one and the same path, not many paths, though it appears to move along several paths' (822A). This statement does not contradict our passage. The proper motion of each planet is confined to one of the seven circles ; it never strays from this orbit into another path. 'It is natural and necessary,' writes Theon,[1] 'that every heavenly body should, like the fixed stars, move uniformly and regularly with one simple *proper* movement. This will be evident if we imagine the universe to be at rest and the planets moving along the Zodiac (which will *ex hypothesi* be at rest). Their movement will then appear no longer variable and irregular, but regular, as we have shown by the construction of Plato's Sphere (σφαιροποιίας).' He goes on to explain that the appearance of variable movement is due to the planets' *proper* movements being twisted into spirals by combination with the movement of the Same in another plane, as the *Timaeus* explains later (39A).[2] As Boeckh pointed out, the unity of the planets' movements in single circles is not supposed in the *Laws*, any more than in the *Timaeus*, to be upset by the fact that the movement of the Same turns them into spirals. Thus, just after the mention of the spiral twist at 39A, Plato speaks of the Moon as describing ' its own circle ' in a month, and of the Sun as describing ' its own circle ' in a year.[3] All that the Athenian asserts is that the planets do not stray about from path to path, but keep to one circular track. This is true of their proper movement. The expression to ' move on several paths ' (πολλὰς ὁδοὺς φέρεσθαι, *Laws* 822A) must not be confused with ' having a movement compounded of more than one motion ' (πλείους φορὰς φέρεσθαι, Aristotle).[4] On Newtonian principles a planet has a

---

[1] p. 244, following Adrastus. The notion is now current that Plato revolutionised his astronomy in his old age, and that this revolution is implied by certain statements in the *Laws* and *Epinomis*. I shall criticise this theory later (122 ff.) ; but I would remark here that the lucid and detailed accounts of Plato's astronomy which Theon took from Adrastus and Dercylides betray no sign that they recognised any contradiction between the *Timaeus* and the later works.

[2] Cf. Pr. iii, 122², ' Each planet has one simple motion, though the combination of more than one revolution—the proper revolution of each one and the revolution shared with the fixed stars—complicates their movement.'

[3] Heath, *Aristarchus*, p. 183.

[4] This confusion invalidates Tr.'s argument (*Class. Rev.* xlix, 54) controverting Shorey's remark (on *Rep.* 530B) that the *Rep.* is consistent on this point with the *Timaeus* and the *Laws*. Tr. says : ' the phrase πολλὰς ὁδοὺς (or φορὰς) φέρεσθαι does not mean to " move irregularly, now this way, now that, but something very different, " to move with several motions at once ", to have a composite movement '. This is not a possible rendering of πολλὰς ὁδοὺς φέρεσθαι.

composite movement, the resultant of two forces acting in different directions, but it keeps to a single elliptical track. Even if we take into account the twisting of the proper circular movement into a spiral by the other component motion, the planet will still be travelling on a single regular track or path. If a man ascends a spiral staircase, he is not straying from one path to another. His position at any moment can be calculated as exactly as if he were moving in a circle or a straight line. The *Epinomis* (982c) gives as a proof of intelligence in the heavenly bodies the regularity of their behaviour; they do not change their places or wander with shifting revolutions. All these statements are directed against the notion popularly entertained by people who knew no astronomy that the term ' planet ' implied irregular and incalculable ' wanderings ' from one track to another.

Another passage in the *Epinomis* [1] has been alleged to contradict the *Timaeus*. After mentioning the seven planets, the author speaks of ' one (divinity), the eighth, which might specially be called the Cosmos on high, who moves in the opposite sense to all those, carrying the others with it—so, at least, it may seem to men who know little of these things. But that of which we are sufficiently well assured we are bound to state and do state ; for to one who has even a small share of right and divine understanding, this appears to be the teaching of true wisdom '. Heath has offered a natural interpretation of this passage. ' It occurs to me,' he wrote, ' that the emphasis is on the word " *men* " (ἀνθρώποις without the article), and that the meaning is " so far as mere human

---

[1] 987B, ἕνα δὲ τὸν ὄγδοον χρὴ λέγειν, ὃν μάλιστά τις ἂν ⟨τὸν ἄν⟩ω (ἄνω libri : ἂν Burnet. I propose τὸν ἄνω κόσμον, to distinguish κόσμος applied to the fixed stars from κόσμος as used of the whole universe) κόσμον προσαγορεύοι, ὃς ἐναντίος ἐκείνοις σύμπασιν πορεύεται, ἄγων τοὺς ἄλλους, ὥς γε ἀνθρώποις φαίνοιτ' ἂν ὀλίγα τούτων εἰδόσιν. ὅσα δὲ ἱκανῶς ἴσμεν, κτλ. Burnet's insertion of οὐκ before ἄγων τοὺς ἄλλους has no authority. Tr. (p. 232) also understands that the outermost circle does *not* really carry the others round with it. He deduces that ' the real motion of the eighth circle, which is still retained in the *Epinomis*, can no longer have anything to do with day and night '. But the *Epinomis* in the context (986B) refers back to an earlier passage mentioning Sun, Moon, and Fixed Stars. There (978D–979A) the Sun is connected with the year, the moon with the months, and the Fixed Stars with night and day : ' When Ouranos ceases not *turning these bodies about for many nights and days*, he never ceases teaching men the lesson of one and two, till even the dullest learns to count well enough. For every one of us who sees the heavenly bodies will go on to form the idea of three and four and higher numbers.' Day and Night—one and two—is the simplest lesson in number, and so is mentioned first ; then the month ; then the year. The lesson is taught by the revolution of the stars in the *Epinomis*, exactly as it is in the *Timaeus* 39C and 47A. Cf. also *Laws* 818 where counting one and two is similarly connected with counting day and night.

beings can judge, who can have little knowledge of these things ''. The words immediately following are then readily intelligible : they would mean '' but if we are reasonably satisfied of a thing we must have the courage to state our view ''.' [1] The view of which the writer is sufficiently well assured to state it as the teaching of true wisdom is that the circle of the fixed stars does carry the others with it—so long as we refrain from inserting the word ' not ' in order to make the *Epinomis* agree with a mistaken interpretation of *Laws* 821. The *Epinomis*, if it be Plato's at all, must anyhow be his latest work ; and he may have wished to hint that, though he still felt sufficiently well assured of the doctrine stated positively in the *Timaeus*, other explanations of the ' appearance ' might be possible. That human beings could know little about the heavenly bodies remained a commonplace long after Galileo had made his telescope. Our knowledge of anything more than their distances and movements dates from the invention of the spectroscope. In any case, whatever the *Epinomis* passage means, it cannot afford proof that Plato did not himself hold the view stated in the *Timaeus* when he wrote that dialogue, perhaps fifteen years earlier. He might have changed his opinion in the meantime.[2]

The conclusion is that the *Laws* (certainly) and the *Epinomis* (quite possibly and, I should say, probably) are perfectly consistent with the theory of the *Timaeus*, which ascribes a compound motion to the seven planets. The conception is fundamental in the system of Eudoxus, who was working at the Academy before the *Timaeus* was written and who died before Plato. It is equally fundamental in Aristotle's adaptation of Eudoxus' system of spheres. The system must have been known to Plato, and the probability is that he incorporated in the *Timaeus* as much of it as he could accept, consistently with his belief that the *proper* motion of each planet keeps to a circular track. It should not be forgotten that

---

[1] *Aristarchus*, 185. (In *Gk. Astron.*, pp. xliii, 61, Heath has adopted a different view.) The above rendering gives its due force to γε and an acceptable meaning to ἀνθρώποις. If this word referred to any individuals it would be slightly insulting. I cannot believe that Plato would have alluded either to his late colleague Eudoxus or (as Tr. suggests, p. 170) to ' Aristotle and his friends ' as ' fellows who know little of these things ', or that such an expression could be characterised as ' urbane irony '. Since Plato had himself made the (alleged) mistake in the *Timaeus*, he might feel that even urbane irony was out of place.

[2] Yet Tr. writes (p. 169) : ' If we turn to the *Laws* and *Epinomis* we further get absolute proof that Plato himself did not hold the theory (of double motion of the planets) in the form in which it is given in the *Republic* and *Timaeus*.' On p. 171 this ' absolute proof ' has become a ' more natural inference ' than the possibility that Plato had changed his view. But, as we have seen, there is no real evidence even for a change of view.

the *Timaeus* is a myth of creation, not a treatise on astronomy. The surprising thing is that Plato should have found room for so many details in his broad picture of rational design in the cosmos, not that he should have simplified by omitting subtleties which would contribute nothing to his main purpose and which might be superseded at any time, as indeed they were very soon afterwards.

36D–E. *The world's body fitted to its soul*

The structure of the World's Soul is now complete. Plato has described its composition out of the three intermediate kinds of Existence, Sameness, and Difference ; its division according to the intervals of the cosmic harmony ; and its rational motions, represented by the two main circles. Nothing has yet been said about the bodies which display these motions and the additional motions of the seven circles. The intention is to emphasise the superior dignity of soul and the truth that the self-moving soul is the source of all physical motions. The next step is to fit the World's body, previously described, into the frame of the soul. This means imparting to the body the motions symbolised by the soul circles.

36D. When the whole fabric of the soul had been finished to its maker's mind, he next began to fashion within the soul all
E. that is bodily, and brought the two together, fitting them centre to centre. And the soul, being everywhere inwoven from the centre to the outermost heaven and enveloping the heaven all round on the outside,[1] revolving within its own limit, made a divine beginning of ceaseless and intelligent life for all time.

The above sentences reiterate the emphasis already laid at 34B on the fact that the soul extends throughout the body of the world from centre to circumference, and communicates its motion to the whole. That is to say, the motions above described are not confined to the stars and planets. The motion of the Same, which is supreme over the seven planetary motions, must affect the entire body of the world, including the Earth at its centre. But we are here concerned not so much with physical movements as with the

---

[1] See note on 34B. Adam compares our passage to *Rep.* 616c, where the light passes through the centre of the universe and round the outer surface of the heavenly sphere, acting as a bond that holds together all the revolving firmament, like the undergirders of a man-of-war. If Chalcidius was right in his interpretation of 36c (p. 77) as referring to the revolution of a meridian circle tracing the circumference of the sphere, this passage may well refer to that enveloping movement of the Same. Compare the language of 34B, where the wrapping of the soul round the body on the outside is immediately followed by mention of the rotation.

motions of the World-Soul as an intelligent being. Hence in the next paragraph 'the circle of the Different' is once more spoken of as representing a single undivided motion.

36E–37C. *Discourse in the World-Soul*

The cognitive activity of the soul's ceaseless and intelligent life is based on the principle that like knows like. As Proclus says, 'Since the soul consists of three parts, Existence, Sameness, and Difference, in a form intermediate between the indivisible things and the divisible, by means of these she knows both orders of things; . . . for all knowing is accomplished by means of likeness between the knower and the known.'[1]

36E.   Now the body of the heaven has been created visible; but
        she is invisible, and, as a soul having part in reason and
37.   harmony, is the best of things brought into being by the
        most excellent of things intelligible and eternal.[2] Seeing,
        then, that soul had been blended of Sameness, Difference,
        and Existence, these three portions, and had been in due
        proportion divided and bound together,[3] and moreover
        revolves upon herself, whenever she is in contact with any-
        thing that has dispersed existence or with anything whose
        existence is indivisible, she is set in motion all through herself
B.   and tells in what respect precisely, and how, and in what
        sense, and when, it comes about that something is qualified as
        either the same or different with respect to any given thing,
        whatever it may be, with which it is the same or from which
        it differs, either in the sphere of things that become or with
        regard to things that are always changeless.[4]

---

[1] Pr. ii, 298. Cf. ii, 135²¹ ff.

[2] Plutarch 1016c (rightly) took τῶν νοητῶν ἀεί τ' ὄντων as depending on τοῦ ἀρίστου. Pr. ii, 294, mentions this as a possible construction, though he suggests, as perhaps preferable, the meaning that soul is the best among those intelligible and everlasting things which are generated, or taking τῶν νοητῶν ἀεί τ' ὄντων with λογισμοῦ καὶ ἁρμονίας (cf. Robin, *Physique de Pl.* 56). That αὐτή means the soul (not 'the heaven itself', Tr.) is plain from 46D, 6. A.-H., Wilamowitz (*Platon* ii, 389), and others are (I think, rightly) inclined to omit ψυχή, though it was read by Plutarch (*loc. cit.*).

[3] Proportion acts as a bond, 31C.

[4] The construction is doubtful. (1) It can be taken (in accordance with the above translation) as follows: 'The soul tells—(ὅτῳ τ' ἄν τι ταὐτὸν ᾖ καὶ ὅτου ἂν ἕτερον) whatever it may be (say B) that something (A) is the same as or different from—in what respect precisely and how and in what sense and when it comes about (ἕκαστα εἶναι καὶ πάσχειν) that it (A) is, or is qualified by, each of these terms (same and different) (πρὸς ἕκαστον) in respect of any such thing (B), either in the sphere,' etc. Grammatically, ἕκαστον (B, 2) is the antecedent of ὅτῳ (A, 7), and the τι of the ὅτῳ clause is the subject of ἕκαστα

37B. Now whenever discourse that is alike [1] true, whether it takes place concerning that which is different or that which is the same, being carried on without speech or sound within the thing that is self-moved,[2] is about that which is sensible, and the circle of the Different, moving aright, carries its message throughout all its soul—then there arise judgments and beliefs that are sure and true. But whenever discourse
c. is concerned with the rational,[3] and the circle of the Same, running smoothly, declares it, the result must be rational understanding and knowledge. And if anyone calls that in

εἶναι καὶ πάσχειν, which I understand (cf. Taylor) as meaning 'is each of these things (same or different) or in other words is qualified by them'. Pr. ii, 304[19], notes that Plato often uses πεπονθέναι for μετέχειν, as at Soph. 245B πάθος ἔχον τοῦ ἑνός and πεπονθὸς ἓν εἶναί πως, mean 'having the attribute or property of unity'.

(2) The words ὅτῳ τ' ἂν . . . ἕτερον might be taken as an interrogative clause depending on λέγει. A parallel occurs at Soph. 262R, ὅτου δ' ἂν ὁ λόγος ᾖ, σύ μοι φράζειν. A grammarian might contend that the full meaning there is : 'Whatever the statement may be about, you are to tell me (what it is about).' So here : 'the soul tells with what thing (whatever it may be) something (τι) is the same'.

The difficult phrase πρὸς ἕκαστον ἕκαστα εἶναι καὶ πάσχειν seems to allude to the ambiguities of the word 'is', explained in the Sophist. 'Is' can mean 'exists' (partakes of Existence) or 'is the same as' (which involves partaking of Sameness or having that property, πάσχειν, as 'is not' involves having the property of Difference). So we can say either that one thing is (εἶναι) the same as, or different from, another, or that it has either of the properties (πάσχει ἕκαστα) with respect to any other (πρὸς ἕκαστον).

[1] κατὰ ταὐτόν 'equally' (A.-H.), for ὁμοίως, which would involve hiatus. The discourse is to be true in either case, whether the judgments are affirmative or negative. Cf. κατὰ ταὐτά, 38D, 5.

[2] The self-moved thing is the Heaven as a whole, which, as a living creature, is self-moved by its own self-moving soul. That an animal (soul and body) is self-moved is a commonplace. Ar., Phys. 265b, 34, 'Witness to this truth (that locomotion is prior to other motions) is borne by those who make soul the cause of motion, for they say that what moves itself is the source of motion and the animal or anything that has a soul does move itself locally'. This explains αὐτοῦ τὴν ψυχὴν below (B, 7) ; and the world (κινηθὲν καὶ ζῶν) is again referred to as αὐτό at c, 6. The passive (κινούμενον ὑφ' αὐτοῦ) is more appropriate to the animal which is moved by its soul than to the soul which moves itself (τὸ ἑαυτὸ κινοῦν). Commenting on the statement (34A) that the Demiurge gave the world 'the motion proper to its body', Pr. (ii, 92[31]) says that it refers to the peculiar constitution of the cosmos, in virtue of which it is so moved by itself (ὑφ' ἑαυτοῦ), ἔχει γάρ τι καὶ αὐτὸς καὶ κατὰ τὴν ζωὴν αὐτοκίνητον καὶ κατὰ τὸ σῶμα σφαιροειδὲς ὂν πρὸς τὴν κύκλῳ κίνησιν οἰκεῖον (where αὐτοκίνητον and οἰκεῖον are both epithets of τι, and the insertion of τὴν after ζωήν is unnecessary).

[3] Pr. ii, 312[11], observes that λογιστικόν here means not, as one might suppose, the subject which reflects, but the object of thought (αὐτὸ τὸ νοητόν), as αἰσθητικόν is used later (61D, 65A, etc.) for αἰσθητόν. Cf. also κινητικόν for εὐκίνητον at 58D.

37C.   which this pair [1] come to exist by any name but ' soul ', his
words will be anything rather than the truth.

Like the earlier description (35A) of the composition of soul out
of the three intermediate kinds of Existence, Sameness, and Difference, this compressed account of the discourse carried on in the
World-Soul can only be understood by reference to the *Sophist*.[2]
There all philosophic discourse is regarded as consisting of affirmative and negative statements about Forms. Discourse is guided
by the science of Dialectic, whose task is ' to divide according to
Kinds, not taking the same Form for a different one or a different
one for the same' (253D). The dialectician discerns the true
structure of the realm of Forms, what each Form is in itself and
how it differs from others—what it *is* and what it *is not*. A false
judgment is described as mistaking one Form for another. Similar
language is used below (44A) : in infancy the motions of the soul-
circles in human beings are perturbed and distorted by the inflow
of nourishment and of sense-impressions, and ' when they meet
with something outside that falls under the Same or the Different
they speak of it as "the same as this" or "different from that"
contrary to the true facts, and show themselves mistaken and
foolish '. When the tide of growth and nutriment flows in less
strongly, the revolutions settle down into their natural course, ' and
giving their right names to what is different and what is the same,
they set their possessor in the way to become rational '. So in our
passage, the true judgment correctly identifies its object (whether
a Form or an individual thing which becomes) with whatever it is
the same as, or distinguishes it from whatever it is different from.

Dialectic is concerned solely with Forms, but here the discourse
of the World-Soul is directed both to the indivisible being of Forms
and to the existence that is ' dispersed ' in the perceptible things
of time and space. The same is, of course, true of human souls,
from which, in fact, the analogy is extended to the Soul of the World.
We have been told that the World's body has no sense-organs,
because there is nothing outside it to be perceived. But the
World's Soul is not pure intelligence ; being united with a perceptible body, it may be imagined as having internal feelings,
which would be covered by the word *aesthesis*.[3] The World's Soul
differs from ours in that its revolutions can never be disordered

---

[1] I incline to think (with A.-H.) that ' this pair ' means rational understanding and knowledge, because Plato thinks it worth while repeatedly to
assert that νοῦς can exist only in soul (30B, 46D, *Soph.* 249A, *Philebus*, 30C),
though the same is true of judgments and beliefs.
[2] 252E ff. See F. M. Cornford, *Plato's Theory of Knowledge*, pp. 260 ff.
[3] Cf. for instance *Theaet.* 156B and the list of feelings at 42A below.

(47C). Hence Plato speaks of its discourse as always true, although it contains, besides rational understanding and knowledge, judgments and beliefs associated with the revolution of the Different —a revolution which is controlled by the superior motion of the Same, but moves in another plane.

Aristotle, after mentioning how Empedocles recognised the principle that like is known by like, continues : ' In the same way Plato in the *Timaeus* fashions the soul out of his elements ; for like, he holds, is known by like, and things are formed out of the principles or elements, so that soul must be so too. Similarly also in his lectures " On Philosophy " it was set forth that the Animal itself is compounded of the Idea itself of the One together with the primary length, breadth, and depth, everything else, the objects of its perception, being similarly constituted. Again he puts the view in yet other terms : Mind is the monad, science or knowledge the dyad (because it goes undeviatingly from one point to another), opinion the number of the plane, sensation the number of the solid ; the numbers are by him expressly identified with the Forms themselves or principles, and are formed out of the elements ; [1] now things are apprehended either by mind or science or opinion or sensation, and these same numbers are the Forms of things ' (*de anim.* 404*b*, 16 ff., trans. J. A. Smith).

37C–38C. *Time, the moving likeness of Eternity*

We turn now from the spiritual motions of the World-Soul—its thoughts and judgments—to the physical motions of perceptible bodies in the Heaven. Planets, stars, and Earth have yet to be created and set in the revolutions symbolised earlier by the eight circles of the celestial mechanism. This work is prefaced by a description of Time, which cannot exist apart from the heavenly clock whose movements are the measure of Time.

37C. When the father who had begotten it [2] saw it set in motion and alive, a shrine brought into being for the everlasting gods, he rejoiced and being well pleased he took thought to make it yet more like its pattern. So as that pattern
D. is the Living Being that is for ever existent, he sought to make this universe also like it, so far as might be, in that respect. Now the nature of that Living Being was eternal, and this character it was impossible to confer in full completeness

---

[1] Not, of course, fire, air, water, earth, but Unity and the Indeterminate Dyad (or Plurality).

[2] αὐτό refers, like αὐτοῦ at B, 7, to τὸ κινούμενον ὑφ' αὐτοῦ, the world as a living and self-moved creature (κινηθὲν καὶ ζῶν).

37D. on the generated thing. But he took thought to make, as it were, a moving likeness of eternity; and, at the same time that he ordered the Heaven, he made, of eternity that abides in unity, an everlasting likeness moving according to number [1] —that to which we have given the name Time.

E. For there were no days and nights, months and years, before the Heaven came into being; but he planned that they should now come to be at the same time that the Heaven was framed. All these are parts of Time, and 'was' and 'shall be' are forms of time that have come to be; we are wrong to transfer them unthinkingly to eternal being. We say that it was and is and shall be; but 'is' alone really belongs to it and describes it truly; 'was' and 'shall be' are properly used of becoming which proceeds in time, for they are motions. But that which is for ever in the same state immovably cannot be becoming older or younger by lapse of time,[2] nor can it ever become so; neither can it now have been, nor will it be in the future; and in general nothing belongs to it of all that Becoming attaches to the moving things of sense; but these have come into being as forms of time, which images eternity and revolves according to number. And besides we make statements like these:[3]

38.

B. that what is past is past, what happens now is happening now, and again that what will happen is what will happen, and that the non-existent is non-existent: no one of these expressions is exact. But this, perhaps, may not be the right moment for a precise discussion of these matters.[4]

---

[1] μένοντος αἰῶνος ἐν ἑνὶ κατ' ἀριθμὸν ἰοῦσαν αἰώνιον εἰκόνα. Even here, where he is contrasting eternal duration (αἰών) with everlastingness in time, Plato will not reserve αἰώνιος for 'eternal' and ἀίδιος for 'everlasting'. ἀίδιος is applied both to the model and to the everlasting gods. But in this particular phrase it is certainly strange that the moving likeness contrasted with abiding duration should be called αἰώνιον. It is tempting to conjecture ἀέναον εἰκόνα, 'ever-flowing likeness', and to compare Laws 966E where the motion of soul gives to Becoming an ever-flowing existence (ἀέναον οὐσίαν), and Critias, Peirithous, frag. 18, ἀκάμας τε χρόνος περί τ' ἀενάῳ ῥεύματι πλήρης φοιτᾷ . . .

[2] Read διὰ χρόνον (F. Eus. Stob. Pr. (lemma): διὰ χρόνου, cett.) οὐδέ, to avoid an intolerable hiatus. See note on 20A.

[3] τὰ τοιάδε, remotely governed by λέγομεν (37E, 5).

[4] The objection is to using the word 'is' in statements about things that become or happen in time or are non-existent. 'Being', in contrast here with Becoming, ought strictly to be reserved for the real unchanging Being of eternal things. Its application to Becoming is at least ambiguous, not 'exact'. The last sentence hints that a discussion of the ambiguity of 'is' will be found in the Sophist. 'The non-existent' means (as in ordinary speech) the absolutely non-existent, of which, as the Sophist shows, nothing whatever can be truly asserted.

38B.   Be that as it may, Time came into being together with the Heaven, in order that, as they were brought into being together, so they may be dissolved together, if ever their dissolution should come to pass ; and it is made after the pattern of the ever-enduring nature, in order that it may

c.   be as like that pattern as possible ; for the pattern is a thing that has being for all eternity, whereas the Heaven [1].has been and is and shall be perpetually throughout all time.

In the first sentence above, ' a shrine brought into being for the everlasting gods ' is a paraphrase of τῶν ἀιδίων θεῶν γεγονὸς ἄγαλμα which calls for some justification. The words are usually translated ' a created *image* of the everlasting gods ', and this expression has troubled commentators, who have assumed that the word *agalma* (image) is simply equivalent to *eikon* (likeness), and that consequently the everlasting gods must be the Forms after whose pattern the world is made, or else (in spite of the plural) the Demiurge himself. But the Demiurge is nowhere in the *Timaeus* identified with his model,[2] and the Forms are nowhere spoken of as gods.

The word *agalma*, however, contains no implication of likeness and is not a synonym of *eikon*. It is true that θεῶν ἀγάλματα is the common phrase for ' images of the gods ', cult-statues ; but the word itself has two main meanings : (1) object of worship, and (2) something in which one takes delight.[3] ' Image ' to our ears suggests a likeness ; ' statue ', a solid and uninteresting effigy in a park. We do not think of a statue as enshrining the spirit of a departed general or politician. It is never an object of worship and seldom a cause of delight. The different associations of *agalma* may be illustrated from other passages in Plato. In the *Phaedrus* (252D) the lover chooses his love (ἔρως) according to his disposition and ' as though that love were a god in his eyes, he fashions and adorns him like an object of worship (οἷον ἄγαλμα), as with the intent to celebrate rites in his honour '. Here the beloved person is worshipped as an incarnation or embodiment of the god answering

---

[1] ὁ δὲ, sc. οὐρανός (Pr. iii, 50²⁹). The existence of the world is spread out all through past, present, and future time. Cf. 31B, οὐρανὸς γεγονὼς ἔστιν τε καὶ ἔτ᾽ ἔσται. Comparison with 37C, 8, and 39E, 1, suggests that οὐρανός is already the subject of ἵν᾽ ὡς ὁμοιότατος αὐτῷ κατὰ δύναμιν ᾖ.

[2] At 92C, 7, εἰκὼν τοῦ νοητοῦ (sc. ζῴου) should be read, not ποιητοῦ.

[3] As object of worship ἄγαλμα is ὅ τις ἀγάλλει (worships) ; in the other sense it is ᾧ τις ἀγάλλεται, a phrase by which ἄγαλμα is frequently glossed. The second appears to be the earlier sense in literature. It is recognised by Proclus with reference to our passage : καὶ γάρ πως τὸ ἄγαλμα παρὰ τὸ ἀγάλλεσθαι τὸν θεὸν ἐπ᾽ αὐτῷ λέλεκται (iii, 6²⁴), and perhaps hinted at by the words ἠγάσθη and εὐφρανθείς in the text.

to his temperament. At *Laws* 931A *eikon* and *agalma* are used side by side: 'Some of the gods whom we honour (the stars) are clearly visible; as likenesses (εἰκόνας) of others we consecrate *agalmata*, and when we worship these, lifeless as they are, we believe that the living gods beyond are gratified and filled with good will towards us.'[1] 'So if a man has parent or grandparent worn out with age laid up as a treasure in his house, let no man think that, so long as he has such a consecrated object set up at his hearth (ἐφέστιον ἴδρυμα), he could have any more efficacious object of worship (ἄγαλμα), if he shall give it due tendance in the true sense . . . In the eyes of the gods we can possess no more precious object of worship than such a parent. Heaven is well pleased when a man worships his progenitors with honours (ἀγάλλῃ τιμαῖς). The consecrated object which is an ancestor (τὸ προγόνων ἴδρυμα, "shrine", Bury) is a marvellous thing, far superior to lifeless ones; for the living ones can join in our prayers when duly tended, or pray against us when neglected. Thus in such parents a man possesses objects of worship most efficacious in securing divine favour.' In this passage the worshipped parent is the *agalma*; 'image' or 'statue' is an inadequate rendering. To the ancient a cult-statue was a thing he worshipped and took delight in because the visible image betokened the presence of the divinity in the shrine. It was set up there in order that the god might come and dwell in it. So the Greek for 'to set up a statue of Hermes' is simply ἰδρύεσθαι Ἑρμῆν. The same word (ἰδρύσατο) is used of the Demiurge setting the planets in the framework of his Sphere (38D). Richard Wilhelm has observed that in Chinese temples the images and pictures of the gods are ordinarily treated with no respect. 'These pictures are not gods at all. They are merely places which they enter if they are called upon in the right way. When the god is there, then the presence in his image is a stern and holy matter. When he is not there, then his image is a piece of wood or clay.'[2] Julian, dwelling on the benefits conferred on the whole world by those visible gods, the heavenly bodies, calls the Sun 'the living *agalma*, endowed with soul and intelligence and beneficent, of the intelligible father (?)'.[3] The Sun is not a statue or a likeness, but a living *embodiment*.

Proclus is fully alive to this mode of thought. Plato, he says

---

[1] τῶν δ' εἰκόνας ἀγάλματα ἰδρυσάμενοι, οὓς ἡμῖν ἀγάλλουσι καίπερ ἀψύχους ὄντας ἐκείνους ἡγούμεθα τοὺς ἐμψύχους θεοὺς πολλὴν διὰ ταῦτ' εὔνοιαν καὶ χάριν ἔχειν. Here the masculine οὓς treats the ἄγαλμα as a god whose life is not in itself but in the living god it portrays.

[2] *The Soul of China*, p. 314.

[3] *Ep.* 51, 434, τὸ ζῶν ἄγαλμα καὶ ἔμψυχον καὶ ἔννουν καὶ ἀγαθοεργὸν τοῦ νοητοῦ πατρός (παντός Osann. The text appears to have a lacuna after this word.)

(iii, 4¹⁸), speaks of the cosmos as an *agalma* of the everlasting gods because it is filled with the divinity of the intelligible gods, although it does not receive those gods themselves into itself any more than cult images (ἀγάλματα) receive the transcendent essences of the gods. The gods in the cosmos (the heavenly bodies) are, as it were, channels conveying a radiance emanating from the intelligible gods. Proclus calls the Demiurge the ἀγαλματοποιὸς τοῦ κόσμου (iii, 6¹⁰), who makes the cosmos as an *agalma* and sets up within it the *agalmata* of the individual gods (iii, 69²).¹ Some of the *agalmata* consecrated by religion are for all to see ; others are hidden within as symbols of the presence of the gods and known only to the initiating priest. In the same way the cosmos is an *agalma* of the intelligible, containing both visible tokens of its Father's divinity and unseen pledges of its participation in reality (i, 273¹⁰). In two places Proclus substitutes the word ' shrine ' (ἱερόν) : ' the cosmos is the holiest of shrines ' (i, 124¹⁷) ; the planetary bodies are set up in it ' as shrines of the gods who together accomplish the perfect year ' (ii, 5²⁷, referring to 38D).²

In our sentence the Demiurge contemplates the cosmos with its body and soul so far as they have yet been organised. The body appears as the celestial Sphere with its turning rings ; animated by soul, whose motions those rings symbolise, it is a living and moving *agalma*, like those statues made by Daedalus which Plato mentions more than once.³ But the everlasting divinities have still to take their places in this vacant shrine. These are the ' heavenly gods ' (οὐράνιοι θεοί, 39E), the stars, the planets, and Earth, all of which are presently to be described as ' living creatures everlasting and divine '.⁴ That the ' everlasting gods ' of our passage are the heavenly bodies is plain from the *Epinomis* 983E, where these are described as divine living beings, which we must either celebrate as

---

¹ This recalls Alcibiades' comparison of Socrates to an image of Silenus which, when opened, is found to contain golden ἀγάλματα of the gods (*Symp.* 216D, E).

² When Aeschylus (*Eum.* 920) describes Athens as φρούριον θεῶν, ῥυσίβωμον Ἑλλάνων ἄγαλμα δαιμόνων, is not ' shrine ' nearer to the true sense than ' bright ornament ' (Weir Smyth) or ' precious jewel ' (Headlam) ? Athens is not a statue or an image, but it is a place wherein the gods delight to dwell.

³ *Euthyphro* 11B, 15B, *Meno* 97D. Curiously enough, Aristotle, just before criticising this part of the *Timaeus*, mentions in the context, dealing with Democritus, the wooden Aphrodite which Daedalus was said to have made to move by pouring quicksilver into it.

⁴ ζῷα θεῖα καὶ ἀΐδια (40B) includes the fixed stars and the planets ; and Earth is ' the most venerable of the gods within the Heaven ' (40C). All these are of the number of τῶν ἐν οὐρανῷ θεῶν (*Rep.* 508). I cannot, therefore, agree with Tr.'s statement that ' all through the story there is only one God who can be called " everlasting ", the Creator himself ' (p. 184).

being actual gods, or consider as likenesses of gods, like *agalmata* which the gods themselves have made. They are not the work of worthless makers, but we must honour them above all other *agalmata* ; for never will there be seen *agalmata* more lovely or more truly a common possession of all mankind, or any set up (ἱδρυμένα) in more excellent regions or of higher purity, majesty, and fulness of life. Here the stars either are actual gods or *agalmata* made by gods for their own habitation.[1] In our passage, the cosmos with its eight moving circles is thought of as an *agalma* which awaits the presence of the divine beings who are to possess the motions symbolised. The addition of the heavenly gods and (later) of the three inferior kinds of living creatures is to complete the resemblance of the copy to its model (92C).

First, however, it must be explained that all these living creatures, even the heavenly gods themselves, are endowed with temporal life that moves in time and lasts throughout all time, but is not the eternal unchanging duration (αἰών) proper to the model. The concept of duration without change, as the attribute of real being, was first formulated by Parmenides. Plato echoes his words about the One Being : ' It never was nor ever will be, since it is now all at once ' (frag. 8, 5). The ' indivisible ' being of Plato's intelligible world demands a duration that ' abides (rests) in unity '. Time is essentially divided into the three ' forms ', past, present, future ; and it ' moves according to number ', being measured by a plurality of recurrent ' parts ', the periods called day, month, year. Nothing that we can call Time can exist without these units of measurement ; and these again cannot exist without the regular revolutions of the heavenly bodies, the motions of the celestial clock. Time, accordingly, is said to ' come into being together with the Heaven ', in the sense that neither can exist without the other.

Plato's treatment of Time presents an important contrast to his treatment of Space. We are apt to speak of Becoming as going on ' in time and space ', as if these two conditions were on the same footing. Plato does not so regard them. Time is here included among the creatures of the divine intelligence which orders the world. It is a feature of that order, not a pre-existing framework. Space, on the other hand, is introduced in the second part of the dialogue, under the heading of ' what happens of Necessity '. The Receptacle of Becoming is there brought into account, as a third factor (besides Being and Becoming) which has hitherto been ignored (48E). This Receptacle, finally identified with Space (52A), is treated as a given frame, independent of the Demiurge and a

---

[1] Cf. Simpl. *Phys.*, 1337, 34, προφανέστατον μὲν θεοὶ καλοῦνται τὰ τῶν οὐρανίων θεῶν περιπολοῦντα ἀγάλματα.

necessary condition antecedent to all his operations.   Time is not
a given frame ; it is ' produced ' by the celestial revolutions (38E),
which are themselves the work of the Demiurge.   It is true that the
existence of Space is implied throughout all this description of the
world's soul and body ; but its existence is due to Necessity,
not to Reason.   Space is a condition without which Reason could
not produce the visible order.   Time is a feature of that order,
inherent in its rational structure.

Plato's view of Time as inseparable from periodic motion is no
novelty, but a tradition running throughout the whole of Greek
thought, which always associated Time with circular movement.
Reviewing popular and philosophic conceptions of Time in connection with his own doctrine, Aristotle remarks that regular circular
locomotion, being most easily counted, provides the best unit of
measurement.   ' Neither alteration nor increase nor coming into
being can be regular, but locomotion can be.   This is why *Time is
thought to be the movement of the sphere* : [1] it is because the other
kinds of change are measured by locomotion and Time by this
(circular) movement.   This also explains the common saying that
human affairs form a cycle, and that there is a *cycle of all other
things that have a natural movement and come into being and pass
away*.   This is because all these things are discriminated by Time
and have their beginning and end as though in a sort of period ;
for *even Time itself is thought of as a sort of circle*.   The reason,
again, is that Time is the measure of this kind of locomotion and is
itself measured by it ; so that to say that things which come into
being form a cycle is to say that there is a circle of Time, which
means that it is measured by the circular movement ' (*Phys.* iv,
223*b*, 13 ff.).

How came it that Time was conceived, not as a straight line, but
as a circle ?   Time is more abstract, unsubstantial, phantom-like,
than Space.   What fills Space is body that we can see and handle ;
what fills Time is movement, and above all the movement of life :
the very word αἰών means both ' time ' and ' life '.   And, as Aristotle
says, there is a cycle of all things that have a natural movement
and come into being and pass away.   The four elements of his
system have a natural movement in the dimensions of Space ; but
they endure for ever, and their motion is straight.   But life, that
comes into being and passes away, moves in the cycle of Time, the

---

[1] At the outset (218*b*, 1) it has been mentioned that some (Plato, according
to Eudemus and Theophrastus) had identified Time with the movement of
the universe ; others (Pythagoreans, Diels, *Vors.* 45B, 33) actually with the
heavenly sphere itself, ' because all things are in Time and also in the sphere '.
Aristotle speaks of this second view as too archaic and naïve for discussion.

wheel of becoming—birth, growth, maturity, decay, death, and rebirth. These words at once suggest the origin of the circular image of Time. It is borrowed from the revolving year—*annus, anulus*, the ring. Hermippus, in his comedy *The Birth of Athena*, thus describes the year, *Eniautos* :

> ' He is round to look at, and he revolves in a circle, containing all things in himself ; and as he runs round the whole earth he brings us men to birth. His name is *Eniautos* ; and being round he has neither end nor beginning, and will never cease wheeling his body round all day and every day ' (frag. 1, Meineke).

The year, says Hermippus, ' contains all things in himself ' (ἐν αὐτῷ). There is an allusion to the derivation of *Eniautos* from ἐν ἑαυτῷ, which we also find in Plato's *Cratylus*. Socrates there explains the two words for ' year '—*eniautos* and *etos*—as significant when taken together : they express that which *seeks within itself* (τὸ ἐν ἑαυτῷ ἐτάζον) and brings forth into the light all things, in turn, that are born and come into being.[1]

In Empedocles' system the old seasonal ' powers ' of summer and winter—the hot, the cold, the moist, the dry—are erected into elements by identification with fire, air, water, and earth. These four ' prevail in turn as the circle of Time comes round ',[2] just as earlier they had prevailed in turn as the seasons came round in the circle of the year. Like Empedocles, Plato speaks here of Time ' revolving ' according to number.[3] Proclus remarks on this that Time revolves as the first among things that are moved ; by its revolution all things are brought round in a circle. He says explicitly that the advance of Time is not like a single straight line of unlimited extent in both directions, but limited and circumscribed.[4] He understands Plato's phrase ' throughout *all time* ' (36E) as meaning the Great Year, the ' single period of the whole ', which embraces all the periods of the planets and contains all Time, ' for this period has as its measure the entire extent and evolution of Time, than which there can be no greater extent, save

---

[1] Cf. Plut., *def. orac.* 12, 416A, ἐνιαυτὸς ἀρχὴν ἐν αὑτῷ καὶ τελευτὴν ὁμοῦ τι πάντων ὧν φέρουσιν ὧραι γῆ δὲ φύει περιέχων. Lydus *de mens.* ii, 4, ἐνιαυτὸς παρὰ τὸ ἐν ἑαυτῷ κινεῖσθαι αὐτόν · κύκλος γάρ ἐστιν ἐφ' ἑαυτὸν εἰλούμενος. Ps.-Hippoc. π. ἑβδ. 16. Soph. *Aj.* 646, ἅπανθ' ὁ μακρὸς κἀναρίθμητος χρόνος φύει τ' ἄδηλα καὶ φανέντα κρύπτεται.

[2] *Vors.* 21B, 17, 29, ἐν δὲ μέρει κρατέουσι περιπλομένοιο χρόνοιο. The same line recurs 26B, 1, with κύκλοιο for χρόνοιο.

[3] 38A, χρόνου . . . κατ' ἀριθμὸν κυκλουμένου.

[4] Pr. iii, 29, ὡρισμένη τε καὶ περιγεγραμμένη.. Contrast Locke (*Essay*, Bk. ii, ch. 15, § 11) : ' duration is but as it were the length of one straight line, extended *in infinitum* '. It is interesting that Locke (in ch. 14) requires a long argument to dissociate Time from the celestial revolutions.

by its recurring again and again ; for it is in that way that Time is unlimited ' (ii, 289). ' The motion of Time joins the end to the beginning, and this an infinite number of times ' (iii, 30³¹).

38C-39E. *The Planets as instruments of Time*

Before proceeding to the creation of all the everlasting heavenly gods who are to be enshrined in the system of revolutions already prepared, Plato takes first those among their number, namely the Planets, whose special utility to mankind lies in their marking off the periods of time and so teaching men to count and calculate. He remarks later (47A) that the observation of these regular periods led to the discovery of number, to all inquiry into nature, and to philosophy itself.

38C.   In virtue, then, of this plan and intent of the god for the birth of Time, in order that Time might be brought into being, Sun and Moon and five other stars—'wanderers', as they are called—were made to define and preserve the numbers of Time. Having made a body for each of them, the god set them in the circuits in which the revolution of the Different was moving ¹—in seven circuits seven bodies :

D.   the Moon in the circle nearest the Earth ; the Sun in the second above the Earth ; the Morning Star (Venus) and the one called sacred to Hermes (Mercury) in circles ² revolving so as, in point of speed, to run their race with the Sun, but possessing the power contrary to his ; whereby the Sun and the star of Hermes and the Morning Star alike overtake and are overtaken by one another. As for the remainder,³ where

---

¹ As Pr. (iii, 59²⁹) remarks, the revolution (περίοδος) of the Different is still spoken of as a single movement of the soul as a whole, going on in all the seven circuits (περιφοραί) among which it is distributed. περιφορά means primarily the circular *motion*, rather than the circular track ; cf. *circuitus*.

² εἰς [τὸν] τάχει μὲν ἰσόδρομον ἡλίῳ κύκλον ἰόντας, Burnet. ' Venus and Mercury are put into circles which have the same period as the sun, but not into one and the same circle. The construction is εἰς (κύκλους) ἰόντας ἰσόδρομον ἡλίῳ κύκλον, κύκλον being an accusative of the internal object after ἰόντας' (Tr.). A.-H. followed Stallbaum in accepting τούς, which appears as a correction in Y and yields the same sense as the omission of τόν. The reading τόν is as old as Albinus, *Didasc.* xiv, φωσφόρον δὲ καὶ τὸν ἱερὸν Ἑρμοῦ λεγόμενον ἀστέρα εἰς τὸν ἰσοταχῆ μὲν ἡλίῳ κύκλον ἰόντα (sic), τούτου δὲ ἀφεστῶτα. It is possible that those who read τὸν understood Plato to have held Heracleides' theory that Venus and Mercury revolve as satellites round the Sun. There would then be only one main circle for all three, the Sun's. But Plato certainly did not hold this. See Heath, *Aristarchus*, pp. 255 ff.

³ The three outer planets, Mars, Jupiter, Saturn. ' Enshrined ' rather over-translates ἱδρύσατο, but the planets are gods and ἱδρύεσθαι θεόν means ' setting up (a statue of) a god ' for cult purposes.

38D. he enshrined them and for what reasons—if one should
   E. explain all these, the account, though only by the way,
would be a heavier task than that for the sake of which it
was given. Perhaps these things may be duly set forth later
at our leisure.

The only difficulty here lies in the statement that Venus and Mercury
(or their circles) ' possess the power contrary to that of the Sun '.[1]
As we have seen (p. 80), the Sun, Venus, and Mercury form a
group with ' similar speed ' (the same angular velocity), which run
their race or finish their course together (ἰσόδρομοι), in the sense
that all complete their journey through the signs of the Zodiac in
a solar year. In contrast with this group, the Moon moves in the
same direction as the Sun, but considerably faster. The three
outer planets had that ' apparent counter-revolution ' mentioned
in the Myth of Er, which we explained by the self-moving power
of their individual souls. Its result was that, relatively to the
circles of the other four, their circles were credited with an additional
contrary movement, slowing down the common motion of the
Different. The effect of this contrary power or tendency, as so
far considered, was that they passed through all angles of divergence
from the Sun, returning into conjunction with him only at intervals
longer than a solar year.

What are we now to make of the statement that Venus and
Mercury ' possess the power contrary to that of the Sun ' ? Evidently not that their behaviour conforms in all respects to that of
the three outer planets. Venus and Mercury do not pass through
all angles of divergence. They keep, as Plato knew, always in the
neighbourhood of the Sun. We are told what phenomenon is
explained by this contrary tendency in the following words : ' whereby the Sun, Mercury, and Venus alike overtake and are overtaken
by one another '. Venus and Mercury, though never far from the
Sun, sometimes get ahead of him and appear as morning stars,
sometimes drop behind, as evening stars.[2] The three are like a
group of racers who reach the goal together (ἰσόδρομοι), but on
the way now one, now another, is in front.

The ancients were not agreed as to the nature of the contrary
power which accounts for this phenomenon, partly because some
were disposed to introduce the complication of epicycles, of which
there is no trace in Plato. But Theon, Proclus, and Chalcidius
all mention the view that, whereas the Sun keeps steadily on at

---

[1] 38D : τὴν δὲ ἐναντίαν εἰληχότας αὐτῷ δύναμιν · ὅθεν καταλαμβάνουσίν τε καὶ καταλαμ-
βάνονται κατὰ ταὐτὰ ὑπ' ἀλλήλων ἥλιός τε καὶ ὁ τοῦ Ἑρμοῦ καὶ Ἑωσφόρος.

[2] Cf. Tim. Locr. 96E. Pr. iii, 66[8].

the same pace, the other two move sometimes faster, sometimes slower.[1] Since Plato nowhere says that each planet moves with a uniform velocity, this view is consistent with the text. I see no reason why it should not be accepted.[2]

Plato has not explained here why the motions of Venus and Mercury have this additional complication, not shared by the Sun. Some ancient interpreters accounted for the variations of speed by the volition of the planets, as living creatures with souls having the power of self-motion.[3] This explanation may be supported

[1] Pr. iii, 66[16]. Theon, p. 222, ' The Sun traverses the signs in a year of about 365½ days. Venus and Mercury *with a movement that is not uniform* (ἀνωμάλως), differing to a small extent in their times, but on the whole running their race with the sun, being always seen in his neighbourhood. Hence they overtake and are overtaken by him.' Chalc., p. 176, ' What he means by these stars having a similar speed, Plato himself explains : they all complete their course in a year, but so that, *moving sometimes slower, sometimes faster*, they now overtake, now are overtaken by, the Sun ; p. 137, *Lucifer* (Venus) *et Stilbon* (Mercury) *imparibus quidem gressibus, isdem tamen paene temporibus quibus sol cursus conficiunt, modo incitato uolatu comprehendentes eum, modo pigro tractu demum ab eodem comprehensi.*

[2] On the question whether and in what sense the motions of the planets are ' uniform ' (ὁμαλής), the ancient commentators are confused. They do not keep distinct (1) what Plato probably thought ; (2) various phenomena which were only discovered later ; (3) later theories of planetary movement, involving concentric spheres, epicycles, eccentrics, etc., which are foreign to Plato's scheme. Tr. (p. 202) concludes : ' Timaeus does not tell us *why* the two planets and the sun in turns gain on one another. No explanation could be offered by a man who assumed all three to be revolving with uniform velocities in the same sense and with the same period in concentric circular orbits.' This seems to me a reason for concluding that Timaeus does not make all these assumptions, which would render the phenomenon not merely inexplicable but impossible.

[3] Pr. iii, 64[8], 117[1], 147[2], δεῖ τὴν ποικιλίαν αὐτὴν ἐξάπτειν τῆς κινήσεως τῶν ψυχῶν, κατὰ τὴν ἐκείνων βούλησιν θᾶττον ἢ βραδύτερον κινουμένων τῶν σωμάτων, ἀλλ' οὐ δι' ἀσθένειαν, ὅπερ οἱ πολλοὶ νομίζοντες, κτλ. Id., *in remp.* ii, 233[2] : according to the *Timaeus* the planets have the following motions : (1) the motion of the Different, a ' single simple motion ' carrying the entire spheres (circles) of all the planets from W. to E. ; (2) axial rotation of each planet (for every divine body must have a circular movement) ; (3) composite movement of the spiral twist ; (4) ' *Some have also a forward and backward movement according to their own will, without ever departing from the movement about their proper centres.*' Chalc., p. 179 : He says (38E) that the heavenly bodies were ' bound with living bonds ', i.e. that the stars are animate and understand the commands of the god, so that not only the planets . . . should possess soul and life, but also the universe with all these should have soul and participate in reason. At *Epin.* 986B the eight circles are actually called ' eight sister powers ' (δυνάμεις), *in* which the heavenly bodies move, either of their own motion or like riders in a chariot. The question where the power resides is left open, as at *Laws* 899A. Cf. also Albinus, *Didasc.* xiv, ὀγδόη δὲ πᾶσιν ἡ ἄνωθεν δύναμις περιβέβληται. πάντες δὲ οὗτοι (stars and planets) νοερὰ ζῷα καὶ θεοί.

by the statement in the *Epinomis* [1] that the revolutions of Venus and Mercury are ' in speed about equal to the Sun, and on the whole neither swifter nor slower. It must needs be that, of these three, the one which has a mind equal to the task leads the way '. The last words indicate that the individual motions of these celestial gods, as distinct from the two motions (of the Same and the Different) to which they are all alike subject, are due to the volition of their own rational souls. The *Laws* (898D) plainly asserts that, besides the Soul which drives the whole heaven round, every one of the heavenly bodies is moved by an individual divine soul. What function can these individual souls have, if not to originate those elements in the motions of stars and planets which are not attributable to the two motions of the World-Soul ? [2] *Laws* 898E suggests three possible ways in which the soul of a star might be related to its body. (1) The soul may reside within the whole spherical body, and move it as our souls move our bodies. (2) Or the soul may provide itself with a body of its own, consisting of fire or air, which envelopes the star's body on the outside and moves it mechanically.[3] (3) Or the soul may have no body at all and guide the star by ' some surpassingly wonderful powers (δυνάμεις) which it possesses '. The ' contrary power ' possessed by Venus and Mercury may be one of these wonderful powers, residing in their individual souls. The Sun leads the whole group because of his superior intelligence, as the *Epinomis* says. The other two possess a power which sometimes counteracts his to some small extent, but on the whole they follow his lead, as he keeps steadily on his course with the actual motion of the Different.

On this view ' *the* power contrary to that of the Sun ' (and to the Different) is, as the words would naturally imply, the power already mentioned in the original account of the planetary circles. The three outer planets exhibited that power constantly, with the result that they passed through all angles of divergence. Venus

---

[1] 986E, δεῖ (Burnet : ἀεὶ libri) τούτων τριῶν ὄντων τὸν νοῦν ἱκανὸν ἔχοντα ἡγεῖσθαι (trans. Harward).

[2] Pr. iii, 70, recognises the two revolutions of the World-Soul as a whole, and seven souls of the planets. In his Platonist period Aristotle maintained that the heavenly bodies (including the planets) were gods and that their motion was voluntary, Π. φιλοσ. fragg. 23, 24.

[3] 898E, ἢ ποθεν ἔξωθεν σῶμα αὐτῇ πορισαμένη πυρὸς ἤ τινος ἀέρος, ὡς λόγος ἐστί τινων, ὠθεῖ βίᾳ σώματι σῶμα. I take this to mean that the star's soul might reside, not in the star's body as a whole, but in an envelope of fire or perhaps of air, ' somewhere on the outside ' (?) of the star's body. The envelope would then be directly moved by its indwelling soul, and would ' push ' the star's body along with it. This seems to be the meaning, even if ποθεν ἔξωθεν be taken with πορισαμένη (which seems most natural) or with ὠθεῖ.

and Mercury exhibit it only intermittently, sometimes dropping behind the Sun, but then quickening their pace to overtake and pass him. Hence their two circles were not reckoned among those which have a motion in the opposite sense to the Sun and Moon. The intermittent dropping behind of Venus and Mercury could not be mentioned in that earlier passage, because it was concerned only with circles representing motions, not with the bodies which have now been created to occupy the circles and possess the motions. Only the main, constant, motions could there be described.

But, it has been objected, if the contrary power here is the same as that mentioned in the account of the soul circles, why is it ascribed only to Venus and Mercury, not also to the three outer planets? The answer is that Plato does not deny it to them. In this passage he mentions the planets in their order from the Earth outwards: first the Moon, then the group of three, Sun, Venus, and Mercury. Of these he notes that, though all three have the same (annual) period, two possess the contrary power which explains why they sometimes drop behind, sometimes get ahead. The remaining three (Mars, Jupiter, Saturn) are dismissed in the last sentence, with the remark that it would take too long to describe their individual motions in detail.[1] It is not denied that they too possess the contrary power which has been already assigned to them for another purpose. The implication is rather that they do possess it, since we are told that their motions are too complicated for description here (cf. 40C), i.e. even more complicated than those of Venus and Mercury. It must be emphasised once more that Plato is not writing a treatise on astronomy, but a myth of creation. The scale of the work demands that the astronomical passages shall be extremely compressed, and we must never assume that some feature which is not explicitly mentioned was unknown to Plato.

In any case, these minor voluntary modifications of planetary motion merely account for changes in the positions of the planets relatively to one another and to the signs of the Zodiac. They do not distort the track of the planet's proper motion, which remains circular. They only counteract, or accelerate, the motion common to them all along their several tracks, as some of our seven passengers on the moving staircase counteract or accelerate its motion by walking in one or the other direction (p. 85).

The upshot, so far, is that the motion of all the planets except

---

[1] In just the same way the *Epinomis* 990B describes the monthly period of the Moon, next, the Sun, who brings the solstices, and ' with him we must group the bodies that keep pace with him ' (Venus and Mercury), and then dismisses ' the remaining paths ' (ὁδούς) as the most difficult to understand.

the Sun is the resultant of at least three components : the motions of the Same and of the Different, which they all share and which are due to the World-Soul as a whole, and individual motions due to the intelligent volition of the planets' own souls, which account for the changes in their relative positions. The Moon alone constantly accelerates the motion of the Different. The remaining five all have the power contrary to the Sun's. Mars, Jupiter, and Saturn exercise it constantly, as we have seen ; Venus and Mercury only intermittently.

There remains the question whether Plato was aware of the phenomenon of retrogradation, as distinct from a mere lagging behind without change of sense in the planet's motion. Against the background of the signs, all the five planets, Venus, Mercury, Mars, Jupiter, Saturn, appear not merely to slow down their main movement, but actually to stand still in their courses, move backwards a certain distance, and then forward again. Proclus [1] held that Plato did recognise actual retrogradation, and there is good reason to believe that this striking phenomenon had been observed. It was provided for in the system of Eudoxus, which must have been familiar to Plato. I suggest that this backward movement, exhibited by all five planets, is here accounted for by the ' contrary power ', explicitly in the case of Venus and Mercury, and implicitly (as I contend) in the case of the other three. Venus and Mercury have the power periodically to reverse the motion they share with the Sun, to come to a stand, and then catch up with him and pass him, though their main annual movement has the same sense and period as his. Mars, Jupiter, and Saturn are always dropping behind the Sun with their constant counter-revolution ; and they have also the power to modify this motion in retrogradation, come to a stand, and then make good the lost ground by speeding up on their normal course. All this might well be thought too complicated to be explained here, without a model of the celestial motions. The accompanying diagram [2] may help the reader to grasp the apparent motions of retrogradation exhibited by Jupiter as seen from a central Earth. The lines marked S^1 and S^4 are those on which the Sun and Jupiter come

---

[1] Pr. iii, 68, ' The Sun does not diminish or augment his speed and has no stations, but Mercury and Venus have advances, stations, and retrogressions ; hence you may say that, according to the observed facts, they possess contrary powers relatively to the Sun. . . . And since they sometimes move quicker, sometimes slower, and do not all move more quickly, or more slowly, at the same time, naturally the more quickly moving overtake the others, and then are overtaken in their turn.'

[2] Adapted, with simplifications, from Bouché-Leclercq, L'astrologie grecque (1899), p. 120.

into conjunction, the Sun moving about twelve times as fast as Jupiter. Between each conjunction and the next Jupiter appears

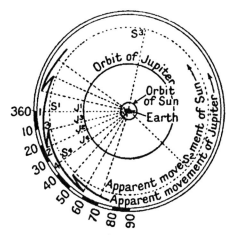

RETROGRADATION OF JUPITER.

The Sun at S^1 and Jupiter at J^1 are in conjunction. When the Sun has moved about 120° to S^2 Jupiter has moved about 22° to J^2 (the first station). While the Sun is moving from S^2 to S^3 Jupiter appears to go back about 10° to J^3 (second station). He then goes forward again and returns at J^4 to conjunction with the Sun at S^4

to stand still, move backwards, stand still again, and then move forwards with accelerated speed. While moving backwards (from J^2 to J^3) Jupiter is exercising his contrary power with more than his normal vigour. Mars and Saturn, having different speeds from his, behave in this way at widely different intervals. All these complications are obviously too intricate for description without models or diagrams.

It may be added that Chalcidius, where he mentions the (correct) view that some of the planetary circles have a movement contrary to others, enumerates the three sets of phenomena which this contrary movement is to account for.[1] They are the three which my explanation covers: (1) the contrast between the steady, forward movement of Sun and Moon and the retrogradations of the other five planets; (2) the contrast between the evening appearance of the Moon (as the fastest) and the morning risings of Mars, Jupiter, Saturn (as the slowest); (3) the peculiar behaviour of Venus and Mercury, appearing sometimes at dawn, sometimes as evening stars. Chalcidius himself does not understand *how* these phenomena are accounted for, being confused by epicycles and eccentrics, which he invokes. But when we find the same sets of

---

[1] Chalc., pp. 167–8.

phenomena in Theon's explanation, it looks as if both were reproducing a tradition which had come down from someone who did
understand what Plato meant.

38E. To resume : when each one of the beings that were to join
in producing Time had come into the motion suitable to it,
and, as bodies bound together with living bonds, they had
become living creatures and learnt their appointed task,[1]
then they began to revolve by way of the motion of the
Different, which was aslant, crossing the movement of the
39. Same and subject to it [2] : some moving in greater circles,
some in lesser ; those in the lesser circles moving faster,
those in the greater more slowly.

So, by reason of the movement of the Same, those which
revolve most quickly appeared to be overtaken by the slower,
though really overtaking them. For the movement of the
Same, which gives all their circles a spiral twist because they
have two distinct [3] forward motions in opposite senses, made
B. the body which departs most slowly from itself—the swiftest
of all movements—appear as keeping pace with it most
closely.

This paragraph explains two consequences of the theory that the
two main factors in the motions of any planet are the motion of
the Same (affecting the whole universe) and the proper motion of
the planet itself in the opposite sense. This proper motion was
represented by one of the seven circles among which the motion
of the Different was distributed. Here, once more, the motion of
the Different is spoken of as a single motion, common to all the
seven circles. We are now concerned only with the effects of these
two main factors, leaving out of account such complications as
retrogradation and the slowing down of the main motion by the
' apparent counter-revolution '. We have only to think of all the
planets revolving at various speeds in the sense of the Different.

(1) Earlier cosmologies, based on the notion of the vortex, had
supposed that all the heavenly bodies were carried round by the
cosmic eddy in one direction only. The apparent backward move-

---

[1] Here, as at *Laws* 898, it is clearly stated that every planet, like the other
heavenly gods, is a living creature with a body and an intelligent soul. So
Pr. iii, 70¹ ; Chalc., p. 179²².

[2] Reading ἰοῦσαν . . . κρατουμένην. The accusatives (read by Cicero and
Chalcidius) are necessary to the sense. Pr's comments (iii, 74²⁷, ἡ θατέρου
περιφορὰ διὰ τῆς ταὐτοῦ τε εἶσι καὶ κρατεῖται ὑπ' αὐτοῦ, 75⁸ τῆς δὴ θατέρου φορᾶς
ἰούσης διὰ τῆς ταὐτοῦ καὶ κρατουμένης ὑπ' αὐτῆς) show that the accusatives should
be read in his lemma, p. 73²².

[3] ' distinct ' (διχῇ), as being in two different planes. Cf. 89E, τρία τριχῇ.

ment of the planets from W. to E. through the signs of the Zodiac was accordingly explained by their being 'left behind' by the more swiftly travelling fixed stars. Lucretius [1] quotes this view from Democritus :

> 'The nearer the different constellations are to the earth, the less they can be carried along with the whirl of heaven ; for the velocity of its force, he says, passes away and the intensity diminishes in the lower parts, and therefore the Sun is gradually left behind with the rearward signs, because he is much lower than the burning signs. And the moon more than the sun : the lower her path is and the more distant she is from heaven and the nearer she approaches to earth, the less she can keep pace with the signs. For the fainter the whirl is on which she is borne along, being as she is lower than the sun, so much the more all the signs around overtake and pass her. Therefore it is that she appears to come back to every sign more quickly, because the signs go more quickly back to her.'

On this theory of the 'leaving behind' (ὑπόλειψις) of the planets, the Moon, overtaken and passed by all the signs in a month, is really moving more slowly than Saturn, who is overtaken and passed by them all only once in about thirty years and so comes much nearer to keeping pace with the outside of the eddy.

Plato's theory reverses the situation. The outermost movement, which alone affects the fixed stars, is still the swiftest of all ; for they complete their circuit in twenty-four hours. But the contrary movement of the planets is now attributed to their own proper motion, at various rates of speed, in the reverse direction. The 'swiftest' of them is the one which completes this journey in the shortest time, namely the Moon. The 'slowest' is the outermost, Saturn, 'the body which departs most slowly from the swiftest of all movements'. Thus the smaller the orbit, the quicker the body.

If we consider only these proper motions of the planets (neglecting the movement of the Same which equally affects them all), the Moon 'overtakes' Saturn. Taking only a month to complete her orbit, she will pass Saturn nearly once every month. But, as Plato adds, 'by reason of the movement of the Same, those which revolve most quickly *seem* to be overtaken by the slower, though really overtaking them'. The movement of the Same carries stars and planets together round the Earth once every day. Suppose that at 10 p.m. to-night the Moon and Saturn are in a line with a certain star in the Zodiac. By 10 p.m. to-morrow the fixed star

---

[1] Lucr. v, 621 ff. (trans. Munro) = Democritus 55A, 88. Cf. Frank, *Plato u. d. sog. Pyth.* 204.

will have come round to the same position. Saturn will have shifted only a very little way eastwards, the Moon a much longer distance. If (as on the vortex theory) we think only of this diurnal movement of the Same and of the planets as trying to keep pace with it, Saturn will have lost much less ground than the Moon and will appear to overtake and pass her. Thus, ' by reason of the movement of the Same those which revolve most quickly appear to be overtaken by the slower, though really overtaking them '. But if we realise that all the planets are trying to make their own way *against* the diurnal movement, Saturn will have gained least ground and be really the slowest ; and the Moon will overtake and pass him.

Plato makes the same point again at *Laws* 822A, where he declares that the planets do not really ' wander ' about, but each one (in respect of its proper motion) ' always travels in a circle one and the same path '. He adds that ' the quickest of them is wrongly supposed to be the slowest, and vice versa '. The false opinion was due to not recognising that each planet has its own proper motion in the reverse direction, and imagining that the planets were merely ' left behind ' by the fixed stars.

(2) The second consequence of the double motion theory is the spiral twist. Martin [1] explains as follows : ' The Sun, for example, which in this system is a planet, describes from the winter to the summer solstice, on the surface of a sphere whose radius is the distance of the Sun from the centre of the Earth, an ascending spiral contained between the two tropics ; then it descends again from the summer to the winter solstice describing on the same sphere a spiral inverse to the former one. The two spirals taken together make up as many turns as there are days in the year. The turns of the two spirals become larger as they approach the equator, but they are all traversed in equal times.' If we imagine a model of the celestial Sphere revolving as a whole towards the right, while the planetary rings inside revolve more slowly to the left, the spiral twist will be the track followed by any one of the planets on a sphere with the same radius as that planet's own circle. It will none the less be true that the planet, in respect of its proper motion, keeps always to one circular track, represented by the ring to which it is fixed.

The only connection between the spiral twist and the other question, which planet is really the swiftest, lies in the point that both involve the recognition of a proper motion opposite in sense to the movement of the Same. Plato is writing with extreme compression, in order to keep this astronomical section within due bounds.

[1] Martin ii, 76. Theon, pp. 324, 329, gives a clear account.

39B. And in order that there might be a conspicuous measure for
the relative speed and slowness with which [1] they moved in
their eight revolutions, the god kindled a light in the second
orbit from the Earth—what we now call the Sun—in order
that he might fill the whole heaven with his shining and that
all living things for whom it was meet might possess number,
learning it from the revolution of the Same and uniform.

c. Thus and for these reasons day and night came into being,
the period of the single and most intelligent revolution.[2]

The purpose of the Demiurge is that mankind shall learn to count
and develop mathematics by the exercise of reckoning periods of
time, days, months, and years. The unit for this reckoning is the
shortest division of time produced by the celestial revolutions, the
period of day-and-night (νυχθήμερον) [3] marked by the daily revolution of the whole heavens in the movement of the Same. Mankind
would not observe this revolution, if the Sun were no brighter than
the other planets. The brilliance of the Sun ' shining through the
whole heaven ', followed by the darkness of night and a new sunrise,
brings it home to man that this daily revolution does occur. The
Sun thus provides a ' conspicuous ' unit of measurement, in terms
of which the other periods can be calculated, with ' their relative
speed and slowness '.[4]

39C. The month comes to be when the Moon completes her own
circle and overtakes the Sun ; the year, when the Sun has

[1] καθ' ἅ, A.-H., Fraccaroli. The subject of πορεύοιτο is easily supplied from
the previous sentence or from πρὸς ἄλληλα. Plut. 1007A, alluding to our
passage, has the phrase μέτρον ἐναργὲς τῆς πρὸς ἀλλήλας βραδυτῆτι καὶ τάχει τῶν
ὀκτὼ σφαιρῶν διαφορᾶς. This might support the conjecture τί⟨νι⟩ πρὸς ἄλληλα
βραδυτῆτι καὶ τάχει [καὶ] τὰ περὶ τὰς ὀκτὼ φορὰς πορεύοιτο ' a conspicuous unit to
measure *with what* relative slowness and speed the bodies involved in the
eight revolutions travel '.

[2] The single (undivided) revolution of the Same, which is the only motion
of translation possessed by the fixed stars.

[3] So in the *Epinomis* 978D the Heaven, causing the stars to revolve ' for
many nights and days ', teaches man to count ' one and two ' and so to
advance to other numbers. Cf. 47A and p. 91, note 1 above.

[4] It is man, not the planets and stars, who is to benefit by this ' conspicuous
measure '. This point involves rejecting the MSS. reading καὶ τὰ at 39B, 3.
Tr. (retaining καὶ τὰ) translates : ' That there might be a plain measure of
their relative slowness and speed, and the eight revolutions go on their way,
God kindled a light. . . .' The planets, he explains (quoting Cook Wilson)
need a light to see their way—' a humorous touch '. The humour, if it can
be detected, is irrelevant. Also the *eight* revolutions include the fixed stars.
Can these need the Sun's help to see their way ? ·Tr.'s suggestion that all
planets (like the moon) reflect the sun's light is supported by no evidence.
Plato's point is that the Sun is the only planet bright enough to make the
difference of day and night conspicuous to mankind.

39c. gone round his own circle. The periods of the rest have not been observed by men, save for a few ; and men have no names for them, nor do they measure one against another by numerical reckoning. They barely know that the wanderings of these others are time at all, bewildering as they are in

D. number and of surprisingly intricate pattern. None the less it is possible to grasp that the perfect number of time fulfils the perfect year at the moment when the relative speeds of all the eight revolutions have accomplished their courses together and reached their consummation, as measured by the circle of the Same and uniformly moving.

In this way, then, and for these ends were brought into being all those stars that have turnings [1] on their journey through the Heaven ; in order that this world may be as

E. like as possible to the perfect and intelligible Living Creature, in respect of imitating its ever-enduring nature.

Men have no names like 'month', 'year', for the periods of planets other than the Moon and Sun. These two are the most conspicuous and they both proceed uniformly on their course. The five remaining planets exhibit apparent irregularities, some of which have been mentioned. The complete analysis of their composite motion involves factors additional to the two great motions of the World-Soul. The result is a 'bewildering' ($\dot{\alpha}\mu\eta\chi\acute{\alpha}\nu\varphi$, not 'incalculable') number of motions of surprisingly intricate pattern. Plato must have been acquainted with the system of Eudoxus, which required for each of these five planets not less than four spheres revolving on different axes, in order to reduce their apparent irregularity to a compound of circular motions. Three spheres each were enough for the Sun and Moon. The total of twenty-seven spheres would certainly make a pattern whose intricacy would bewilder a layman. Plato does not commit himself to Eudoxus' system, which may have been recognised at the time as only giving an approximate picture, and was soon to be still further complicated by Callippus and Aristotle. If the 'contrary power' of the five planets has been rightly explained above as causing variations in speed without change of track, Plato's own system is different, and an armillary sphere representing the planetary movements, if it were not required to work mechanically, would be of much simpler construction.

Though the readers of the *Timaeus* would be bewildered by these complications, 'none the less it is possible to grasp' the notion of

---

[1] τροπαί. The Sun, for instance, 'turns back' at the top of its spiral when it touches the *tropic* of Cancer at midsummer.

a Great Year, completed when all the heavenly bodies come back to the same relative positions. This notion was an ancient one, going back to the earliest attempts to arrive at a period of years which would coincide with a number of complete months. Plato extends it to include the periods of the remaining planets. He gives no estimate of its length.[1] There is, as Taylor remarks, no suggestion that the end of the period is marked by any cosmic cataclysm. Such a catastrophe is, in fact, out of the question. The hands of a perfect clock would regain at every moment the position at which they were twelve hours before. Since the celestial clock was never set going at any moment of time, there was never any original position to serve as starting-point. The period, whatever it may be, is beginning and ending at every moment of time. This perpetual recurrence, as the concluding sentence remarks, is the nearest approach that the visible world can make to the eternal duration of the unchanging model. If the language of our passage suggests a period beginning at some one date and ending at another, that is only because the myth speaks as if Time and its instruments had been created at some moment which would mark the beginning of such a period.

39E–40B. *The four kinds of living creature. The heavenly gods*

So far, the planets are the only living creatures, within the universal frame, whose creation has been described. Among the everlasting gods who were to take up their positions in that frame, the planets were singled out because they are, in a special way, the ‘ instruments of Time ’ ; and Plato wished first to define Time in order to contrast the temporal existence of even the everlasting gods with the unchanging duration of the eternal model. Time cannot exist without the clock. Plato, accordingly, had to anticipate the creation of the heavenly gods by mentioning the planets. He now repeats the statement (37C, 38D) that the Demiurge designed to make his image as like as possible to the model. This is to be done by making all the four chief families of living creature, corresponding to the four regions of fire, air, water, and earth.

39E. Now so far, up to the birth of Time, the world had been made in other respects in the likeness of its pattern ; but it was still unlike in that it did not yet contain all living creatures brought into being within it. So he set about accomplishing this remainder of his work, making the copy after the nature of the model. He thought that this world must possess all the different forms that intelligence discerns contained in the Living Creature that truly is. And there

---

[1] See Tr., pp. 217 ff. Heath, *Aristarchus* 172.

39E. are four : one, the heavenly race of gods ; second, winged
40   things whose path is in the air ; third, all that dwells in the
       water ; and fourth, all that goes on foot on the dry land.

The Demiurge himself, however, makes only the living creatures
of the first class, the gods within the heaven.[1] These are the fixed
stars, the planets, and Earth. Since the planets and some of their
motions have already been mentioned, the following sentences
refer specially to the fixed stars. But the planets are brought in
at the end of the paragraph.

40.   The form of the divine kind he made for the most part of
       fire, that it might be most bright and fair to see ; and after
       the likeness of the universe he gave them well-rounded [2]
       shape, and set them in the intelligence of the supreme to
       keep company with it, distributing them all round the heaven,
       to be in very truth an adornment (cosmos) for it, embroidered
       over the whole. And he assigned to each two motions : one
       uniform in the same place, as each always thinks the same
B.   thoughts about the same things ; the other a forward motion,
       as each is subjected to the revolution of the Same and uniform.
       But in respect of the other five motions he made each motion-
       less and still, in order that each might be as perfect as possible.
          For this reason came into being all the unwandering stars,
       living beings divine and everlasting, which abide for ever
       revolving uniformly upon themselves ; while those stars that
       having turnings and in that sense [3] ' wander ' came to be in
       the manner already described.

The stars have spherical bodies, mostly composed of fire, but
containing some portions of the other primary bodies. Without
earth, as Proclus says, they would not be ' solid ' masses resistant
to touch ; and the other two primary bodies are the ' means '
which hold fire and earth together (31B). Their composition is
similarly described in the *Epinomis* (981D) in a passage which
refers to all the heavenly bodies. There is no reason to doubt that
the statement here applies to the planets, as Proclus held.
   ' The intelligence of the supreme ', in which the stars are set,
is a short expression for the revolution of the Same, that rational

[1] At *Rep.* 508 the heavenly bodies are called ' the gods in the heaven '
(τῶν ἐν οὐρανῷ θεῶν).
[2] εὔκυκλον for ' spherical ' is reminiscent of Parmenides 8, 43. εὐκύκλου
σφαίρης, quoted by Plato at *Soph.* 244E.
[3] τοιαύτην. But only in that sense. They are not really ' wanderers ',
but keep to their regular paths, though they ' turn ' back at the limits of
their spiral tracks.

118

motion of the World-Soul which was described (36c) as having the supremacy over the interior motion and in fact affects the whole universe.[1] The circle symbolising the plane of that motion is the equatorial circle of the sphere, over the whole of whose surface the stars are scattered. All the fixed stars move together in the daily revolution, as if they were set in a solid sphere. But there is no material sphere ; the stars move freely, though they keep their relative positions. The rotation of the heaven thus becomes for each individual star an imparted motion of translation : the star moves 'forward' along its circular track parallel to the equator. Every star has also, we are now told, a second motion, rotation on its own axis. The reason is that 'each always thinks the same thoughts about the same things'. Here, for the first time in the *Timaeus*, it is explained why axial rotation is regarded as 'that one of the seven motions which above all belongs to reason and intelligence' (34A).

Every star has its own intelligent soul, 'and accordingly its own proper motion ; for the soul is the source of motion' (Pr. iii, 119). The same is true of the planets, as Proclus remarks. They also must have axial rotation ; and, in fact, the Moon is the only heavenly body whose rotation could actually be observed. She must rotate on her axis in order to keep the same face always towards the Earth. This is a consequence of the free movement of stars and planets. If they were set rigidly in material spheres or rings which carried them round, they would, of course, all have the same face always turned towards the Earth, but it would be possible to deny (as Aristotle does) that they have an independent motion of rotation. Since Plato's circles symbolise movements only and are not material rings, he recognises this rotation as an independent proper movement, due to the individual soul of star or planet.

The last sentence is intended to convey that the statements about the composition and proper movements of the heavenly gods cover the planets, which are, just as much as the 'unwandering stars', divine and everlasting living beings, and must have the movement proper to their intelligent souls.[2] Earlier the planets were treated merely as the instruments of Time, and the periodic motions relevant to this function were alone described. Their axial rotation was not there relevant ; we are to understand that it is added here, as the movement of intelligence, which they possess equally with the fixed stars.

---

[1] Cf. 47B, 7, 'the circuits of intelligence (τοῦ νοῦ) in the heaven'.

[2] So also Albinus, *Didasc.* xiv, 'All these (stars and planets) are intelligent living beings and gods and spherical in shape.' The *Laws* and *Epinomis* leave no doubt on this point.

40B–C. *Rotation of the Earth.*

The Earth is now included, with the stars and planets, as ' the most venerable of all the gods within the heaven '. She, too, is a ' living being, divine and everlasting ' ; as such, she must possess a soul as well as a body, and Soul being defined as ' the self-moving thing ', she may be expected to possess a proper movement of axial rotation, in the same right as the stars and planets. But is this consistent with the rest of Plato's astronomical scheme in the *Timaeus* ? The question has been debated by ancient and modern critics without reaching any agreement. It turns on the interpretation of the word ἰλλομένην (' winds ') in the following sentence :

40B. And Earth he designed to be at once our nurse and, as she
c. winds [1] round the axis that stretches right through, the guardian and maker of night and day, first and most venerable of all the gods that are within the heaven.

The problem is this : (1) Day and Night have been described at 39C as ' the period (or circuit, περίοδος) of the single and most intelligent revolution ', namely the revolution of the Same, which carries round with it (συνεπόμενον, 40A) all the fixed stars. Everything that has been said about this revolution clearly implies that it is a real movement, due to the self-moving Soul of the World,

[1] Burnet reads ἰλλομένην δὲ τὴν περὶ τὸν διὰ παντὸς πόλον τεταμένον, with the note : ' τὴν A P : om. FY Plut '. If τὴν is sound, we must supply ὁδόν, and some movement is certainly intended. But τὴν is omitted not only by Plut. 1006C, ἰλλομένην περὶ τὸν διὰ πάντων πόλον τεταμένον, but also by Aristotle, *de caelo* 293b, 30, ἔνιοι δὲ καὶ κειμένην ἐπὶ τοῦ κέντρου φασὶν αὐτὴν ἴλλεσθαι καὶ κινεῖσθαι περὶ τὸν διὰ παντὸς τεταμένον πόλον, ὥσπερ ἐν τῷ Τιμαίῳ γέγραπται ; by Simplic., *de caelo* 517, 8, ἡ μὲν ἐν Τιμαίῳ ῥῆσις τοῦ Πλάτωνος οὕτως ἔχει · 'γῆν δὲ τροφὸν ἡμετέραν, ἰλλομένην δὲ περὶ τὸν διὰ παντὸς τεταμένον πόλον,' κτλ. (τροφὸν Ab : τροφὸν μὲν Fc), cf. *ibid.* 532, 5, 12 ; by Proclus 133[11] (lemma) ; and, as we may infer, by all who took ἰλλομένην to mean ' packed ' or ' globed ', or maintained that the Earth has no movement, including, e.g. Albinus, *Didasc.* xv, περὶ τὸν διὰ παντὸς τεταμένον σφιγγομένη πόλον, Pr. iii, 136, Ἴλλεσθαι λέγεται περὶ τὸν διὰ παντὸς τεταμένον πόλον, διότι δὴ περὶ τὸν ἄξονα τοῦ παντὸς συνέχεται καὶ συσφίγγεται ; Chalcid. *constrictam limitibus per omnia uadentis et cuncta continentis poli* (trans. p. 41), which he says (p. 187) may mean ' *medietati mundi adhaerentem quiescere terram* ' (impossible with τὴν) ; Theon (representing Adrastus and Dercylides), p. 212, φερομένης δὲ τῆς οὐρανίας σφαίρας περὶ μένοντας τοὺς ἑαυτῆς πόλους καὶ τὸν ἐπιζεύγνυντα ἄξονα, περὶ ὃν μέσον ἐρήρεισται ἡ γῆ (the last words paraphrase our passage, which is not discussed elsewhere in Theon) ; Iamblichus (Pr. iii, 139). Since no ancient authority betrays any knowledge of the reading τὴν, which *must* imply motion, I cannot believe in its antiquity, though I hold that ἰλλομένην does mean motion, and the presence of τὴν would not invalidate my view. At *Phaedo* 108E the Earth is said to be at the centre of the heaven and to stay there because equidistant from the extremity in all directions. There is nothing to show whether or not it is regarded as rotating.

not an apparent movement to be explained by saying that the stars really stand still while the Earth rotates daily. This appears to me indisputable. (2) It follows that, if Plato is consistent, the Earth must stand still, relatively to the diurnal revolution of the stars. If it had an actual daily rotation in either sense, then day and night would not be produced, as they are, by that revolution. Earth would be anything but ' the guardian and maker of day and night '. As Proclus says (iii, 139),[1] she must ' guard ' day and night by not moving, and ' make ' night by her shadow. (3) The chief objection to supposing that the Earth is absolutely at rest is a very serious one. Aristotle (*de caelo* ii, 13), intent on proving that the Earth must be at rest and at the centre of the universe, discusses two other views. (*a*) ' The Italian philosophers known as Pythagoreans hold that there is a fire at the centre, and that the Earth is one of the heavenly bodies (ἄστρα, i.e. planets), creating night and day as it revolves about the centre. They further provide another Earth, in opposition to ours, which they call " Counterearth ".' Some added yet more revolving bodies, which the Earth hides from our sight, to account for eclipses. (*b*) ' Some, again, say that the Earth, though situated at the centre, " winds ", i.e. moves, " round the axis which stretches right through ", as it is written in the *Timaeus*.' It is beyond question that Aristotle interprets our passage as meaning that the Earth is situated at the centre, not a planet revolving round a central fire ; and that it has a ' winding ' motion round the axis of the universe.[2] What sort of motion he understood will appear later.

Modern critics have been driven to suppose either (as some ancients thought) that Aristotle misunderstood the word ἴλλεσθαι, or that he deliberately misrepresented Plato's doctrine. Others think that neither Plato nor Aristotle noticed that an axial rotation was inconsistent with the earlier statement that day and night are the period of the revolution of the Same. But if Aristotle had wished to misrepresent Plato, he would have done better to make out that the phrase means planetary revolution at a distance from the centre and to class Plato with the Pythagoreans, instead of carefully distinguishing his view from theirs ; for the Pythagorean view, which removes Earth from the centre, is to Aristotle the more objectionable and bears the main brunt of his attack. That neither

---

[1] Following Plut. 1006E.

[2] Cf. the summary of the *Timaeus* in Diog. L. iii, 75, οὖσαν δὲ ἐπὶ τοῦ μέσου κινεῖσθαι περὶ τὸ μέσον. The author of *Tim. Locr.* (97D) understood that the earth is at the centre (ἐν μέσῳ ἱδρυμένα, a word which does not exclude motion). So did Albinus, *Didasc.* xv. κεῖται δὲ ἡ μὲν γῆ τῶν ὅλων μέσῃ, adding a paraphrase of the rest of our sentence.

Plato nor Aristotle should notice the discrepancy is to me incredible. Eudoxus and his students had been working at the Academy on the problem of the celestial motions, and surely someone would have pointed out the contradiction, if Plato and Aristotle were too stupid to see it. These suggestions are desperate expedients, which ought to be cheerfully abandoned, if it could be shown that the Earth can have some circular motion without upsetting the explanation of day and night as due to the diurnal revolution of the Same.

If there is any such possible motion, the choice lies between (*a*) planetary revolution at a distance from the centre and (*b*) some 'winding' motion at the centre. These are the only alternatives, known to Aristotle, to an absolutely stationary Earth. '*All*,' he says, 'who deny that the Earth is situated at the centre think that it revolves (as a planet) about the centre.' (These are the Pythagoreans previously mentioned.) Some admit that the Earth is at the centre, but assign to it a 'winding' motion round the axis, 'as is written in the *Timaeus*'.[1]

(*a*) The difficulties supposed to be involved in any axial rotation have led some critics to obliterate this clear distinction and to identify the winding motion of the *Timaeus* with planetary revolu-

---

[1] Aristotle's 'all' formally excludes Burnet's suggestion of 'a motion up and down (to speak loosely) on the axis of the universe itself' (*E.G.P.*³ 303). It is hard to take this explanation of ἴλλεσθαι seriously. Even if it were true that 'the only clearly attested meaning of the rare word ἴλλομαι is just that of motion to and fro, backwards and forwards: Cf. Soph., *Ant.* 340, ἰλλομένων ἀρότρων ἔτος εἰς ἔτος', it may be remarked that ploughs do not go backwards and forwards *in the same furrow*, but *wind* to and fro in a serpentine track. This is not oscillation, and cannot be supported by the oscillation (αἰώρα) of water inside the Earth at *Phaedo* 111E. Aristotle (*Meteor.* 356a, 5) describes this up and down movement as 'oscillation about the centre' (περὶ τὸ μέσον εἰλεῖσθαι), but 'about the centre' is not the same thing as 'about the axis'. Oscillation *along* the axis is not compatible with 'the only admissible translation: Earth, our nurse, going to and fro on its path *round* the axis' (Burnet, *Gk. Phil.* 348), as Heath observes (*Gk. Astr.* xli). There is no trace in the history of Greek astronomy or, so far as I know, anywhere else, of this grotesque notion that the Earth jumps up and down along the axis. Such a motion would upset the whole theory of the Spiral Twist. It certainly never occurred to any ancient commentator that Plato meant this or that Aristotle was arguing against this view. His description of the Earth as 'situated' or 'lying' (κειμένην) at the centre excludes oscillation to and from the centre. If the most venerable of the gods within the heaven has any motion, it can only be the circular motion of reason, not any of those rectilinear motions which are expressly excluded for all the other gods. Mr. F. H. Sandbach has pointed out to me that in Sext. Emp. *math* x, 93, αἱ περὶ τοῖς κνώδαξιν εἰλούμεναι σφαῖραι unquestionably means 'spheres rotating on pivots'. Armillary spheres may be meant. Cf. also Lydus *de mens.* ii, 4, κύκλος ἐφ' ἑαυτὸν εἰλούμενος. I cannot pursue the intricacies resulting from Tr.'s theory that Timaeus puts forward a view which Plato did not hold himself.

tion at a distance from the centre. The *Timaeus*, in fact, is, in spite of Aristotle,[1] to be interpreted as stating the Pythagorean theory. The objections to this view are, to my mind, overwhelming.

(1) Plato clearly implies that the effect of the movement, if movement it be, is that Earth is the guardian and maker of day and night.[2] The inference is that the period of the alleged planetary revolution is twenty-four hours. In accomplishing this daily revolution, does the Earth also rotate on her axis, like the moon, so as to keep the face on which we live always turned away from the centre of the universe? So the Pythagoreans held. But if so, her revolution will, according to its sense, either cancel the effect of the daily revolution of the stars, and there will be no day and night, or else the period of day and night will be forty-eight hours. If she does not rotate on her axis, the face we live on will be turned once every day towards the centre of the universe. This, according to these Pythagoreans, is occupied by the Central Fire. Why, then, do we never see this Central Fire crossing the skies?

(2) If the Earth revolves as a planet, why was not the circle of the Different divided into eight (not seven) planetary circles? Why was not the Earth reckoned among the planets where they were described as the instruments of Time? Why was her period not counted, as a ninth, with the eight others whose consummation makes up the period of the Great Year? I can see no answer to these questions.[3]

[1] And Proclus, who does not doubt that Earth is at the centre (iii, 133). On 41D (sowing of the souls into Earth and into the planets as instruments of time) he remarks : οὔτε γὰρ ἡ γῆ ἄστρον . . . οὔτε τὰ ἀπλανῆ ὄργανα εἴρηται χρόνου . . . μόνα δὲ τὰ πλανώμενα καὶ ἄστρα ἐστὶ καὶ ὄργανα χρόνου. Indeed, a statement of Theophrastus (which we shall consider later) that Plato in his old age repented of having given earth the central position, stands alone. All other ancient authorities either state or assume that Plato's Earth was at the centre.

[2] This has been denied, e.g. by Tr. (p. 240), but Tr., like other translators, ignores the effect of μὲν and δὲ in the sentence. The god gives Earth two functions : he makes her (1) τροφὸν μὲν ἡμετέραν, (2) ἰλλομένην δὲ . . . φύλακα καὶ δημιουργὸν νυκτὸς καὶ ἡμέρας : ' to be *at once* our nursing-mother, *and*, as winding round the axis, the guardian and maker of day and night'. There is no proper contrast between τροφόν (μὲν) and ἰλλομένην (δὲ). The translation : ' But earth, our foster-mother, that goes to and fro on her path about the axis of the universe, he contrived for a guardian,' etc., simply ignores the existence of μὲν and δὲ.

[3] Pr. iii, 138[11], urges the last point as an argument against any movement of the Earth. Tr. (p. 239) admits a contradiction, but attributes it to a ' want of adaptation ' of Timaeus' views about the Great Year and about the movement of the Earth; Timaeus has ' no finished system '; he is ' engaged on the working out of a science which is progressive '. On this principle, no statement in the *Timaeus* can be used to determine the meaning of any other; the science may always have progressed in the interval.

(3) If Earth is not at the centre of the universe, what is at the centre ? The only alternatives are : the Pythagorean Central Fire and no solid body at all. The second is entirely incredible. No ancient system of astronomy ever contemplated the possibility that the centre of the world should be unoccupied. Aristotle, writing before the heliocentric theory was propounded, says that *all* who regarded the entire universe as finite [1] held that Earth was at the centre, with the exception of the Pythagoreans, who had their Central Fire. We must, then, assume (as the adherents of the planetary theory do) that the Earth is to revolve round the Central Fire. But the *Timaeus* says nothing whatever about any Central Fire. Can anyone believe that, if Plato had thought of the Earth as a planet, he would have made no mention at all of the body round which Earth, planets, and stars all revolve ? [2] No writer with that picture in his mind could describe the motion of the Earth as ' winding *round the axis that stretches all through* '. Moreover, the very existence of a free body of fire at the centre contradicts the whole theory of the natural motions of the primary bodies. We learn later that the main body of fire is at, or towards, the circumference (63B ff.), and that every primary body has a natural tendency towards its like. The so-called ' lightness ' of fire is explained by this tendency. If we can imagine someone stationed aloft in the region of fire and trying to force fire ' downwards ' into the alien region of air, he would find that fire resisted his efforts and he would have to call it ' heavy '. We shall later come to an elaborate account of the interaction of the primary bodies, explicitly designed to explain why all the fire in the universe has not escaped to the main body on the outside (58A). All this flatly contradicts the notion of a free body of fire properly situated at the centre.

I conclude that, when Plato said that the Earth ' winds round the axis ', he did not mean that it revolves at a distance from the axis round a body which he never mentions and which cannot

[1] As distinct from the Atomists, who believed in an unlimited plurality of worlds scattered over infinite space. Infinite space has no centre. But even the Atomists held that our Earth is at the centre of its own world.

[2] Burnet (E.G.P.³ 304) says : ' We know from the unimpeachable authority of Theophrastus, who was a member of the Academy in Plato's later years, that he had then abandoned the geocentric hypothesis, *though we have no information as to what he supposed to be in the centre of our system* ' (my italics). Unless this is an oversight, Burnet's Earth must bounce up and down at some distance from the centre. If there were any body there, the collision that would result from the Earth reaching the centre or attempting to cross it would have frightful consequences. Tr. (p. 235) recognises that on this view (which he attributes to Timaeus, but not to Plato) the centre must be empty except when the Earth happens to be just passing the centre in its excursions.

exist in his physical system; or that it revolves round nothing at all. Finding no support whatever in the *Timaeus* itself, the adherents of the planetary theory fall back on a statement which Plutarch attributes to Theophrastus: that 'Plato, when he had grown old, repented of having assigned to earth the central position, which did not properly belong to it'.[1] Theophrastus does not often disagree with Aristotle, and the two could be reconciled, if we could suppose that Plato's repentance took place after he had written the *Timaeus*. But then we should expect to find the Central Fire and the planetary motion in the *Laws* and the *Epinomis*. Neither of these works ever hints at the existence of the Pythagorean Central Fire, and the passages (*Laws*, 822c and *Epin.* 987B) alleged to support planetary motion are at least capable of other interpretations.[2] If Aristotle had known of Plato's repentance, he had no motive for not mentioning his master's adoption of the Pythagorean scheme.

So far the weight of evidence seems to be against Theophrastus, but perhaps a reconciliation is possible on somewhat different lines. It has been suggested that Aristotle himself alludes to the repentant Plato in his opening passage (*de caelo* ii, 13). He first mentions the Pythagoreans as the only philosophers with a finite universe who do not place the Earth at the centre.

'At the centre, they say, is fire, and the Earth is one of the heavenly bodies, which makes day and night as it revolves round

---

[1] Plut., *Plat. Qu.* viii, 1006c; *Life of Numa* xi. Tr. (p. 228) says, it would be most natural to suppose that the statement of Theophrastus occurred in his Περὶ φυσικῶν δοξῶν, 'where as we see from Aetius, *Placita* iii, 11–13, the questions whether the earth is at the centre and whether it moves were discussed'. But Aetius iii, 11, attributes the doctrine that Fire, not Earth, is at the centre to Philolaus only, without mentioning Plato at all; and 13 attributes a stationary earth to all except Philolaus (planetary motion), Heracleides and Ecphantus (axial rotation at the centre), and Democritus. Aet. ii, 7 (περὶ τάξεως τοῦ κόσμου), says that Plato arranged the elements in the order: 'fire first, then aether, next air, next water, and last earth; though sometimes he connects aether with fire'. Further on, Philolaus' system is described, with 'fire in the midst about the centre', which is 'primary by nature'. If Aetius represents Theophrastus' *Physical Opinions* correctly, Plutarch must have had some other source. That Theophrastus cannot have attributed the planetary theory to Plato in that work may be inferred from Simplic., *de caelo* 513: Alexander said the question who these 'others' were was ἐκ τῆς ἱστορίας ζητητέον. If the answer had been in Theophrastus' history, it would have been found by Alexander and reproduced by Simplicius; but he can suggest no one earlier than Aristotle who had agreed with the so-called Pythagoreans.

[2] See above, p. 89 ff. I may appeal to the fact that so careful and judicious an authority as Sir Thomas Heath has interpreted both passages differently at different times.

the centre. They further provide another Earth in opposition to ours, which they call "Counter-earth", not looking for theories and explanations to account for observed facts, but rather attempting to force facts into agreement with certain theories and opinions of their own.

Many others, however, might agree that it is wrong to give earth the central position, looking for confirmation to theory rather than the facts of observation.[1] *They think that the most honourable place fittingly belongs to the most honourable thing (τιμιωτάτῳ), that fire is more honourable than earth, and the limit more honourable than the intermediate, and centre and circumference are limits. Reasoning from these premisses they think it is not earth that lies at the centre of the sphere, but rather fire.* Besides, the Pythagoreans, at any rate, have the further reason that the most important part (κυριώτατον) of the world ought to be most strictly guarded, and this is the centre, which they call the Guardhouse of Zeus (Διὸς φυλαχήν)—the fire which occupies this position.'

The sentences in italics referring to the 'many others' who might agree that earth ought not to hold the most honourable place, certainly recall Theophrastus' statement and it has been inferred that 'many others' means the elderly Plato and perhaps Speusippus and other members of the Academy. But it is important to observe precisely what these others 'might agree to'. Aristotle has said that the Pythagoreans are *alone* in holding the planetary motion of Earth and Counter-earth round a Central Fire; and later he adds that in the *Timaeus* the Earth is not a planet but at the centre. The 'others' are not said to agree to planetary motion round a Central Fire [2] but to the Pythagoreans' estimate of the element, fire, as more honourable than the element, earth. The most honourable element ought to occupy both centre and circumference because both these are 'limits', and limits are more honourable than what lies between them. That is all. Aristotle then returns to the Pythagoreans.

Now we know from Simplicius that the doctrine of a central fire existed among the Pythagoreans in another form. Some, whom

---

[1] By these 'facts of observation' Aristotle may mean the fact that any piece of earth, and therefore (as he argues) Earth as a whole, has a natural tendency to seek rest at its proper region, the centre. He insists on this in his criticism.

[2] This observation disposes of Tr.'s argument leading to the conclusion that Plato (whom Tr. has earlier identified with these 'others') 'had consistently taught that the earth is a planet during the twenty years of Aristotle's connexion with him in the Academy' (p. 231).

he describes as 'more genuine' adherents of the school, 'mean by fire at the centre the creative power which gives life to the whole Earth from the centre and revives warmth in that part of her which has grown cold. Hence some call it the Tower (πύργον) of Zeus, as Aristotle himself says in his account of the Pythagoreans, some the Guardhouse (φυλακήν) of Zeus, as here, some the Throne of Zeus, as others report. They spoke of the Earth as a 'star' (ἄστρον) in the sense that she is herself too an instrument of time as the cause of days and nights, making day on the side illuminated by the sun and night by her conical shadow. The Pythagoreans gave the name 'Counter-earth' to the Moon (as also 'heavenly Earth'), both as intercepting the Sun's light, which is a peculiarity of Earth, and as marking the limit of the heavenly bodies, as the Earth marks the limit of what is beneath the Moon'.[1] Hilda Richardson [2] used this passage among others to support her view that 'the earliest generations of the Pythagorean school conceived of fire as existing at the heart of their central, spherical earth. It was only the separation of this fire from the earth and the conversion of the earth into a planet that was late'. She claims that this passage in Simplicius shows 'that some Pythagoreans at some period held the doctrine of a central fire hidden in the bowels of the earth and that the doctrine was considered a piece of genuine Pythagoreanism. Simplicius gives no indication of date, but it has been shown above that the doctrine need not necessarily be late. It may quite well have been early.'

It seems to be, at least, a not improbable view that the 'more genuine' Pythagoreans adhered to the primitive doctrine of a fire in the heart of the central Earth. Rejecting the Central Fire of the planetary theory, they transferred its peculiar terminology to established features of the older system. 'Tower of Zeus' becomes another name for their own fire, which may already have been known as the 'throne' or 'guard house' of Zeus; 'Counter-earth' is transferred to the Moon as a 'heavenly Earth'. Since Simplicius

---

[1] Simplic., *de caelo* 512.

[2] *Class. Qu.* xx (1926), p. 119. She writes that this form of the doctrine was 'regarded by Zeller (I⁴, 420) as a late modification of the central fire system described by Aristotle in the *De Caelo* on the ground that the doctrine of the earth's revolution on its axis is only found among the Pythagoreans of the fourth century. But it is not necessary to suppose that the earth in the system described by Simplicius rotated on its axis. (This is pointed out by Sir T. Heath *Aristarchus of Samos*, p. 250.) Rather it is exactly like the central earth of Plato's *Timaeus* which, while possessing no rotatory motion on its axis, yet is called φύλακα καὶ δημιουργὸν νυκτός τε καὶ ἡμέρας, because by remaining fast in its central position on the axis of the cosmos it creates night by casting its shadow. . . .'

actually gives Aristotle's books *On the Pythagoreans* as his authority
for 'Tower of Zeus' as applied to the fire in the centre of the
Earth, I see no reason to doubt that he took the whole account of
both forms of the doctrine from the same source.[1]   In the *de caelo*
itself Aristotle mentions the Pythagoreans who hold that the
centre, as the most important part of the world, needs to be
guarded by a fire called the Guardhouse of Zeus, immediately
after those 'others' who think that the most honourable element
should hold the most honourable place, and that there should
consequently be fire at the centre as well as at the circumference
of the sphere.[2]

If we put all this together, it is a reasonable conclusion that those
'others' did not hold the planetary theory (as indeed Aristotle
implies), but were quite content with the perhaps older doctrine
of a fire in the heart of a central Earth.   If the 'others' are Plato
and Speusippus, the repentance of the elderly Plato may be traced
back to some remark of his, which Theophrastus had heard of, to
the effect that in the *Timaeus* he had wrongly spoken as if the
element, earth, had its proper place in the centre, and the element,
fire, were naturally situated at the circumference.   He had, indeed,
recognised the presence of fire and of the other primary bodies
inside the Earth, both in the *Timaeus* and in the *Phaedo*, where the
central Earth contains rivers of fire, air, and water ; but he ought
to have acknowledged that fire, as the most honourable element,
was not merely entrapped in the Earth but had its rightful place
at the core of the Earth and of the universe.   The last sentence of
the *Critias* describes Zeus as summoning all the gods ' to their most
honourable habitation (τιμιωτάτην οἴκησιν) which stands at the
midst of the universe and surveys all that has part in becoming.'
This is, of course, mythical language; it recalls the procession of
the gods in the *Phaedrus*, where 'Hestia alone stays in the house
of the gods'.   If Hestia there is the Earth, the name at least suggests that Earth is the central hearth of the world.   The *Politicus*
myth (272E) leaves doubtful the situation of that 'place of outlook' (περιωπή), to which the Governor of the universe retires
when he abandons control.   But all these passages suggest that
Plato was familiar with that 'Tower of Zeus' which the more

---

[1] See Ar. frag. 204R.

[2] Proclus (who assumes as a matter of course that Plato's Earth is at the
centre) mentions that ' the Pythagoreans called the centre of the universe
Ζανὸς πύργον, ὡς δημιουργικῆς φρουρᾶς ἐν ἐκείνῳ τεταγμένης ', and says that this
Tower of Zeus is inside the Earth (iii, 141[11], 143[26]).   In the context he refers
to the *Phaedo* as authority for the Earth containing all the elements—rivers
of fire, water, and air—and so being a sort of microcosm.

genuine Pythagoreans identified with the fire at the centre of the Earth.[1]

What is certain is that Theophrastus' statement is, in any case perfectly consistent with the repentant Plato's recognising a fire properly situated at the centre of the Earth. It provides no ground for rejecting Aristotle's plain assertion that the Earth in the *Timaeus* is not a planet but situated at the centre.[2] In the history of astronomy the planetary theory was an aberration, confined, according to Aristotle, to a section of the Italian philosophers who called themselves Pythagoreans, in the early fourth century. As he remarks, they were not trying to account for observed facts, but constructing a system to fit preconceived notions. They did not stick at inventing two non-existent bodies which could never be observed without visiting the antipodes—the Central Fire and the Counter-earth—in order to give fire the most honourable position and to raise the number of circles to the sacred number ten. Plato, we know, had set his own school the task of working out a scheme which should best account for the observed facts ; and Eudoxus, among others, took up the challenge. Plato's attitude towards astronomy had become more scientific since the *Republic*, which recommends the student to dispense with the starry heavens. I cannot believe that in his old age he repented of this attitude and adopted a system which had no future among serious

[1] Hilda Richardson (loc. cit.) developes further the connection between the πόλος διὰ παντὸς τεταμένος of *Tim.* 40C, the διὰ παντὸς τοῦ οὐρανοῦ καὶ γῆς φῶς εὐθύ, οἷον κίονα of *Rep.* 616B, and the World-Soul of *Tim.* 34B, ψυχὴν εἰς τὸ μέσον αὐτοῦ θεὶς διὰ παντός τε ἔτεινεν. She suggests that ' the epithet Ζηνὸς πυργός for the central fire, for which we have the excellent evidence of Aristotle (frag. 204) has some connection with the pillar of the sky-god. At any rate, both this epithet and those which correspond to it, such as Διὸς φυλακή (Ar., *de caelo* 293b, 2), Διὸς θρόνος (Simplic.), and Διὸς οἶκος (Aet. ii, 7, 7, Philolaus) point to connections of the central fire with the sky-god as well as the earth ; and these connections lend some support to the theory that the central fire may have been regarded as flaming upwards and outwards from the earth and may have eventually come to be shaped into the form of a cosmic axis.'

[2] So little foundation is there for Frank's assertion (*Plato u. d. sog. Pyth.* 207) that Theophrastus explicitly attributes the planetary theory to Plato in his old age, and that this remained the system of the Academy after his death : ' fast alle unmittelbaren Schüler Platos haben es gelehrt, Speusipp (*Fr.* 41 *Lang*) ebenso wie Philippus von Opus (V.S. 45B, 36) und Heraklides vom Pontus (*Fr.* 49–59 Voss).' Let us look at the evidence adduced. Speusippus, frag. 41, reads : εἰκῇ γὰρ οἱ περὶ τῆς ὅλης οὐσίας λέγοντες ὥσπερ Σπεύσιππος σπανιόν τι τὸ τίμιον ποιεῖ τὸ περὶ τὴν τοῦ μέσου χώραν, ᾽τὰ δ' ἄκρα καὶ ἑκατέρωθεν᾽. V.S. 45B, 36 (Aet. ii, 29, 4) says that ' certain ' Pythagoreans, κατὰ τὴν Ἀριστοτέλειον ἱστορίαν καὶ τὴν Φιλίππου τοῦ Ὀπουντίου ἀπόφασιν account for eclipses by the Counter-earth. Heraclides, as Heath (*Arist. of Samos* 275ff.) has proved, did not anticipate Copernicus ; there is clear and detailed testimony that he held that the Earth rotates *at the centre*, while the heavens stand still.

astronomers. Neither the *Laws* nor the *Epinomis* has the faintest suggestion of a Central Fire or of a Counter-earth or of a ninth circle for the planetary Earth.[1]

(*b*) So we come back to the question : Can the Earth have an axial rotation compatible with the doctrine that day and night are due to the daily revolution of the fixed stars ? The answer is that she must rotate on her axis relatively to the stars, in order to preserve the effect of that daily revolution.

Some writers have failed to notice that the revolution of the Same is a movement of the World-Soul, which, ' everywhere inwoven *from the centre to the extremity* of heaven and enveloping the heaven all round on the outside, revolving upon itself, made a divine beginning of ceaseless and intelligent life for all time ' (36E). Physically, this is that rational movement whereby the entire spherical body of the world rotates upon its axis (34A). This movement must not only carry the planets with it (as we have seen), but extend from the circumference to the centre and therefore include the Earth.[2] If this movement alone existed, it would be indistinguishable from rest. There would be no change in the relative positions of any parts of the world's body, and there would be no day and night.

In the account of the other heavenly gods, Plato has just added, for the first time, the individual motion of axial rotation, due to the self-moving souls of stars and planets. The Earth is mentioned last. Earth too is a god, ' a living being divine and everlasting ', with a self-moving soul, as well as a body. She ought to have the same property of axial rotation.[3] And she needs it, precisely for

---

[1] On the strength of Theophrastus' statement, Mondolfo (*L'infinito nel pensiero dei Greci*, 329) says that the elderly Plato embraced with the ardour of a neophyte the system which made the Earth revolve round a Central Fire, the source of light, of heat, and of motion to the whole universe, and regarded as of dimensions perhaps greater than those of the earth. Is it credible that Plato should never mention by far the most important body in the universe or explain that the Sun no longer held the position, as source of light and life, which he has in the *Republic* ?

[2] Some modern critics are obsessed by the Aristotelian division of the world into the heavens above the Moon where the celestial bodies have the circular movement proper to the ether, and the sublunary region of the four simple bodies which move in straight lines. This distinction is foreign to Plato.

[3] A passage in the *Epinomis* mentions a movement of the Earth : ' Nothing can receive a soul in any other way than by the action of God, as we have proved. And since God can do this, it is the easiest of things for him, first to put *life* into any body and the whole of any bulk, and then to *make it move* as he has thought best. Now with regard to *all these bodies* [Sun, Earth, and all the stars have just been mentioned] I hope that we may truthfully lay down one conclusion. It is not possible for the *earth* and heaven and all the

the purpose here mentioned—in order that 'winding round the axis' she may be 'the guardian and maker of day and night'. She must rotate on her axis daily in order not to be carried round by the movement of the whole. The effect is that in relation to absolute space she stands still, while in relation to the other makers of day and night, the fixed stars, she rotates once every twenty-four hours in the reverse sense. In the planetary theory of the Pythagoreans Earth rotates on her axis, like the Moon, in order to keep the same face always turned towards the Central Fire. In Plato's theory she rotates so as not to keep the same face always turned towards the same quarter of the revolving heaven. The two notions lie not far apart. It was easy for Heraclides to take the next step and make the Earth's rotation an absolute movement, not merely relative to the fixed stars. The stars can then stand still, while Earth rotates absolutely.

This solution of the problem was all but discovered by Martin, who saw that the Earth must be involved in the revolution of the whole. 'In Plato's system,' he wrote, 'in order that Earth may produce the succession of days and nights, she must resist the diurnal movement of the universe. To an impulse which would make her turn upon herself in a day, she must constantly oppose an equal force in the contrary sense, and remain motionless.' 'If Earth had not an individual soul, a Platonist should say, she would yield without effort to the diurnal motion imparted by the World-Soul to the entire heaven, and then the succession of days and nights would not take place. But she has a soul, whose circles, turning on themselves, give her body a force of rotation contrary and equal to that which she receives from the Soul of the World, whose centre she occupies. The complete immobility of the terrestrial globe is, consequently, the result of two forces of rotation, whose physical effects annul one another, and one of which belongs to her intelligent soul.'[1] It is surprising that, in the same breath, Martin should dispute Ideler's assertion that 'the present participle εἰλλομένην (sic.) indicating a continuous action, ought to express the rotation of the Earth', and should argue at length that ἰλλομένην does not mean any sort of movement, but only that the Earth is closely wound round the axis, to which she clings (as it were) in order to resist the movement that would otherwise carry her

stars with all their solid bodies, unless there is a *soul attached to each*, or actually in each, to carry out accurately their yearly, monthly and daily *movements* . . .' (983B, trans. Harward). Whoever wrote this must have thought of Earth as a living creature with a soul and a movement due to that soul.

[1] Martin, ii, 88, 137.

round.[1] On this point, Aristotle's opinion that ἴλλεσθαι means movement is to be preferred to that of any modern critic.

It remains to ask, what kind of movement Aristotle understood by ἴλλεσθαι. It is unfortunate that in his criticism (*de caelo* ii, 14) he attacks both the Pythagorean planetary theory and Plato's winding motion at the centre simultaneously. Two of his arguments turn on his own doctrine that the natural movement of earth is in a straight line towards the centre of the universe ; from which it follows that both planetary movement and rotation would be 'unnatural', and therefore could not be eternal. Plato, who attributes axial rotation to all the other heavenly gods by virtue of the self-moving power of their individual intelligent souls, and denies them any rectilinear motion, would be unmoved by an argument which assumes that the divine Earth as a whole must behave like any 'clod' that is lifted and falls to the ground. Also in his view the clod falls to Earth because like moves to like ; it does not fall towards the central point of the universe, as such.

The remaining argument is explicitly aimed at both theories :

'Again, everything that moves with the circular movement, except the first sphere, is observed to be passed, and to move with more than one motion' (i.e. all the planets have two motions : the Same, shared with the fixed stars, and the Different, their proper motion in the contrary sense about the axis of the Zodiac, the result of which is that they are passed, or left behind, by the fixed stars). 'The Earth, then, also, whether it move about the centre or as situated at it, must necessarily move with two motions. But if this were so, there would have to be passings and turnings of the fixed stars. Yet no such thing is observed. The same stars always rise and set in the same parts of the Earth' (296*a*, 34 ff.).

As directed against planetary motion, the argument is that Earth, as a planet, ought to have both the contrary motions. But since one of these is oblique to the other, the effect would be that the

---

[1] It may be that the recent neglect of Martin's explanation is due to his denial that ἰλλομένην means motion. A.-H. followed Martin in this (' globed round ') and holds that the Earth must be ' absolutely motionless '. He quotes with approval Martin's view that Earth's soul enables her to *resist* rotation on her axis, which would occur if she were lifelessly carried round with the rotation of the whole. It was the notion that, in order to stay still relatively to the fixed stars, Earth must ' cling ' to a stationary axis, that led so many ancient authorities (like Martin) to paraphrase ἰλλομένη by σφιγγομένη and to imagine the axis as if it were, not a mathematical line, but a solid rod which the Earth could be ' packed round ' and ' cling ' to.

pole of the sphere of the fixed stars would appear to describe a circle in the sky, and the stars would not rise and set as they do.

Simultaneously Aristotle uses this same argument against Plato's movement at the centre. On this theory, too, he says, the Earth must have two motions. Let us disentangle this application in the following dialogue :

ARISTOTLE. The Earth, you say, is situated at the centre and has a winding motion round the axis of the universe.

PLATO. Yes.

AR. But this motion must be a compound of two motions. First, there is the motion of the Same, which the Earth will share as part of the whole body of the world rotating on its own axis.

PL. Clearly.

AR. But if that were all, there would be no day or night. You must have a second motion to counteract the Same and restore day and night.[1]

PL. Exactly. That is why I wrote that the Earth, winding round the axis, is the guardian of day and night.

AR. The Earth, then, has two contrary motions. But the second motion you invoke in the case of the planets is the Different, and that is oblique to the Same.

PL. True.

AR. But if you give the Earth this oblique motion to counteract the other, the compound of the two will have this effect. The actual motion of the Earth will be that of a globe fastened to the axis of a sphere rotating eastwards in the plane of the ecliptic (the motion of the Different), and having its poles fixed in an outer sphere (the fixed stars) rotating westwards in the plane of the equator. The centre of the Earth will always be at the centre of the universe and of both axes. But the poles of the Earth on the axis of the ecliptic will describe circles round the axis of the universe. The effect would be the same as in planetary motion : the pole of the universe (the pole star) would *appear* to describe a circle in the sky, and the fixed stars would not rise and set where they do. This compound movement of rotation on two axes must be the ' winding ' motion you meant by ἰλλομένην. It is certainly hard to find a suitable word.

PL. Your argument is sound, but for one flaw. It rests on the assumption that the second motion is oblique to the first.[2] But I meant by the second motion, not the motion of the Different

---

[1] Observe that Aristotle does not raise the modern objection that rotation of the Earth would upset day and night. He must have understood that Plato gave the Earth two motions in order to preserve day and night.

[2] This is remarked by Simplicius, *de caelo* 537, 20–26.

(which I explicitly limited to the seven planetary circles), but a self-motion of the Earth, whom I regard as a living creature. It is, like the first, rotation 'round the axis of the universe', not round the axis of the Zodiac. So it takes place in the same plane as the first motion—the plane of the equator—and exactly cancels it. I am sorry that my word ἰλλομένην has misled you. But if I had written στρεφομένην or any of the other usual expressions for rotation or revolution, that would have suggested that the Earth, like the planets and stars and the world as a whole, has an absolute rotation. Then people less acute than yourself would have supposed me stupid enough not to see that day and night would be upset. So I chose this word ἰλλομένην—the best I could think of—to describe the Earth 'winding' or 'curling' round the axis. Perhaps it is really more appropriate to that armillary sphere I mentioned in the next sentence. Imagine the machine which my Demiurge made out of his strips of soul-stuff. The axis of the universe is a vertical rod attached at its ends to a vertical (meridian) circle, which serves to support the horizontal equator and the oblique circle of the Zodiac or ecliptic. Suppose that all this part of the apparatus revolves on a pivot in the stand. The Earth is a globe at the centre. It will be kept stationary by being separately supported on a hollow pillar fixed to the stand—hollow, so that the axis rod may turn inside it. The rod passes through a hole in the Earth globe, so as not to carry the Earth round with it. Now as the machine revolves, the axis rod turns round inside the hole through the stationary Earth. But the axis of the universe is really a mathematical line, which cannot turn round. So I looked at the thing from the other standpoint and spoke of the Earth globe as 'winding' or 'curling' round a stationary axis. After all, the Earth *has* a rotatory movement relatively to the fixed stars; and when I added that the purpose of the movement was to preserve day and night, I assumed that no one could misunderstand.

At this point Aristotle would perhaps have admitted that his mind had been confused, partly by his own picture of concentric spheres, partly by the attempt to criticise simultaneously two different views of the Earth's motion. He would, however, have thought Plato sufficiently refuted by his first and third arguments, resting on his own dogma that the only natural motion of earth is rectilinear, towards the centre. When earth is actually at the centre, it can have no motion at all. Plato held that it was at the centre. It cannot, according to Aristotle, rotate there, because rotation, being an 'unnatural' movement, could not be eternal. So, whatever motion ἰλλομένην might mean, Plato was wrong.

40C–D. *The further movements of the heavenly bodies are too complicated for description here*

With the creation of Earth the list of the heavenly gods is complete. The astronomical chapter is now closed with the remark that, without a visible model, all the complicated movements cannot be described.

40C.   To describe the evolutions in the dance of these same gods, their juxtapositions, the counter-revolutions of their circles relatively to one another, and their advances ; to tell which of the gods come into line with one another at their conjunctions, and which in opposition, and in what order they pass in front of or behind one another, and at what periods of time they are severally hidden from our sight and again

D.   reappearing send to men who cannot calculate panic fears and signs of things to come—to describe all this without visible models of these same [1] would be labour spent in vain. So this much shall suffice on this head, and here let our account of the nature of the visible and generated gods come to an end.

With this conclusion Plato breaks off his account of the motions of the heavenly gods. A sphere or orrery would be needed to illustrate all the complications that result, in particular, from the changes in the relative positions of the planets, due to their composite motions and differences of speed. ' Juxtapositions ' (or ' comings along-side one another', παραβολαί) is explained by Proclus as the ' rising and setting together ' of two heavenly bodies. The ' counter-revolutions (ἐπανακυκλήσεις) of the (planetary) circles relatively to one another ' I understand to refer to (1) the additional constant movement, contrary to the Different (and to the Moon and the Sun group), possessed by the outer planets, and (2) the intermittent retrograde movements of all the planets, except the Sun and Moon.[2] ' Advances ' (προχωρήσεις) describes the accelerated

---

[1] ἄνευ ⟨τῶν ?⟩ δι᾽ ὄψεως τούτων αὐτῶν μιμημάτων. αὐτῶν F has the support of Pr. iii, 145¹⁶ (lemma), though Diehl has altered it there to αὖ τῶν. For the insertion of τῶν A.-H. appeals to Pr. iii, 145¹³, τὸ γὰρ λέγειν περὶ τούτων ἄνευ τῶν δι᾽ ὄψεως μιμημάτων μάταιός ἐστι πόνος, ὥς φησιν αὐτός. Cf. also 149¹¹, and Theon p. 238, αὐτός φησιν ὁ Πλάτων ὅτι τὸ ἄνευ τῶν δι᾽ ὄψεως μιμημάτων [τῶν] τὰ τοιαῦτα ἐθέλειν ἐκδιδάσκειν μάταιος πόνος. αὖ τῶν yields no tolerable sense (Tr. is not convinced by his own suggestion) and should be dismissed as simply a case of wrong division. The opposite error (αὐτῶν for αὖ τῶν) appears to occur at 66A, 1.

[2] See pp. 88, 110. The phrase is sometimes understood as if ἐπανακυκλεῖσθαι were the same as ἀνακυκλεῖσθαι (πρὸς αὑτήν, 37A). So Heath (*Gk. Astr.* 55) translates : ' the returnings of their orbits upon themselves ' (cf. *Aristarchus*, p. viii). But a model would not be needed to explain that a circle ' returns upon itself '. Pr. (iii, 145², 146¹⁰) read ἀνακυκλήσεις, but understood it as equivalent to ὑποποδισμοί (retrogradations). Cicero has *conuersiones*.

forward movement (προποδισμός) of a planet after retrogradation, whereby Venus and Mercury overtake the Sun once more, and the outer planets resume their main proper movement.[1]   The phrase ' coming into line with one another ' (κατ' ἀλλήλους γιγνόμενοι) refers to the cause of eclipses of the Sun and Moon, with which the rest of the sentence is concerned.   Occultations and transits of other planets are not noticed at all by ' men who cannot calculate ', and they did not cause panics in the Greek world.   The Sun is eclipsed when he and the Moon are ' at their conjunction ' ; the Moon, when she is ' in opposition ' ; in both cases the three bodies, Sun, Moon, and Earth, are ' in a line with one another ', but in different orders : the Moon at her own eclipse passes ' behind ' the Earth, at the eclipse of the Sun, ' in front ' of it.

It will be convenient here to give a table of all the celestial motions mentioned in the *Timaeus.*

TABLE OF CELESTIAL MOTIONS

A. MOTIONS OF THE WHOLE :

Self-motions of the World-Soul :
  (1) The Same (37C), imparted as axial rotation to the whole spherical body from centre to circumference (34A, B, 36E).
  (2) The Different, a single motion (36C, 37B, 38C), imparted to the planets (only) by distribution among seven circles (36C, D).

B. MOTIONS OF PARTS :

(a) *Individual Stars :*
  (1) The Same, imparted to each star as a ' forward ' motion of diurnal revolution (40B).
  (2) Self-motion : axial rotation (40A).
(b) *The Seven Planets :*
  (1) The Same, imparted to each planet by the ' supremacy ' of the Same (36C, 39A).
  (2) The Different, imparted to each planet as a constituent of its proper motion on a circular track (the seven circles, 36C, D).
  The composition of these two motions results in the Spiral Twist (39A).

---

[1] The variant ' approachings ' (προσχωρήσεις) might apply to one planet *nearing* another as it hastens to overtake it ; but προχωρήσεις (= προποδισμοί) makes a better antithesis to ἐπανακυκλήσεις (= ὑποποδισμοί).

(3) Self-motions :
    (a) Axial rotation of each planet (implied at 40A, B).
    (β) Differences of speed of the several planets (36D) :
        The Moon accelerates the movement of the Different. The Sun, Venus, Mercury, as a group, move with the actual speed of the Different, completing their course in a year. The Sun alone has the actual motion of the Different unmodified ; Venus and Mercury modify it by intermittent retrogradation (38D). Mars, Jupiter, Saturn slow down the movement of the Different by an additional motion of counter-revolution (ἐπανακύκλησις 40C). These are the three circles with a motion contrary to the Different and to the remaining four (36D).
    (γ) Retrogradation of all planets, except Sun and Moon : This is the ' contrary tendency ' (ἐναντία δύναμις, 38D) explicitly ascribed to Venus and Mercury, but also shared by Mars, Jupiter, Saturn. It involves variations in the speed of each planet, and intermittent counter-revolution accelerated to the point of bringing the main motion to a stand and temporarily reversing its sense.
        (None of these self-motions distorts in any way the circular track of the planet's proper motion. So the planets do not ' stray ' from one path to another, *Laws* 821, *Epin.* 982C.)

(c) *Earth :*
    (1) The Same, imparted to Earth as part of the whole body of the world rotating on its axis (34A, 36E).
    (2) Self-motion : axial rotation at the centre, relatively to the fixed stars, counteracting the imparted motion of the Same (40B).

THE HUMAN SOUL AND BODY

40D–41A. *The traditional Gods*

THE celestial gods, living beings whose intelligent souls have voluntary motions, are now enshrined in the system of circular movements provided by the self-moving power of the World-Soul. The celestial mechanism is finished ; but there remain three other classes of living creatures, ' which intelligence discerns contained in

137

the Living Creature that truly is ' (39E). These are neither gods nor everlasting, but subject to birth, change, and death, in the inferior regions of air, water, and earth. The making of them is, accordingly, now to be delegated to the created gods, whose handiwork will not be indissoluble, like that of the Demiurge himself. Before proceeding to this next stage, Plato finds it necessary to make some mention of the anthropomorphic gods of traditional religion.

40D.   As concerning the other divinities, to know and to declare their generation is too high a task for us ; we must trust those who have declared it in former times : being, as they said, descendants of gods, they must, no doubt, have had certain knowledge of their own ancestors. We cannot, then, mistrust the children of gods, though they speak without

E.   probable or necessary proofs ; when they profess to report their family history, we must follow established usage and accept what they say. Let us, then, take on their word this account of the generation of these gods. As children of Earth and Heaven were born Oceanus and Tethys ; and of these Phorkys and Cronos and Rhea and all their company ;

41.   and of Cronos and Rhea, Zeus and Hera and all their brothers and sisters whose names we know ; and of these yet other offspring.

Plato has given his own ' likely account ' of the creation of the celestial gods. The authors of the theogonies attributed to Orpheus, Musaeus, and other descendants of the Olympian gods, had professed to speak with knowledge, but had not given even probable, much less necessary, proofs of their assertions.[1] In an earlier dialogue Plato had not hesitated to make Socrates echo the famous saying of Protagoras in the remark : ' We know nothing about the gods— neither about the gods themselves nor about the names they may call one another by ' (*Crat.* 400D). If Protagoras had scandalised the contemporaries of Pericles, the Athenians of fifty years later, who had assimilated the plays of Euripides, were perhaps no longer to be shocked. But Plato stops short at the agnostic position which may well have been taken up by Socrates himself ; he does not flatly deny that the traditional gods exist. In the *Phaedrus* again (246c) Socrates says that to speak of an ' immortal living creature ', compact of soul and body, has no ground in any principle of reason. ' We have never seen a god or adequately conceived

---

[1] The Theogonies are again dismissed at *Laws* 886c as hard to censure because of their antiquity, but certainly false and unhelpful with respect to the honour due to parents. The same view is expressed at *Epin.* 988c.

one, but we imagine (πλάττομεν) him as a kind of immortal living creature possessing both a soul and a body combined in a unity which is to last for ever.'[1] This does not apply to the celestial gods of the *Timaeus*, whom we can see ; it means that we have no evidence in reasoning or in perception for the existence of gods in human form. But if we reject the human form and the mythical genealogies, it does not follow that we must deny altogether any invisible beings answering to the divinities of recognised belief. The *Epinomis* (984D), like the *Timaeus*, lays emphasis on the divinity of the visible celestial gods ; but it adds invisible spirits in the air and spirits sometimes visible in water, so that the heaven may be completely filled with living beings. Mankind has come into touch with these real beings, perhaps in visions, dreams, prophecies, or clairvoyance at the hour of death ; and hence have arisen beliefs in individuals and in States and widespread forms of worship. No wise lawgiver will wish to innovate here or ' turn away his own State to a form of piety which has no certainty ; he will not prevent men from obeying traditional laws about sacrifices, seeing that he has no knowledge at all about them, as in fact it is not possible for our mortal nature to have knowledge about such matters '. He ought, however, to insist on the worship of the visible gods as well. The attitude towards the traditional gods is still that of an agnostic, not of an atheist. There is no reason to question its sincerity or to suggest that Plato is hedging in order to escape a criminal charge of impiety. The irony in our passage is aimed, not at the pious beliefs of the common man, but at the pretensions of ' theologians ' to know the family history of anthropomorphic deities.[2]

41A–D. *The address to the gods*

The speech in which the Demiurge now delegates the task of making inferior living creatures, is addressed to all the visible gods as well as to those invisible powers which reveal themselves, in so far as they will, and thereby occasion the current beliefs in the deities of tradition.

41A. Be that as it may, when all the gods had come to birth—both all that revolve before our eyes and all that reveal themselves in so far as they will—the author of this universe addressed them in these words :

[1] Cf. *Laws* 904A ἀνώλεθρον δὲ ὄν γενόμενον, ἀλλ' οὐκ αἰώνιον, ψυχὴν καὶ σῶμα, καθάπερ οἱ κατὰ νόμον ὄντες θεοί.
[2] Cf. the judicious remarks of Mr. W. K. C. Guthrie in his excellent book, *Orpheus and Greek Religion* (1935), p. 240.

41A.   'Gods,[1] of gods whereof I am the maker and of works the father, those which are my own handiwork are indissoluble, save with my consent. Now, although whatsoever bond [2]

B.   has been fastened may be unloosed, yet only an evil will could consent to dissolve what has been well fitted together and is in a good state ; therefore, although you, having come into being, are not immortal nor indissoluble altogether, nevertheless you shall not be dissolved nor taste of death, finding my will a bond yet stronger and more sovereign than those wherewith you were bound together when you came to be.

'Now, therefore, take heed to this that I declare to you. There are yet left mortal creatures of three kinds that have not been brought into being. If these be not born, the Heaven will be imperfect ; for it will not contain all the

C.   kinds of living being, as it must if it is to be perfect and complete. But if I myself gave them birth and life, they would be equal to gods. In order, then, that mortal things may exist and this All may be truly all, turn according to your own nature to the making of living creatures, imitating my power in generating you. In so far as it is fitting that something in them should share the name of the immortals, being called divine and ruling over those among them who at any time are willing to follow after righteousness and after you—that part, having sown it as seed and made a beginning,

D.   I will hand over to you. For the rest, do you, weaving mortal to immortal, make living beings ; bring them to birth, feed them, and cause them to grow ; and when they fail, receive them back again.'

If the slight correction I have proposed in the first sentence of this address be accepted, the sense is satisfactory. ' Gods and works whereof I am father and maker ' means the whole universe, of which the Demiurge has been called maker and father at 28c and just above (41A). Among all these creatures, those which have so far been described—the body and soul of the living world and the heavenly gods—are ' my own handiwork ' ; and these, we are now told, are indissoluble save with their maker's consent. That consent, it is added, will never in fact be given ; hence the created

---

[1] Reading θεοί, θεῶν ὧν ἐγὼ δημιουργὸς πατήρ τ' ἔργων τὰ (for ἃ) δι' ἐμοῦ γενόμενα ἄλυτα ἐμοῦ γε μὴ ἐθέλοντος.. This conjecture and other interpretations are discussed in the Appendix (p. 367).

[2] The ' living bonds ' connecting the souls and bodies of the celestial gods, mentioned at 38E.

gods are everlasting and can never die.[1] But the world, as a living creature that must embrace all kinds of lesser living creatures, is not yet complete. The mortal kinds must now be added, and since they are to die, they must be made indirectly through the agency of the created gods. The Demiurge himself will supply only the immortal element of the human soul.

This delegation of the rest of the work to the celestial gods may perhaps be connected with the notion that the heavenly bodies, especially the Sun, are active in generating life on the Earth. The male, says Aristotle, is that which generates in another, the female that which generates in itself; hence in the universe also men call the Earth female and mother, and speak of the Heaven and the Sun or some other such thing as begetters and fathers[2] (*de gen. anim.* 716a, 14). In the *Republic* vi the Sun is singled out among the heavenly gods as 'the offspring of the Good which most resembles his parent'. He is the cause of the birth, growth, and nourishment of things in the visible world (509B). Aristotle elaborates the doctrine that the cause of coming to be and passing away is not the revolution of the First Heaven, but the annual movement of the Sun in the ecliptic or zodiac circle. This motion of 'the generator' is a compound of two motions. It includes the motion imparted by the revolution of the First Heaven (Plato's motion of the Same): this secures that coming to be shall be perpetual. The other motion in the reverse sense along the ecliptic, by causing the Sun to approach and retreat alternately, provides that generation shall alternate with decay, birth with death. If we were right in supposing that the annual motion of the Sun actually is the motion of the Different, unmodified in the Sun's case and variously retarded or accelerated by the other planets, Aristotle's explanation fits Plato's scheme. The activity of the created gods in making perishable things can be associated with the combined motions of the fixed stars (the Same) and of the planets (the Different).

The only mortal creatures whose making will be described in detail are human beings. Timaeus' task was at the outset defined as 'ending with the birth of mankind'. Even the plants on which man is to feed are not mentioned till far on at 77A. The lower animals are dealt with very briefly at the end (91D) and treated

---

[1] The *Epinomis* 982A says that 'opinion' must assign to the stars one of two destinies: either they are wholly indestructible and divine by all necessity, or each has a length of life sufficient to him and of such duration that no longer span could ever be required.

[2] Cf. Soph., frag. 752P, Ἥλιε . . . ⟨ὃν οἱ⟩ σοφοὶ λέγουσι γεννητὴν θεῶν ⟨καὶ⟩ πατέρα πάντων.

only as degraded forms suitable for the reincarnation of men who have lived unwisely. The physical differences between men and women are postponed to the same context (90E ff.), because they are irrelevant to the whole account of our common human nature which fills most of the remaining discourse. Plato does not mean that men ever existed without women and the lower animals.

41D–42D. *The composition of human souls. The Laws of Destiny*

The Demiurge next fulfils his promise to fashion with his own hands the immortal part of the individual souls which are to be incarnated first in human form. They are composed of what was left of the original ingredients used to compound the World-Soul, namely the intermediate kinds of Existence, Sameness, and Difference (35A).[1]

41D. Having said this, he turned [2] once more to the same mixing bowl wherein he had mixed and blended the soul of the universe, and poured into it what was left of the former ingredients, blending them this time [3] in somewhat the same way, only no longer so pure as before, but second or third in degree of purity. And when he had compounded the whole, he divided it into souls equal in number with the stars, and

E.   distributed them, each soul to its several star.

The human soul, no less than the World-Soul, must be so composed as to be like the objects it is to know, and it must possess the faculties of intelligence and knowledge, opinion and belief (37A–C). It is assumed later (43D), though not mentioned here, that its substance is divided into the ratios of the same *harmonia*, and given the motions of the Same and the Different. Human souls

---

[1] So Pr. here (iii, 254[19]) : ' Soul is a substance intermediate between the substance that has real Being and Becoming, being a compound of the intermediate kinds.'

[2] Reading καὶ πάλιν ἐπὶ τὸν πρότερον ⟨ἰὼν or τρεπόμενος⟩ κρατῆρα. Anyone reading the words as they stand in the MSS. would expect τρεπόμενος or its equivalent to follow, not κατεχεῖτο ; κατεχεῖτο ἐπὶ τὸν κρατῆρα is not Greek for ' poured *into* the bowl '. Cf. above τρέπεσθε ἐπὶ τὴν τῶν ζῴων δημιουργίαν (41C). Pr. evidently felt this (though he had our text), for he writes εἰπόντα γὰρ τὸν δημιουργὸν εὐθὺς ἐπὶ τὸν κρατῆρα τρέπει (ὁ λόγος). I conjectured ⟨ἰὼν⟩, but Professor Robertson points out to me that τρεπόμενος has many letters in common with πρότερον and might easily disappear after it.

[3] I suggest κατέχει (cf. Ar., *Plut.* 1021, ἐνέχεις) τό⟨τε⟩ μίσγων. I can see no sufficient justification for the middle καταχεῖσθαι, which is correctly used at *Laws* 637E, κατὰ τῶν ἱματίων καταχεόμενοι, ' *letting it pour down* over *their* garments '. The active occurs at *Rep.* 398A, μύρον κατὰ τῆς κεφαλῆς καταχέαντες and Ar., *Ach.* 1127, κατάχει σύ, παῖ, τοὔλαιον. For τότε = ' now ', cf. 37E, 2, 43C, 7.

are inferior, because they can do wrong of their own wills. ' Second or third in degree of purity,' if it does not mean ' second or even worse ', may refer to the superiority of man's soul over woman's (42A).

The souls are equal in number to the stars, among which they are distributed, one to each star. (The ' sowing ' into the planets comes later.) There is no reason to doubt the obvious meaning of these words : that there are just as many individual'souls as there are stars, whose number must be finite. But in all this section of the dialogue the veil of myth grows thicker again, and it is useless to discuss problems that would arise only if the statements were meant literally.

41E.   There mounting them as it were in chariots, he showed them the nature of the universe and declared to them the laws of Destiny.[1] There would be appointed a first incarnation one and the same for all, that none might suffer disadvantage at his hands ; and they were to be sown into the instruments of time, each one into that which was meet for it, and to

42.   be born as the most god-fearing of living creatures ; and human nature being twofold, the better sort was that which should thereafter be called ' man '.

Whensoever, therefore, they should of necessity [2] have been implanted in bodies, and of their bodies some part should always be coming in and some part passing out, there must needs be innate in them, first, sensation, the same for all, arising from violent impressions ; second, desire blended with pleasure and pain, and besides these fear and

B.   anger and all the feelings that accompany these and all that are of a contrary nature : and if they should master these passions, they would live in righteousness ; if they were mastered by them, in unrighteousness.

---

[1] νόμους τοὺς εἱμαρμένους.   Cf. *Laws* 904C (referring to the promotion and degradation of souls according to character) : Whatever has soul contains in itself the cause of change and in changing moves from place to place according to the disposition and law of Destiny ' (κατὰ τὴν τῆς εἱμαρμένης τάξιν καὶ νόμον).

[2] A.-H. notes the recurrent references to Necessity in this sentence : ἐξ ἀνάγκης . . . ἀναγκαῖον εἴη . . . βιαίων παθημάτων, echoed in the parallel passage (69C, D) where the created gods, after the long intervening section on ' What happens of Necessity ', fashion the mortal soul : τὸ θνητόν, δεινὰ καὶ ἀναγκαῖα ἐν ἑαυτῷ παθήματα ἔχον . . . συγκεραοάμενοι ταῦτα ἀναγκαίως . . . ὅτι μὴ πᾶσα ἦν ἀνάγκη. All the feelings and emotions mentioned come under the term *aesthesis* in its widest sense (*Theaet.* 156B), and have bodily concomitants. *Aesthesis* in the narrower sense was not present in the World-Soul, whose body has no organs of sense or nourishment and cannot be attacked by any ' strong powers ' from without (33A–D).

42B.   And he who should live well for his due span of time should journey back to the habitation of his consort star and there live a happy and congenial life [1]; but failing of this, he should shift at his second birth into a woman;

c.   and if in this condition he still did not cease from wickedness, then according to the character of his depravation, he should constantly be changed into some beast of a nature resembling the formation of that character, and should have no rest from the travail of these changes, until letting the revolution of the Same and uniform within himself draw into its train [2] all that turmoil of fire and water and air and earth that had later grown about it, he should control its irrational turbulence

D.   by discourse of reason and return once more to the form of his first and best condition.

The souls are set in the stars ' as it were in chariots ', an image intended to recall the procession of the gods in the *Phaedrus*, where the soul-chariots are taken round the outside of the heaven, and the charioteers are vouchsafed a vision of the realm of Forms. Here they are shown ' the nature of the universe '.   Such knowledge of reality as they will acquire in earthly life will be gained by Recollection (*Anamnesis*).   They are also taught the laws of their own destiny, as the souls in the Myth of Er, between their incarnations, hear the discourse of Lachesis, daughter of Necessity.   The chief lesson, here as there, is that the soul is responsible for any evil that it may suffer.   Proclus reproduces the genuinely Socratic doctrine that moral evil is the only real evil : ' neither disease nor poverty nor any other such thing is really an evil, but only wickedness of the soul, intemperance, cowardice, and vice in general ; and we are responsible for bringing these upon ourselves ' (iii, 313[18]).

---

[1] In Pindar, *Ol.* ii, and *Phaedrus*, 249A, the soul which has kept pure for three lives finally escapes from the wheel of reincarnation.   The present passage might mean this, or that the soul waits on its star before being reincarnated as man.   So Pindar provides a paradise where good souls, between their incarnations, ' spend a life free from tears in the presence of gods high in honour ' (*Ol.* ii, 65).   The hiatus συνήθη ἕξοι suggests that καὶ συνήθη should be omitted with FY. Stob.   Cf. note on 20A.

[2] συνεπισπώμενος.   The rational revolution in the human soul's movements is to establish its supremacy over the irrational motions, as the Same in the World-Soul has supremacy (κράτος) over the circles of the Different (36c).   Cf. 44A, where the revolutions assailed by sensations from without, which ' draw in their train ' (συνεπισπάσωνται) the whole vessel of the soul, only seem to be in control (κρατεῖν).   Plut., *Pl. Qu.* 1003A, ἐπεὶ δὲ ἡ ψυχὴ νοῦ μετέλαβε καὶ ἁρμονίας, καὶ γενομένη διὰ συμφωνίας ἔμφρων μεταβολῆς αἰτία γέγονε τῇ ὕλῃ καὶ κρατήσασα ταῖς αὑτῆς κινήσεσι τὰς ἐκείνης ἐπεσπάσατο καὶ ἐπέστρεψεν . . . The word προσφύντα recalls the comparison of the incarnate soul to the image of Glaucus encrusted with shells and seaweed (προσπεφυκέναι, *Rep.* 611D).

144

In the *Phaedrus* (248D), it is a law of Adrasteia that no soul shall be implanted in the form of a beast ' at its first birth '. So here all the souls are to start on their course in human form, the better as men, the worse as women.[1] We need not understand that there were no women until the bad men of the first generation began to die and to be reincarnated in female form, but only that a bad man will be reborn as a woman, a bad woman presumably as a beast. In the *Laws* (721C) the Athenian says that ' the race of man is twin-born with all Time, which it accompanies and shall accompany all through,[2] being in this way immortal : by leaving children's children and existing always one and the same, it partakes of immortality by means of generation '. Since Time itself has no beginning or end, the human race must have always existed. Proclus[3] took this to be Plato's view. He appeals to *Laws* 676B, where the Athenian speaks of the unlimited length of time, in which ' myriads upon myriads ' of States must have come into existence and perished, and no one could ascertain any date at which mankind began to live in cities. The world, says Proclus elsewhere (iii, 282), had no beginning in time. If it had had a beginning, then some soul would have been the first to descend to its incarnation. But since there was none, male and female must always exist, and all that is meant is that every soul that is at any time incarnated for the first time, is incarnated in male form. The soul, mankind, and the universe are all ' ungenerated ' in the sense of having no beginning in time, though ' generated ' in the sense of being in the realm of temporal becoming (i, 287 ; iii, 294). At *Laws* 781E,

---

[1] There is nothing in the text here to suggest that the first living creatures are ' without sex-differences, the differentiation of the sexes and the infra-human species coming about later by a kind of " evolution by degeneration ' " (Tr., p. 258). The latter statements are founded on 90E ff. where Plato says that those who were born as *men* (not sexless creatures), if they lived ill, were reborn as women *at their second incarnation* (as he says here, 42B). ' Also at that time they fashioned Eros,' and the physiological apparatus of sex in both men and women is described. In our passage the first generation of men have ἔρως (42A, 7), an element in the mortal soul which the created gods proceed to make at 69C. There is nowhere in the *Timaeus* any mention of sexless creatures. As I have suggested, the physical differences of the sexes are postponed to a sort of appendix at the end because all that will be said in the interval applies equally to men and women.

[2] This passage illustrates Norden's remark : Die Vorstellung, dass der χρόνος, als Begleiter des Menschen gedacht, mit ihm geboren wird und mit ihm altert, ist in dem Hellenentum geläufig (*Die Geburt des Kindes* (1924), p. 44).

[3] i, 288, εἰ δὲ ἀεὶ γένος ἐστὶν ἀνθρώπων, καὶ τὸ πᾶν ἀναγκαῖον ἀΐδιον ὑπάρχειν. So Oc. Luc. iii argues that the main parts of the cosmos must always exist, including man : ἀνάγκη τὸ γένος τῶν ἀνθρώπων ἀΐδιον εἶναι. Diodorus i, 6, 3, remarks that those physicists who make the world ungenerated and imperishable say that the human race has existed from all eternity.

however, Plato leaves open the alternatives that either the human race always has been and always will be, or it must have existed for an incalculable length of time. In any case, the details of the mythical story here are not to be taken literally.

42D-E. *Human souls sown in Earth and the planets*

After the journey in their star chariots, the immortal souls are next sown like seed in the planets and committed to the care of the created gods. Only the immortal element in the soul, as the immediate creation of the Demiurge, is indissoluble. The subordinate divinities must add the body and those mortal parts of the soul which temporary association with the body entails.

42D.   When he had delivered to them all these ordinances, to the end that he might be guiltless of the future wickedness of any one of them, he sowed them, some in the Earth, some in the Moon, some in all the other instruments of time. After this sowing he left it to the newly made gods to mould mortal bodies, to fashion all that part of a human soul that there was still need to add and all that these things entail, and

E.   to govern [1] and guide the mortal creature to the best of their powers, save in so far as it should be a cause of evil to itself.

In the machinery of the myth, it is natural to suppose that the first generation of souls is sown on Earth, the rest await their turn, unembodied, on the planets.[2] The sowing of the immortal souls in the Earth and the planets, the instruments of Time, may symbolise that the soul possesses that intermediate kind of existence which partakes both of real being and of becoming. The soul is subject to Time and change; and her earthly life is spent in the region where the government of Reason is conditioned by Necessity. She

---

[1] The comma after ἄρχειν should be omitted. A.-H. prints it, but rightly ignores it in his translation.

[2] So Chalcidius, p. 241. I cannot see why this notion is ' foolish ', as Tr. calls it (p. 259). Some of the ancients who thought the moon was composed of earth imagined that it might be inhabited (or at least habitable, as Anaxagoras said: οἴκησις does not necessarily mean actually inhabited). Tr. produces no evidence that anyone regarded any other planet as habitable by men, except a statement by Chalcidius that Pythagoras believed that men exist on all the planets, though Plato does not. (At Pr. iii, 280, στοιχεῖα does not mean ' planets ' but ' elements ', as elsewhere in the commentary.) Plato, who speaks of all the heavenly gods (including all the planets, as I have argued, p. 118) as mainly composed of fire, was not likely to think of men living on them. Did any ancient ever hold that men lived in the Sun ? Cf. Guthrie, *Orpheus and Gk. Relig.*, pp. 232, 247, note 10, for Anaxagoras and the Orphic belief in an inhabited Moon.

will be subject to the 'violent' assaults of the corporeal environment. If she does not reduce to order the consequent turbulence in the bodily members, the fault will be her own. Her will is free, to follow after righteousness and the created gods (δίκη καὶ ὑμῖν ἕπεσθαι, 41C), whose guidance is revealed to her eyes in the orderly revolutions of the heavens.

42E–44D. *The condition of the soul when newly incarnated*

How the gods established the mortal parts of the soul and framed the body it was to inhabit will be described in detail later, in the third section of the dialogue (69A ff.). The whole account, in the second section, of the structure and behaviour of the primary bodies and of the physical processes of sensation and perception will have intervened. For the present we are concerned only with the picture of the immortal principle of reason, made by the Demiurge himself, plunged for the first time into the turbulent tide of bodily sensation and nutrition. The mythical machinery of the soul circles is woven into an account of infant psychology with an imaginative power that few other writers could equal. The whole leads up to the central problem of human life, the establishment of rational control over the bodily nature.

We are here approaching the stage at which the works of Reason will give place to 'what happens of Necessity'. The 'errant cause' begins to come into view, with factors in the economy of the visible world that are not the creatures of divine purpose but limit the conditions under which Reason must operate. The language hints at a certain analogy between the task of the human reason and the task of the Demiurge himself, who 'took over all that was visible, not at rest but in discordant and unordered motion, and brought it from disorder into order' (30A). But the World-Soul was not exposed to the invasion of violent affections from without, such as beset every new-born soul of man.

42E. When he had made all these dispositions, he continued to abide by the wont of his own nature [1]; and meanwhile his sons took heed to their father's ordinance and set about obeying it. Having received the immortal principle of a mortal creature, imitating their own maker, they borrowed from the world portions of fire and earth, water and air, on
43.   condition that these loans should be repaid, and cemented

---

[1] ἔμενεν is hard to render. The word does not mean rest or cessation of activity (contrast *Gen.* ii, 1, κατέπαυσε τῇ ἡμέρᾳ τῇ ἑβδόμῃ ἀπὸ πάντων τῶν ἔργων αὐτοῦ): 40B, the stars στρεφόμενα μένει. The meaning seems to be that the Demiurge left these further operations to the created gods, confining himself to his own proper activity.

43.   together what they took, not with the indissoluble bonds
        whereby they were themselves held together, but welding
        them with a multitude of rivets too small to be seen and so
        making each body a unity of all the portions. And they
        confined the circuits of the immortal soul within the flowing
        and ebbing tide of the body.

            These circuits, being thus confined in a strong river,
        neither controlled it nor were controlled, but caused and
        suffered violent motions ; so that the whole creature moved,
B.   but advanced at hazard without order or method, having all
        the six motions ; for they went forward and backward, and
        again to right and left, and up and down, straying every way
        in all the six directions.[1] For strong as was the tide that
        brought them nourishment, flooding them and ebbing away,
        a yet greater tumult was caused by the qualities [2] of the
        things that assailed them, when some creature's body chanced
C.   to encounter alien fire from outside, or solid concretion of
        earth and softly gliding waters, or was overtaken by the
        blast of air-borne winds, and the motions caused by all these
        things passed through the body to the soul and assailed it.
        (For this reason these motions were later called by the name
        they still bear—' sensations ').[3] And so at the moment we
        speak of, causing for the time being a strong and widespread
        commotion and joining with that perpetually streaming
D.   current in stirring and violently shaking the circuits of the
        soul, they completely hampered the revolution of the Same
        by flowing counter to it and stopped it from going on its
        way and governing [4] ; and they dislocated the revolution of

---

[1] πάντῃ κατὰ τοὺς ἐξ τόπους πλανώμενα. Contrast the World's spherical body
which the Demiurge made without ' all the other six motions, giving it no
part in their wanderings ' (ἀπλανές, 34A). The stars (ἀπλανῆ) have orbital revolution ' forwards ', but not the other five motions (40B). Later we shall hear
of the Errant Cause (πλανωμένη αἰτία) and of the motions it causes (48A).
The human soul, as self-moved, has its own revolutions ; but these are dis-
organised by motions originated by bodies acting on it, through its own body,
from outside. We next hear of two elements of cognition—sensations and
false judgments, which did not occur in the World-Soul (37A–C). Not only
are the creature's bodily movements erratic, but all processes of rational
thought are thrown out of gear.

[2] παθήματα can mean ' affections ' of the sentient body, causing sensation
in the soul, as at 42A, αἴσθησιν ἐκ βιαίων παθημάτων, or the perceptible ' qualities ' of external bodies, as here and at 61C.

[3] Pr. iii, 332, suggests that Plato connected the word αἴσθησις either with
ἀίσθω (θυμὸν ἀίσθειν, ' breathe out ', Hom. So Etym. Mag.) or with ἀίσσω
' rush '. The latter seems more probable. In Plato's view both sensations
and qualities are movements, Theaet. 156.

[4] The higher faculty of reason is put completely out of action.

148

43D.   the Different. Accordingly, the intervals of the double and
the triple,[1] three of each sort, and the connecting means of
the ratios, $\frac{3}{2}$ and $\frac{4}{3}$ and $\frac{9}{8}$, since they could not be completely
dissolved save by him who bound them together, were

E.   twisted by them in all manner of ways, and all possible
infractions and deformations of the circles were caused ; so
that they barely held together, and though they moved, their
motion was unregulated, now reversed, now side-long, now
inverted. It was as when a man stands on his head, resting
it on the earth, and holds his feet aloft by thrusting them
against something : in such a case right and left both of the
man and of the spectators appear reversed to the other
party.[2] The same and similar effects are produced with
great intensity in the soul's revolutions ; and when they

44.   meet with something outside that falls under the Same or
the Different, they speak of it as the same as this or different
from that contrary to the true facts, and show themselves
mistaken and foolish.   Also [3] at such times no one revolution
among their number is acting as governor or guide ; but
whatever revolutions are assailed by certain sensations
coming from without, which draw in their train at the same
time the whole vessel of the soul,[4] at such times only seem to
be in control, whereas really they are overpowered. It is,
indeed, because of these affections that to-day, as in the
beginning, a soul comes to be without intelligence at first,

B.   when it is bound in a mortal body.[5]

[1] The first mention of the harmonic intervals as present in the individual
soul. They stand for that harmony and κοσμότης which need to be re-estab-
lished by contemplation of the kindred harmony of the World-Soul, revealed
in the heavenly revolutions (47B, c).

[2] Correctly translated and explained by A.-H. (except that ἄνω should be
taken with ἔχῃ) : ' if A and B stand face to face, B's right is of course opposite
A's left. But if A stand on his head, still facing B, then B's right will be
opposite A's right ; the normal relation being inverted.' προσβαλὼν πρός τινι
can only mean ' thrusting his feet (so that they rest) against some support '.

[3] This clause goes with what follows ; it refers to lack of control over
behaviour. An infant's earliest actions are determined not by judgment or
thought or will, but mechanically by ' motions ' of sensation rushing in from
without and sweeping with them the motions of the soul. Its behaviour only
looks like voluntary self-motion. Martin and others forget that all this refers
to infancy, not to the enslavement of reason by passion in later life.

[4] καὶ τὸ τῆς ψυχῆς ἅπαν κύτος, the whole body which contains the soul, as
well as (καὶ) the revolutions of the soul itself.

[5] The whole description applies to every new-born baby's soul, not only
to the first generation of mankind. Contrast the World-Soul, which, as soon
as it was joined with its body, began an ' intelligent life ' (ἔμφρων βίος, 36E),
not being exposed to external assaults.

44B.   But when the current of growth and nutriment flows in less strongly, and the revolutions, taking advantage of the calm, once more go their own way and become yet more settled as time goes on, thenceforward the revolutions are corrected to the form that belongs to the several circles in their natural motion ; and giving their right names to what is different and to what is the same, they set their possessor in the way to become rational. And now if some right nurture lends help towards education,[1] he becomes entirely

c.   whole and unblemished, having escaped the worst of maladies ; whereas if he be neglectful, he journeys through a life halt and maimed and comes back to Hades uninitiate and without understanding.[2]

These things, however, come to pass at a later stage. Our present subject must be treated in more detail ; and its preliminaries, concerning the generation of bodies, part by part, and concerning soul, and the reasons and forethought of

D.   the gods in producing them—of all this we must go on to tell, on the principle of holding fast to the most likely account.

44D–45B. *Structure of the human body : head and limbs*

The matter in hand, to which Timaeus now returns, is the implanting of souls in bodies possessed of sense-organs and of all the feelings and emotions that accompany sense (42A). The first duty of the gods is to provide a residence for the immortal part of the soul, which they have just received from the hands of the Demiurge. We have not yet come to the addition of the two mortal parts of the soul (69C). So the body is here regarded as consisting of the head, which houses the immortal, rational part, and an apparatus of limbs to carry the head about, together with the organs of sight to direct its movements.

44D.   Copying the round shape of the universe, they confined the two divine revolutions in a spherical body—the head, as we now call it—which is the divinest part of us and lord over all the rest. To this the gods gave the whole body, when they had assembled it, for its service, perceiving that it

---

[1] Cf. 47C : the observation of the unperturbed revolutions of the heavens will lead to philosophy, and we shall learn ' to reproduce the perfectly unerring (ἀπλανεῖς) revolutions of the god (the Heaven) and reduce to settled order the wandering (πλανωμένας) motions in ourselves '. Cf. 90D, and 87B, διὰ τροφῆς καὶ δι' ἐπιτηδευμάτων μαθημάτων τε.

[2] Plato uses terms borrowed from Mystery ritual. A.-H. compares *Phaedrus* 250C, *Laws* 759C (ὁλόκληρος), and Dem., *de cor.* 259, ἔφυγον κακόν, εὗρον ἄμεινον. Cf. also *Phaedrus* 248B, ἀτελὴς τῆς θέας ; *Gorg.* 469B, τοὺς ἀνοήτους ἀμυήτους.

44D. possessed all the motions that were to be.[1] Accordingly, that the head might not roll upon the ground with its heights

E. and hollows of all sorts, and have no means to surmount the one or to climb out of the other, they gave it the body as a vehicle for ease of travel; that is why the body is elongated and grew four limbs that can be stretched out or bent, the god contriving thus for its travelling. Clinging and supporting itself with these limbs, it is able to make its

45. way through every region,[2] carrying at the top of us the habitation of the most divine and sacred part. Thus and for these reasons legs and arms grow upon us all.[3] And the gods, holding that the front is more honourable and fit to lead than the back, gave us movement for the most part in that direction. So man must needs have the front of the body distinguished and unlike the back; so first they set the face on the globe of the head on that side and fixed in

B. it organs for all the forethought of the soul, and appointed this, our natural front, to be the part having leadership.

This description of the human body has the same oddly archaic character as that of the World's body at 33A–34A; but it is hard for a modern reader to gauge the effect. Many passages in Sir Thomas Browne strike us as 'quaint' or funny, that may not have seemed so to his contemporaries. The evidences of design in the human body were a serious matter to Plato. A more systematic account of the body's structure will be given in the third section of the dialogue. This paragraph is mainly intended to compare and contrast the human body and its motions with the body and motions of the universe.

45B–46A. *The eyes and the mechanism of vision*

Plato singles out the sense of sight, first because it is useful for locomotion, and secondly because sight and hearing, which will presently be added, are the two senses which above all reveal the

---

[1] The bodies of the universe and of the created gods possessed only rotation and orbital revolution—the rational motions. Inferior creatures have all the six rectilinear motions proper to the primary bodies, portions of which are 'assembled' to compose their bodies.

[2] The six regions (τόποι) of 43B, answering to the six motions (34A) 'up and down', 'forward and backward', 'right and left', which the World's body has not.

[3] προσέφυ πᾶσιν. πᾶσιν is at least superfluous: why 'all'—as if some of us might be expected to do without arms and legs? It is, accordingly, tempting to conjecture προσπεφύκασιν, which removes the very unusual construction of the singular προσέφυ. Chalcidius ignores πᾶσιν: *addita est crurum quoque et brachiorum porrigibilis et flexuosa substantia*; but his version is loose.

harmony of the world.[1] He begins with the bodily mechanism of vision, for the sake of leading up to the contrast between these ' secondary causes ' and the true reason or purpose, which is that man may learn number by seeing the heavenly bodies and so pass on through the sciences of number to all philosophy.

The mechanism of vision involves three kinds of ' fire ' or light. (Several varieties of fire will be enumerated at 58c.) These are : (1) Daylight, a body of pure fire diffused in the air by the Sun. This (like (2)) is ' pure ', not admixed with other primary bodies. At 58c it is contrasted with flame ($\varphi\lambda\delta\xi$) as ' that which flows off from flame, and does not burn but gives light to the eyes '. (2) The visual current, a pure fire of the same kind as daylight, contained in the eye-ball and capable of issuing out in a stream directed towards the object seen. At 67D it appears that the visual current or ray is not composed of the very smallest grade of fire. (3) The colour of the external object, defined at 67c as ' a flame ($\varphi\lambda\delta\xi$) streaming off from every body, having particles proportioned to those of the visual current, so as to yield sensation '.

Plato begins by describing (1) Daylight.

45B. First of the organs they fabricated the eyes to bring us light, and fastened them there for the reason which I will now describe. Such fire as has the property, not of burning, but of yielding a gentle light, they contrived should become the proper body of each day.[2] For[3] the pure fire within us is akin to this, and they caused it to flow through the eyes, making the whole fabric of the eye-ball, and especially the central part (the pupil), smooth and close in texture,[4]

c. so as to let nothing pass that is of coarser stuff, but only

[1] So Ar., *Eudemus*, frag. 47, 48, speaks of sight and hearing as heavenly and divine senses, revealing the harmony to mankind with sound and light. The other senses are for the sake of mere existence, these for well-being.

[2] Taking οἰκεῖον ἑκάστης ἡμέρας σῶμα together (with Madvig and A.-H.). Each day, as it follows night, has a ' body of its own ' (οἰκεῖον), consisting of sunlight diffused in the air, which ' withdraws ' at nightfall (45D), following the sinking sun. This body actually *is* daylight, not ' *similar* ' to daylight or ' akin ' to it (as A.-H. renders). But οἰκεῖον contains the suggestion that a ' gentle ' (ἥμερον) light is naturally appropriate to day (ἡμέρα, a word which some modern authorities agree with Plato in connecting with ἥμερος ; cf. *Crat.* 418D). Tr.'s translation, ' a gentle light proper to day ', ignores ἑκάστης.

[3] The connection of thought (' for ') is : the gods made daylight (essentially a visible thing) of a suitable kind of fire, *for* they wanted us to see and so arranged that the fire within the eye should be similar and capable of coalescing with daylight.

[4] Empedocles (84B), whom Plato is following, compares the eye to a horn lantern, and explains that the fire confined in the eyeball is so fine as to pass through tissues impervious to water.

45C.   fire of this description to filter through pure by itself.   Accordingly, whenever there is daylight round about, the visual current issues forth, like to like, and coalesces with it and is formed into a single homogeneous body in a direct line with the eyes, in whatever quarter the stream issuing from within strikes upon any object it encounters outside.   So the whole, because of its homogeneity, is similarly affected and passes on the motions of anything it comes in contact with or that

D.   comes into contact with it, throughout the whole body, to the soul, and thus causes the sensation we call seeing.[1]

But when the kindred fire (of daylight) has departed at nightfall,[2] the visual ray is cut off ; for issuing out to encounter what is unlike it, it is itself changed and put out, no longer coalescing with the neighbouring air, since this contains no fire.   Hence it sees no longer, and further induces sleep.   For when the eyelids, the protection devised by the

E.   gods for vision, are closed, they confine the power of the fire inside, and this disperses and smooths out the motions within, and then quietness ensues.   If this quiet be profound, the sleep that comes on has few dreams ; but when some stronger motions are left, they give rise to images answering

46.   in character and number to the motions and the regions in which they persist—images which are copies made inside and remembered when we awake in the world outside.[3]

[1] What is transmitted along this sympathetic chain is *motion* partly originated by qualitative changes (ἀλλοιώσεις) in the object, as the *Theaetetus* explains.   This motion reaches the bodily organ and causes qualitative changes there, which when they penetrate to the soul (but not before) are called ' sensations ' (43C).   There is no ground for Tr.'s notion of a pencil of light, a temporary extension of my body which may be miles long and ' is *sensitive throughout*, and so " transmits " *sensation* from one extremity to the other '.   Sensation, as Plato clearly says, occurs in the soul, not at the surface of a mountain ten miles distant and throughout the interval.

[2] εἰς νύκτα, *sub noctem*, as at Xen., *Hell.* 4, 6, 7 ; not ' into night '.   Albinus, *Didasc.* xviii, paraphrases : τοῦ φωτὸς νύκτωρ ἀπιόντος.   Plato seems to imagine the ' proper body of each day ' moving away, following the sinking sun and superseded by the night air with little or no fire in it.   He was probably thinking of Empedocles' two hemispheres of night and day ' revolving round the earth, the one altogether composed of fire, the other of a mixture of air and a little fire ' (Ps.-Plut., *Strom.* 10).   The night-air, being damp, ' puts out ' the fire issuing from the eye.

[3] The last words may mean ' when we have emerged into the waking world ', or that, when we recall a dream, the persons and things we dreamt of appear to be outside us, as they do in the dream itself.   The latter interpretation is perhaps favoured by *Rep.* 476C (cited by Beare, *Gk. Theories of Elementary Cognition* 46) : Dreaming, whether we are awake or asleep, consists in taking an image for the real thing it resembles.   I am not convinced that Plato could not write ' made inside and remembered outside ' in this sense.

46A–C. *Mirror images*

A short appendix on mirror images is added here, seemingly for its own sake rather than as contributing to the main argument. It has, however, the effect of emphasising the purely mechanical processes of vision, which will presently be contrasted with its rational purpose.

46A. There will now be little difficulty in understanding all that concerns the formation of images in mirrors and any smooth reflecting surface. As a result of the combination of the two fires inside and outside, and again as a consequence of the formation, on each occasion, at the smooth surface, of a single fire which is in various ways changed in form, all

B. such reflections necessarily occur, the fire belonging to the face (seen) coalescing, on the smooth and bright surface, with the fire belonging to the visual ray. Left appears right because reverse parts of the visual current come into contact with reverse parts (of the light from the face seen), contrary to the usual rule of impact.

In interpreting this short account of mirror images we must beware of ascribing to Plato too much knowledge of optics. There is no reference to the lens or the retina. He knew that the angles of incidence and reflection of a ray are equal. This proposition is assumed in Euclid's *Optics*, where Def. 1 ' embodies the same idea of the process of vision as we find in Plato, namely that it is due to rays proceeding from our eyes and impinging upon the object, instead of the other way about : " the straight lines (rays) which issue from the eye traverse the distances (or dimensions) of great magnitudes "; Def. 2 : " The figure contained by the visual rays is a cone which has its vertex in the eye, and its base at the extremities of the object seen "; Def. 3 : " And those things are seen on which the visual rays impinge, while those are not seen on which they do not." ' [1]

Plato speaks first of ' the combination of the two fires inside and outside '. As above, this means ' inside and outside the eye '. He has just been explaining that such combination of the visual ray with the sunlight does not occur at night, and how in sleep the visual fire confined inside gives rise to dream images. He now returns to the case where combination does occur, resulting in coalescence of the internal fire with the external into one homogeneous body which can transmit the motions from object to eye. That is the first condition of all vision.

[1] Heath, *Gk. Math.* 1, 441.

In the special case of reflections, there is a second condition: 'the formation at the smooth surface of a single fire which is in various ways changed in form'. At the reflecting surface the visual ray which has coalesced with the daylight encounters a stream of fire from the object, and the two now form 'a single fire', extending from the object to the mirror and from the mirror to the eye. The object taken as illustration is 'the face', which may be the face of someone else standing beside the observer and facing the mirror (as in the diagram), or the observer's own face. The single fire is said to be 'in various ways changed in form'. This probably refers forward to the transposition of right and left mentioned in the next sentence, and also to the distortions due to the mirror having a curved surface. The transposition of right and left is mentioned in an earlier dialogue, the *Sophist* 266c: a reflection

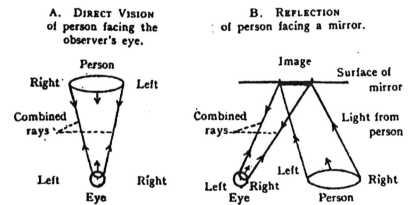

A. DIRECT VISION of person facing the observer's eye.

B. REFLECTION of person facing a mirror.

occurs 'when the light belonging to the eye meets and coalesces with light belonging to something else on a bright and smooth surface and produces a form yielding a perception that is the reverse of the ordinary direct view'.

Finally, in the next sentence, there is the case of a mirror whose two sides curve forward so that the surface becomes cylindrical, with the curvature horizontal. The effect is that the rays 'change sides', and right again becomes right as in direct vision. If the mirror is turned through a right angle so that the curvature becomes vertical, the image will appear inverted.

46B. On the contrary, right appears right and left left, when the visual light changes sides in the act of coalescing with the light with which it does coalesce; and this happens when

c. the smooth surface of the mirror, being curved upwards at either side, throws the right part of the visual current to the

46c. left, and the left to the right. The same curvature turned lengthwise to the face makes the whole appear upside down, throwing the lower part of the ray towards the top and the upper part towards the bottom.

This disquisition on optics will seem less intrusive if we remember that the whole apparatus of vision was peculiarly significant to Plato because of the analogy between the bodily eye and the eye of the soul, and between the sunlight and truth. Dream images, shadows, and reflections occupied in the *Republic* (510A) the lowest section of the Divided Line. The relation of these *eidola* to the actual visible things whose images they are was there used to illustrate the relation not only of the lower objects of intelligence to the higher, but also of the whole visible world to its intelligible pattern. In the *Sophist* (266) a parallel is drawn between divine and human production. Divine production covers the same field as the work of the Demiurge in the *Timaeus*: it is the creation of the whole visible world, divided into (1) originals, ' ourselves and all other living creatures and the elements of natural things—fire, water, and their kindred', and (2) images which attend on all these products: dream images, shadows, reflections. In human production the two classes have their analogues in (1) the production of useful things by crafts such as the builder's, who makes an actual house, and (2) images, such as the painter's, who makes, as it were, ' a man-made dream for waking eyes '. In this lower class rank all the fine arts, political rhetoric, and sophistry. Thus the relation of dreams and reflections to their originals was associated with what may be called the metaphysical problem of the *eidolon*, a problem raised but not answered in the *Sophist*: How can there be such a thing as a visible world, which is not perfectly real (ὄντως ὄν) and yet has some sort of existence (ὄν πως).[1] The problem was there consciously shelved ; if Plato meant to deal with it in the *Philosopher* that dialogue was never written. We must look for the answer, if anywhere, in the *Timaeus*. We are now approaching the second section of the dialogue, which brings into account a hitherto neglected factor in Becoming—the Receptacle. This, we shall find, plays a part analogous to the mirror holding the reflections of actual things (52B, C).

46c–47e. *Accessory causes contrasted with the purpose of sight and hearing*

The account of eyesight has brought us to the point of contact between the knowing soul and the external world of visible bodies.

---

[1] Cf. F. M. Cornford, *Plato's Theory of Knowledge*, pp. 199 ff., 320 ff.

The form in which it was cast was designed to serve another purpose. It leads to the transition from the first section of the dialogue to the second, from the works of Reason to what comes about of Necessity. We have been told about the mechanism of vision, what happens to the rays of light and colour in the commerce between the sense-organ and its object outside. All such physical transactions we need to study; but they will not reveal the true reason or explanation (αἰτία) of vision, the purpose it is rationally designed to serve. They tell us 'how' we see, but not 'why'.

46C.   Now all these things are among the accessory causes which the god uses as subservient in achieving the best result that is possible. But the great mass of mankind regard them,

D.   not as accessories, but as the sole causes of all things, producing effects by cooling or heating, compacting or rarefying, and all such processes. But such things are incapable of any plan or intelligence for any purpose. For we must declare that the only existing thing which properly possesses intelligence is soul, and this is an invisible thing, whereas fire, water, earth, and air are all visible bodies; and a lover of intelligence and knowledge must necessarily seek first for the causation that belongs to the intelligent nature,[1] and

E.   only in the second place for that which belongs to things that are moved by others and of necessity set yet others in motion. We too, then, must proceed on this principle: we must speak of both kinds of cause, but distinguish causes that work with intelligence to produce what is good and desirable, from those which, being destitute of reason, produce their sundry effects at random and without order.

    Enough, then, of the secondary causes that have contributed [2] to give the eyes the power they now possess; we must next speak of their highest function for our benefit,

47.   for the sake of which the god has given them to us. Sight, then, in my judgment is the cause of the highest benefits to us in that no word of our present discourse about the universe could ever have been spoken, had we never seen stars, Sun, and sky. But as it is, the sight of day and night, of months and the revolving years, of equinox and solstice, has caused the invention of number and bestowed on us the notion of time and the study of the nature of the world; whence we

---

[1] τῆς ἐμφρονος φύσεως, i.e. τῆς ψυχῆς, possessive genitive. For ὅσαι (αἰτίαι) γίγνονται, cf. Soph. 265C, θείας (sc. αἰτίας) ἀπὸ θεοῦ γιγνομένης, 'causation which has its origin in deity'.

[2] συμμεταίτια recalls Soph., Antig. 537, καὶ συμμετίσχω καὶ φέρω τῆς αἰτίας, 'I take my share with you in the burden of the accusation (or responsibility).'

47B. have derived all philosophy, than which no greater boon has ever come or shall come to mortal man as a gift from heaven. This, then, I call the greatest benefit of eyesight ; why harp upon all those things of less importance, for which one who loves not wisdom, if he were deprived of the sight of them, might 'lament with idle moan' ?[1] For our part, rather let us speak of eyesight as the cause of this benefit,[2] for these ends : the god invented and gave us vision in order that we might observe the circuits of intelligence in the heaven and profit by them for the revolutions of our own thought, which are akin to them, though ours be troubled and they

c. are unperturbed ; and that, by learning to know them and acquiring the power to compute them rightly according to nature, we might reproduce the perfectly unerring revolutions of the god and reduce to settled order the wandering motions in ourselves.

Of sound [3] and hearing once more the same account may be given : they are a gift from heaven for the same intent and purpose. For not only was speech appointed to this same intent, to which it contributes in the largest measure,

D. but also all that part of Music that is serviceable with respect to the hearing of sound is given for the sake of harmony ;[4] and harmony, whose motions are akin to the revolutions of the soul within us, has been given by the Muses to him whose commerce with them is guided by intelligence, not for the sake of irrational pleasure (which is now thought to be its utility), but as an ally against the inward discord that has come into the revolution of the soul, to bring it into order and consonance with itself. Rhythm also was a succour

---

[1] ὧν governed by τυφλωθείς. Stallb. compares Xen., *Symp.* iv, 12, τυφλὸς δὲ τῶν ἄλλων ἀπάντων μᾶλλον ἂν δεξαίμην εἶναι ἢ ἐκείνου ἑνὸς ὄντος. The last words quote Eur., *Phoenissae* 1762, τί ταῦτα θρηνῶ καὶ μάτην ὀδύρομαι;

[2] Taking τούτου (like τοῦτο above) to mean philosophy, and ἐπὶ ταῦτα as referring forward to the rest of the sentence. Cf. c, 5, ἐπὶ ταὐτὰ τῶν αὐτῶν ἕνεκα.

[3] φωνή, as opposed to ψόφος (noise), covers articulate speech and musical sound.

[4] Reading φωνῆς χρήσιμον πρὸς ἀκοήν, φωνῆς being governed by ἀκοήν. 'Music' is a wide term, including poetry and the thought conveyed in it. That part which 'is serviceable with respect to the hearing of sound' is vocal and instrumental music in our sense. φωνῇ χρήσιμον can hardly mean 'vocal'; and why should instrumental music be excluded ? Nor can it mean 'expressed in sound'; and 'useful to the voice' is irrelevant. ἕνεκα ἁρμονίας ἐστὶ δοθέν must be taken as predicate, to give ἕνεκα ἁρμονίας the necessary emphasis. Ἁρμονία is not the 'harmony' of simultaneous concordant sounds (συμφωνία), but strictly the adjustment of notes in the concordant ratios of the scale. But 'harmony' (tunefulness) is the nearest English equivalent.

47D. bestowed upon us by the same hands to the same intent,
  E. because in the most part of us our condition is lacking in
  measure and poor in grace.

II. WHAT COMES ABOUT OF NECESSITY

THE distinction drawn in the last paragraph between subsidiary
causes and rational purpose has provided the transition to the
second part of the dialogue, which begins here. The opening
sentence describes the contents of the first part as the works wrought
by the craftsmanship of divine intelligence (τὰ διὰ Νοῦ δεδημιουργημένα). We have traced, in the structure of the visible universe and
of man, the manifestations of benevolent purpose; but we have
been perpetually reminded that the work of the most ungrudging
benevolence cannot be perfect; it can only be ' as good as possible '.
The Demiurge has been operating all through under certain given
conditions, which he did not originate and which set a limit to the
goodness of his work. We have now to bring into account that
' other principle ' concerned in the production. It is introduced
under the names of Necessity and the Errant Cause.

If we consider the plan of the whole discourse, we see that Plato,
who has hitherto been looking at the world, as it were, from above,
and following the procedure of intelligence as it introduces order
into chaos, now shifts to the opposite pole and approaches the
world from the dark abyss that confronted its maker. Step by
step he analyses those elements which were pictured at the outset
as ' taken over ' by the Demiurge—' all that was visible, not at
rest, but in discordant and unordered motion ' (30A). These
factors are gradually distinguished, until we reach the fundamental
factor, Space. Space being given, Plato can then proceed to
discover elements of rational design even in the ' tumultuous welter
of fire, air, water, and earth '. The geometrical shapes of the
primary bodies are constructed; and once they are formed into
regular particles of determinate size and shape, the transformation
of one into another, which had bulked so large in earlier physical
systems, can be translated into terms of the disintegration and
reformation of these solids. In some degree, the sensible qualities
(or ' powers ') which act upon our sense-organs can then be cor-
related with the peculiarities of geometrical shape; and so we shall
come back once more, at the end of this second part, to the mechanism of sensation and perception—that point of contact between
the knowing soul and the external world, to which the first part
has brought us here.

47E–48E. *Necessity. The Errant Cause*

The opening paragraph is of fundamental importance for the understanding of the whole discourse. It describes the relations between Reason and Necessity, and how they co-operate to produce the visible world.

47E. Now our foregoing discourse, save for a few matters,[1] has set forth the works wrought by the craftsmanship of Reason ; but we must now set beside them the things that come about

48. of Necessity. For the generation of this universe was a mixed result of the combination of Necessity and Reason. Reason overruled Necessity by persuading her to guide the greatest part of the things that become towards what is best ; in that way and on that principle this universe was fashioned in the beginning by the victory of reasonable persuasion over Necessity. If, then, we are really to tell how it came into being on this principle, we must bring in also the Errant Cause—in what manner its nature is to cause motion.[2] So we must return upon our steps thus, and taking, in its turn,

B. a second principle concerned in the origin of these same things, start once more upon our present theme from the beginning, as we did upon the theme of our earlier discourse.

We must, in fact, consider in itself the nature of fire and water, air and earth, before the generation of the Heaven, and their condition [3] before the Heaven was. For to this day no one has explained their generation, but we speak as

---

[1] Namely, the account of the physical processes of vision, which are only secondary causes, subservient to the true ' reason ' for the gift of sight.

[2] τὸ τῆς πλανωμένης εἶδος αἰτίας, ᾗ φέρειν πέφυκεν. " Literally ' how it is its nature to set in motion ' " (A.-H.). For this use of φέρειν cf. *Epin.* 983B, ὅτε δὲ τοῦτο οἷός τέ ἐστιν θεός, ἅπασα αὐτῷ ῥαστώνη γέγονεν τοῦ πρῶτον μὲν ζῷον γεγονέναι πᾶν σῶμα καὶ ὄγκον σύμπαντα, ἔπειτα ᾗπερ ἂν διανοηθῇ βέλτιστα, ταύτῃ φέρειν, ' And since God can do this, it is the easiest of things for him, first to put life into any body and the whole of any bulk, and then to *make it move* as he has thought best' (trans. Harward). Cf. also 43A, where the soul-circles ' cause and suffer violent motions ' (βίᾳ ἐφέροντο καὶ ἔφερον), ' straying (πλανώμενα) every way in all the six directions ', and note there (p. 148). The meaning will be further discussed below. (2) Some critics have followed Stallbaum in taking φέρειν to mean ' endure ' and so ' admit ', *ratione qua ipsius natura fert* ; ' comme la nature des choses le comporte ' (Martin) ; ' so far as its own nature admits ' (Tr.). It may be questioned whether φέρειν with no expressed object can bear this sense. (3) Robin (*Phys. d. Plat.* 14) : *et la suivre distinctement* ' par où sa nature est de porter '. This is impossible, because *la suivre* is not in the Greek.

[3] πάθη is vague. It might cover the chaotic *condition* and *behaviour* of the ' powers ' before the elementary bodies received geometrical form, and ' *what happened to them* ', namely the construction of those bodies, which no one has yet explained.

48B.   if men knew what fire and each of the others is, positing
       them as original principles, elements (as it were, letters) of
       the universe ; whereas one who has ever so little intelligence
  C.   should not rank them in this analogy even so low as syllables.[1]
       On this occasion, however, our contribution is to be limited
       as follows.   We are not now to speak of the ' first principle '
       or ' principles '—or whatever name men choose to employ—
       of all things, if only on account of the difficulty of explaining
       what we think by our present method of exposition.   You,
       then, must not demand the explanation of me ; nor could
  D.   I persuade myself that I should be right in taking upon
       myself so great a task ; but holding fast to what I said at
       the outset—the worth of a probable account—I will try to
       give an explanation of all these matters in detail, no less
       probable than another, but more so, starting from the
       beginning in the same manner as before.[2]   So now once
       again at the outset of our discourse let us call upon a pro-
       tecting deity to grant us safe passage through a strange and
       unfamiliar exposition to the conclusion that probability
  E.   dictates ; and so let us begin once more.

In this prefatory passage the word ἀρχή (' beginning ', ' principle ',
' starting-point ') is reiterated many times, with a certain fluctuation
of sense.

The discourse needs a fresh *starting-point*.   The previous part
started from the question, for what reason (purpose, motive, αἰτία)
the world was made (29D).   The answer was found in the maker's
desire that all things should be as like himself, that is to say, as
good, as possible.   This was the ' supremely valid principle ' (or
starting-point, ἀρχή) to be accepted from men of understanding ;
and we have followed its guidance to the point where rational
design came into contrast with factors in the visible world that are
' incapable of any plan or intelligence for any purpose ' (46D).
We must now start afresh upon a study of these irrational factors.

They are at once connected with ' the nature of fire and air,
water and earth '.   These four so-called ' elements ', or some one

---

[1] στοιχεῖα, letters of the alphabet, first used in extant literature of the
physical elements at *Theaet.* 201E.   It is, however, not unlikely that Leucippus
or Democritus illustrated the infinitely various combinations of atoms by the
rearrangement of the same set of letters to form a tragedy or a comedy (Diels,
Elementum 13).

[2] καὶ ἔμπροσθεν seems untranslatable.   I suggest ⟨ᾗ⟩ καὶ ἔμπρ.   Cf. 48B, 2,
καθάπερ περὶ τῶν τότε . . . πάλιν ἀρκτέον ἀπ' ἀρχῆς.   But, just below at E, 2,
it is added that the new starting-point must be a fuller classification than the
one we started from ' before ' (τῆς πρόσθεν).

or more of them, had been regarded by Ionian science and by popular thought as the *original principles* (ἀρχαί) of all things. The earliest Ionians had chosen water or air as the one *original condition* (ἀρχή) from which a manifold world had emerged, and also as the *fundamental form* (φύσις) of which all things at all times ultimately consist. Empedocles had taken all four and clearly endowed them with the status of *elements*, irreducible and immutable factors which are merely mixed and rearranged in space to yield all the variety of compounds. The unexplained existence of the four elements had been taken as the *starting-point* for cosmogony, their properties and behaviour assumed, ' as if men knew what fire and each of the others is '. Plato at once denies them the status of elements, and promises to ' explain their generation ' from prior and simpler beginnings. He intends to construct the geometrical shapes of the four primary bodies from triangles which he takes as elementary. Only he adds that even this analysis will not claim to have reached ' the first principle or principles of all things '. This warning may mean that the elementary triangles themselves are reducible to numbers, and number perhaps to be derived from unity ; but he will not here push the analysis so far. Or it may mean that no one can ever really know the ultimate constitution of body, because there can be no such thing as physical science, but only a ' probable account '.

There was, however, another and more objectionable sense in which the elements had been called ἀρχαί : they had been taken as the original *source of motion* (ἀρχὴ κινήσεως), ' producing effects by cooling or heating, compacting or rarefying, and all such processes ' (46D). These effects were produced blindly by things incapable of any rational plan or forethought ; and from their casual interplay the world-order was believed to have emerged. In this way the elements and the physical processes due to their properties or ' powers ' (δυνάμεις) were made responsible as the true and only *causes* of all things (αἴτια τῶν πάντων, 46D). Plato intends to maintain that they are not original causes of motion and so of world-formation. The only source of motion is the self-moving soul, ' the causation of the intelligent nature ' (46D). These bodies hold only the second rank, as ' things that are (passively) moved by others and of necessity set yet others in motion '.

Reason and Necessity. With all this in mind, Plato opens this second part of the discourse with the contrasted powers of Reason and Necessity. Both, he says, contribute their part to the formation of the world of Becoming. Reason, aiming at the best, must use persuasion to win over Necessity, inducing her ' to guide the greatest part (but not all) of the things that become towards what is best '.

Immediately afterwards, he speaks of this second factor, Necessity, as an Errant Cause, whose manner of causing motion must be taken into account.

This central utterance has been much misunderstood, because the conceptions are foreign to the modern mind. How can Reason overrule Necessity by persuasion ? Is not Necessity precisely the inexorable, which can listen to no persuasion ? Necessity, in association with the material, suggests to us the unbroken and unbreakable chain of cause and effect, determining the whole course of events. What opening is left for persuasion ? Moreover, we connect Necessity with the element of intelligible order and regular sequence in becoming ; and we look to that quarter for the objects of knowledge, of natural science, whose aim is to formulate laws of necessary causation. How can Plato speak of Necessity as the errant or wandering cause, as something essentially irregular and unintelligible, needing to be brought, so far as possible, into order and persuaded to subserve, in some measure, the intelligent direction of Reason ?

In interpreting this passage some modern commentators are, perhaps unconsciously, influenced by the desire to bring Plato into conformity with the Jewish-Christian doctrine of an omnipotent Creator. They are reluctant to admit any factor in the visible world that does not owe its existence to God, who, having called all things into being out of nothing, must himself be the author of Nature's inexorable laws, and responsible for every defect in his handiwork. Archer-Hind's interpretation goes to the extreme in this direction, though he substitutes for the Christian God an idealistic equivalent—an absolute Spirit evolving everything out of itself by a timeless process of thought (whatever that may mean). By identifying the Demiurge with the Form of the Good, the World-Soul, and the sovereign Reason, he finds that Plato's system is ' a form of pantheism ' and ' an absolute idealism '. Matter is reduced to extension, and extension ' exists only subjectively in our minds ' (p. 45). In this view there is really nothing left but God, who must accordingly be the author of Necessity ; and Necessity is identified with natural law. It ' signifies the forces of matter originated by νοῦς, the sum total of the physical laws which govern the material universe : that is to say, the laws which govern the existence of νοῦς in the form of plurality ' (p. 166). The forces of nature ' are themselves expressly designed by Intelligence for a good end. . . . Necessity persuaded by Intelligence means in fact that necessity is a mode of the operation of intelligence '. The phrase ' Errant Cause ' implies no uncertainty or caprice in the operation of necessity, but only that necessity, though working strictly in obedience to a certain law, is for the most part as *inscrutable to us* as if it acted from arbitrary caprice (p. 167).

In all this Archer-Hind has pushed too far (and in the wrong direction) his principle of ' stripping off the veil of allegory ' from Plato's myth. By pursuing that principle the Neoplatonists discovered in the *Timaeus* a hierarchy of divinities that would have astonished Plato. It is no less easy for a modern critic to unveil the outlines of Christian theology or of the Hegelian absolute. We must pause to ask whether there is any sense in speaking of Reason as ' persuading ' a Necessity which has emanated wholly from Reason itself, or of an ' Errant Cause ' which is only an unerring cause that happens to be inscrutable to us and may become less and less inscrutable as knowledge advances.

By assuming that Necessity means the laws of nature and identifying these laws with a mode of the operation of Reason, Archer-Hind has eliminated one of Plato's two factors and left Reason in complete control. Professor Taylor reaches the same result by a different route. We are not, he remarks, to confuse Plato's *Ananke* with ' scientific necessity ' or ' the reign of law ' for she is expressly called the ' rambling ' or ' aimless ' or ' irresponsible ' cause (πλανωμένη αἰτία). ' Thus it is not the " necessary " but the " contingent ", the things for which we do not see any sufficient reason, the *apparently* arbitrary " collocations " in nature which are the contribution of that which Plato here calls ἀνάγκη . . . We must not take ἀνάγκη to represent anything inherently lawless and irrational, and yet must not take the word to mean necessity in the sense of conformity to law.' If we speak of ' mechanical causality ', it must be with reservations. Mechanism is entirely subordinate to intelligent purpose ; and, as the term ' errant cause ' implies, ' this " mechanism ", if we are to call it so, is supposed to be most prominent in the apparently anomalous, exceptional, and singular. I take it this means that where we can see a rational connection in nature we are dealing with what Timaeus calls a creation of νοῦς . . . But there is in the world a good deal of what we may call " brute " fact. We know it is there but we do not see " what the good of it " is, though, if we think with Timaeus and Plato, we feel satisfied that it subserves *some* good end. .... If we could ever have complete knowledge, we should find that ἀνάγκη had vanished from our account of the world. But since the sensible world itself is an ἀεὶ γιγνόμενον and never complete, there can be no complete knowledge of it ' (pp. 300–1).

As against Archer-Hind, Professor Taylor seems right in refusing to identify Necessity with natural law, which is neither an errant cause nor open to persuasion. But it is impossible to dispose of

Necessity as a mere residuum of hitherto unexplained fact, which complete knowledge (if man could ever attain to it) would reduce to nothing. Consider the effect of substituting this notion for Plato's Necessity. Could he have written that the generation of the universe was a mixed result of a combination of Reason and a certain amount of brute fact which dwindles as we come to see the reason for it ? Is there any sense in saying that Reason overruled this residuum of facts which we cannot yet account for and persuaded it to guide most things that become towards what is best ? Professor Taylor seems to have explained away the name Necessity as completely as Archer-Hind explained away the name Errant Cause. Both are influenced by the desire to make Plato's Demiurge really omnipotent.

Now, in discussing the Demiurge (p. 36), we have already remarked that the omnipotent Creator is foreign to ancient Greek thought, which unanimously denied the possibility of creating anything out of nothing. Plato's Demiurge, whatever he stands for, is represented as like the human craftsman, who must have materials to work upon. His task is to bring some intelligible order into a disorder which he ' takes over ', not to create the material before he fashions it. The material may have properties of its own, which he can, within limits, turn to his purpose, but which he did not institute. This possibility should be kept open, not foreclosed by the gratuitous assumption that the Demiurge must possess unrestricted omnipotence. In this respect the difficulty, as Professor Field remarks, is rather to conceive a purpose that is *not* restricted by given conditions. It is the familiar experience of every craftsman that his material limits the scope of his design and may hinder it from reaching a perfection he can imagine but never achieve. So far, there is really nothing but modern prejudice against accepting Plato's picture of the divine Reason as confronted by something which partly thwarts his benevolent purpose and needs to be persuaded, because it is not wholly under his control. The difficulty for us lies rather in a different quarter, in the seemingly contradictory notion of a Necessity which is also an Errant Cause, and associated, not with order and intelligibility, but with disorder and random chance.

We may start from a passage where Aristotle, discussing 'necessity' in contrast with final causation in Nature, associates necessity with accident, coincidence, chance, and spontaneity, because they are all contrasted with design. He puts the opponent's case in this way :

' Why should not nature work, not for the sake of something, nor because it is better so, but just as the sky rains, not in order

to make the corn grow, but of necessity (ἐξ ἀνάγκης)? What is
drawn up must cool, and what has been cooled must become
water and descend, the result of this being (συμβαίνει) that the
corn grows. Similarly if a man's crop is spoiled on the threshing-
floor, the rain did not fall for the sake of this—in order that the
crop might be spoiled—but that result just followed (συμβέβηκεν).
Why then should it not be the same with the parts in nature,
e.g. that our teeth should come up *of necessity* (ἐξ ἀνάγκης)—the
front teeth sharp, fitted for tearing, the molars broad and useful
for grinding down the food—since they did not arise for this end,
but it was merely a coincident result (συμπεσεῖν); and so with
all other parts in which we suppose that there is purpose (τὸ
ἕνεκά του)? Wherever then all the parts came about (συνέβη) just
what they would have been if they had come to be for an end,
such things survived, being organised spontaneously (ἀπὸ τοῦ
αὐτομάτου) in a fitting way; whereas those which grew otherwise
perished and continue to perish, as Empedocles says his " man-
faced ox-progeny " did.'[1]

In this passage the idea of necessity is opposed to purpose, and
linked with spontaneity, coincidence, chance. If we toss a coin
and it comes down heads up, it would not occur to us to call that
a ' necessary ' result, because (we should feel) there is no law that
coins must always come down so. Aristotle would call it indiffer-
ently a ' chance ' result or a ' necessary ' result: it ' comes about '
by causes that cannot act otherwise than they do and are not
directed by purpose. Empedocles' oxen with men's heads and
other such monstrous creatures were formed by the chance con-
currence of limbs which came into existence separately and were
never intended to fit together. The monsters perished because
they could not reproduce their kind. Others, more fortunately
composed, were able to survive. In the minds of Plato and
Aristotle this Empedoclean theory stood for the view of nature
which they condemned. The two alternatives, as they saw the
question, were that the order of the world should be due either to
intelligible purpose or to the undirected play of necessity and
chance. At *Philebus* 28D Socrates asks: ' Which are we to say,
Protarchus—that everything, this " whole " as we call it, is at the
disposal of a force that works without plan, at random, and just
as it may chance,[2] or on the contrary, as our predecessors said,
that it is an ordered system, guided by some admirable reason or
intelligence? ' Protarchus replies that it seems impious to doubt

[1] Ar., *Phys.* B, viii, 198b, 17 (Oxford trans.).
[2] τὴν τοῦ ἀλόγου καὶ εἰκῇ δύναμιν καὶ τὸ ὅπῃ ἔτυχεν.

that all things are directed by a mind worthily manifest in the whole appearance of the cosmos and in the revolutions of the heavenly bodies. Socrates concludes that we shall not agree when some clever person tells us that all things are in a disorderly condition (ἀτάκτως ἔχειν). There is a similar passage in the *Sophist*, where the alternative to divine craftsmanship is ' the belief commonly expressed that Nature (Φύσις) gives birth to things as a result of some spontaneous cause that generates without intelligence ' (265c). Here, as in the *Physics*, we find, in contrast with design, a spontaneous power of generation ascribed to a vaguely personified ' Nature '.

The earliest cosmogonies were of the evolutionary type and led to what Plato regarded as the atheistic materialism of which he draws à generalised picture in the *Laws*. Some, says the Athenian, assert that all things come into being partly by nature (φύσει), partly by chance (τύχη), and partly by design (art, τέχνη). Fire and water, earth and air, they say, all exist by nature and chance, not by design ; and these inanimate things then bring into existence the Sun and Moon, the Stars, and the Earth. ·They all move ' by the *chance* of their several powers (active properties, δυνάμεως), and according as they clash and fit together with some sort of affinity— hot with cold, dry with moist, soft with hard, and in other mixtures that result, *by chance, of necessity* (κατὰ τύχην ἐξ ἀνάγκης), from the combination of opposites—in that way they have generated the whole Heaven, animals and plants, and the seasons, not owing to intelligence or design or some divinity, but by *nature and chance*' (φύσει καὶ τύχη). Art (design, τέχνη) is a later product, mortal and of mortal origin. There are the fine and useful arts, and the art of statesmanship. All law is artificial, not natural ; so religion and morality are matters of convention, which vary from place to place and can be altered at human pleasure. This leads to the belief that might is right, to impiety and faction · (888E–890B). The Athenian himself denies that fire and air, water and earth are the primary things and deserve, in that sense, the name of ' nature '. Soul is really ' the first cause of the becoming and perishing of all things '. Soul is prior to all bodies, and governs their change and rearrangement. Judgment, forethought, intelligence, design, law (νόμος), are prior to ' hard and soft, heavy and light '. If ' nature ' means the generation of primary things, then soul has the best right to be described as existing ' by nature ' (891C–892C).

In this passage of the *Laws*, as in the *Physics*, we find necessity linked with chance, while law (νόμος) and order are connected with design. Chance and necessity, moreover, are associated with

'Nature', which is credited by the materialist with some spontaneous power of generation. This idea had survived from the earliest cosmologies, which had conceived the primary element or 'nature of things' as living. In consequence, the first physical philosophers had felt no difficulty about an original cause of motion. The divine and immortal substance of the world moved and gave birth to individual things, because it was alive. It was only later that this substance came to be reduced to the level of the bodily, which needs some external force to move it about. At that stage separate moving powers emerged: the Mind of Anaxagoras, the Love and Strife of Empedocles. These forces, however, remained part of Nature; they were not what we should call immaterial, but were extended in space. They retained that power of self-motion which had originally resided in the primary substance; but their motion was not directed by purpose towards any ideal of perfection in an ordered world. Even Anaxagoras' Mind, in spite of its name, had not been represented as working with conscious design for any good end, but only as giving the first impulse of mechanical motion to the revolution, or cosmic eddy, in which the world takes shape.

In the last of these physical systems, the atomism of Leucippus and Democritus, the cause of motion seems to have entirely disappeared. Matter or body has now been reduced to tiny impenetrable particles of solid 'being'. These and the void or 'not-being' in which they move are the sole realities in the universe. Rational design or purpose had no part in the formation of the world. The atoms move unceasingly in all directions. As they collide and fly off to a new quarter, they form vortices here and there in the field of unlimited space. In these vortices atoms of similar size and shape tend to drift together, like the sticks and straws in the eddies of a stream; and so finally worlds are always being formed, innumerable worlds scattered throughout the void, holding together for a time and then shattered and dispersed.

Why do the atoms move? Aristotle complains that the atomists merely declared motion to be everlasting; they did not explain what motion is, or how it occurs, or why it should be in one direction rather than another. He accuses them of indolence in ignoring these questions; but the truth was that, by reducing all the contents of the universe to solid bodies with no qualitative differences, they had left nothing that could possibly originate motion. The atoms collided and inflicted shocks on one another, so as to be constantly changing the direction of their movements. The process had no beginning; atoms have always been moving in all directions, aimlessly and at random. The only principle governing their motion is the tendency of like atoms to come together in the vortices. This is assumed as an unanalysed axiom, supported only by superficial analogies and proverbial maxims : ' birds of a feather flock together '. It is the last remnant of that spontaneous moving power in Nature which had originally animated the living substance. ' Like tends to move towards like ' is now taken as a bare unexplained fact ; but the principle is evidently akin to the more concrete images of Love and Strife in Empedocles, though his Love is the attraction between unlikes. It is not for nothing that Love and Strife reappear in the poem of Lucretius as Venus and Mars, though these mythical figures seem to have no right to any place in the arid universe of atoms and void. The principle 'Like moves towards like ' is important for our purpose ; for we find it, still as an ultimate unexplained assumption, at work in the chaos of the *Timaeus*.

A world in the atomists' system can thus be described as a product of chance or, as Aristotle calls it, spontaneity. ' There are some,' he writes,[1] ' who ascribe this Heaven of ours and all the worlds to spontaneity ($\tau\dot{o}\ a\dot{v}\tau\acute{o}\mu a\tau o\nu$). They say that the vortex, that is, the motion which separated and arranged in its present order all that exists, arose spontaneously.' From another point of view the result may be called necessary, in the sense that every motion takes place ' under constraint ' ($\dot{v}\pi'\ \dot{a}\nu\acute{a}\gamma\varkappa\eta\varsigma$) of some previous motion : an atom receives a shock and blindly passes it on. But the ancients had not discovered the laws of motion : to say that a movement happens ' by constraint ' is not to say that it conforms to any law. Necessity, in fact, did not carry with it the associations of law and order, at any rate in the earlier phases of atomism. The system might develop later towards a complete determinism, threatening to exclude any freedom of the will ; but Democritus shows no trace of having perceived this implication in the moral sphere.[2] The reason, I suspect, is that he had not arrived at what

---

[1] *Physics*, B, 4, 196a, 25. The reference to ' all the worlds ' shows that the atomists are meant.

[2] This has been pointed out by Dr. Bailey. See *The Greek Atomists*, p. 122. In his paper on Fate, Men, and Gods (Proc. Class. Assoc., 1935, p. 16), Dr. Bailey writes : ' It is in Democritus that we find for the first time anything like a consistent theory of Ethics, yet it is strange that there is no trace of any link between it and his physical theory of the world. The problem was really fundamental, for if the rule of " necessity " is absolute, then men's actions must be determined like everything else, and it is no good telling them what they ought to do, if they are not free to do it. Yet of this difficulty there is no sign ; the figure of " chance " now and then raises its head in Democritus' aphorisms, but never the thought of " fate " or of an inexorable " necessity ". The scientific view of the world has been laid down, but its implications have not been worked out.'

we should call a strictly mechanical or ' scientific ' conception of the world. His necessity was compatible with spontaneity.[1]

The thought of the fifth century in general was still farther removed than atomism from any closed system of determinism.[2] An attempt to arrive at the philosophy implied in Thucydides' conception of the course of history [3] led me to the conclusion that Thucydides, like his contemporaries, did not conceive nature as a domain of causal law. He believed in Fortune, defined as ' any non-natural agency which breaks in, as it were, from outside and diverts the current of events, without itself being a part of the series or an effect determined by an antecedent member of it. Human actions are not to be fitted into such a series. Their only causes—if we are to speak of causes at all—are motives, each of which is itself uncaused by anything preceding it in time ; all human motives are absolute " beginnings of motion ". A view of

[1] The statement which most clearly attributes a complete determinism to Democritus is in [Plutarch] *Strom.* 7 (*Vors.* 55A, 39) : He declared the universe to be unlimited, because it had never been fashioned by any design. . . . The causes of what now happens had no beginning (ἀρχήν), but all things absolutely " both past, present, and future " were determined by necessity (constraint, τῇ ἀνάγκῃ) without any beginning in time. The words in inverted commas are the only ones recognised by Diels as Democritus' own, and we cannot be sure that the doxographer's statement was not based, for example, on the view attributed to Democritus by Aristotle (*Phys.* 252a, 34) : ' Thus Democritus reduces the causes that explain nature to the fact that things happened in the past in the same way as they happen now : but he does not think fit to seek for a first principle (ἀρχήν) to explain this " always ".' Aristotle makes this remark in connection with the doctrine that ' there never was or will be a time at which motion did not or will not exist '. If Democritus was only affirming that principle, he might easily be understood to mean what the doxographer states. In other testimonies we are told that he actually *identified* ' necessity ' or ' constraint ' with the whirl or vortex of atoms (55A, 1) or with ' the collision, motion, and shock of matter ' (55A, 66). ' Atoms hold together until some more powerful constraint present in the environment shakes them apart and disperses them ' (55A, 37, Simplicius). This is not the ' necessity ' of causal law.

[2] It has been remarked that in Greece oracular predictions were normally hypothetical. ' It is extremely common for an oracle to answer : if you act in such and such a way, the result will be such and such. . . . The oracle foretells the future subject to certain conditions ; it can predict the consequences of a certain course of action. Such prophecies presuppose the existence of an order, a regularity in what happens, which yet leaves some scope for the individual. Life is not foreordained except in so far as its events are the effects of definite causes,' E. Ehnmark, *The Idea of God in Homer*, Uppsala (1935), p. 75. Even this statement is, perhaps, expressed in too modern terms.

[3] *Thucydides Mythistoricus*, London (1907), ch. vi. My excuse for quoting my own words at length is that the book is out of print. I can only reproduce here the conclusions without the supporting evidence.

the universe in which this irruption of free human agency is tacitly assumed is at any rate illogical if it denies the *possibility* of similar irruptions into the course of Nature by non-human agencies.' Thucydides, like the Socrates of Xenophon,[1] contrasts ' the field of ordinary human foresight (γνώμη) with the unknown field, which lies beyond it, of inscrutable non-human powers, whether we call these gods or spirits or simply Fortune. This antithesis is more frequently in Thucydides' thoughts than any other, except the famous contrast of word and deed. The two factors—γνώμη human foresight, purpose, motive, and Τύχη unforeseen non-human agencies —divide the field between them. They are the two factors, and the only two, which determine the course of a series of events such as a war ; neither Socrates nor Thucydides thinks of natural law. One speaker after another in the History dwells on the contrast between a man's own γνώμη, over which he has complete control, and Fortune, over which he has no control at all. . . . An examination of all the important passages where this contrast occurs has convinced me that Thucydides does not mean by Fortune " the operation of unknown (natural) causes ", the working of ordinary causal law in the universe. He is thinking of extraordinary, sudden interventions of non-human agencies, occurring especially at critical moments in warfare, or manifest from time to time in convulsions of Nature. It is these irruptions, and not the normal sway of " necessary and permanent laws ", that defeat the purposes of human γνώμη, and together with γνώμη are the sole determinant factors in a series of human events. The normal, ordinary course of Nature attracts no attention and is not felt to need explanation or to be relevant in any way to human action. When Thucydides speaks of the future as uncertain, he means not merely that it is unknown, but that it is undetermined, and that human design cannot be sure of completely controlling human events, because other unknown and incalculable agencies may at any moment intervene.' No one will deny that the outlook of Thucydides was as scientific as any to be found in the fifth century, and more scientific than that of any later historian before Polybius. The above account of his philosophy was written without any reference to Plato's ; but it now appears that there is a certain analogy between the two. Thucydides sees the field of human action divided between human foresight and chance ; Plato sees the world of physical events divided between divine purpose and chance associated with necessity.

That Necessity in Plato was the very antithesis of natural law was clearly seen by Grote. ' This word (necessity) ', he wrote, ' is

[1] *Mem.* i, 1.

171

now usually understood as denoting what is fixed, permanent, unalterable, knowable beforehand. In the Platonic *Timaeus* it means the very reverse : the indeterminate, the inconstant, the anomalous, that which can be neither understood nor predicted. It is Force, Movement, or Change, with the negative attribute of not being regular, or intelligible, or determined by any knowable antecedent or condition—*vis consili expers*' (*Plato*, iii, ch. 36). Grote, however, attempted no explanation of this factor in Plato's system. We may seek further light from the manner in which Plato approaches the subject, where he distinguishes between two types of causation, the divine and the necessary. At the end of the first part, he has just described the mechanical processes involved in the act of seeing—what happens to the rays of light and colour in their commerce with the visual fire that streams out from the eye. These physical transactions he then contrasts with the true reason or explanation (αἰτία) of sight, the purpose it is rationally designed to serve, namely to reveal to man the order and harmony of the visible heavens. Thus the manner ' how ' is contrasted with the reason ' why '. Most men, he adds, imagine that bodily processes, producing their effects without plan or purpose, are the sole causes of everything. But the lover of wisdom will seek first for the causation whose source lies in a self-moving and intelligent soul, and only in the second place for the causation characteristic of ' things that are moved by others and *of necessity* (ἐξ ἀνάγκης) set yet other things in motion '. ' Causes that work with intelligence to produce what is good and desirable ' must be distinguished from ' those which, being destitute of reason, produce their sundry effects *at random and without order* ' (τὸ τυχὸν ἄτακτον ἐξεργάζονται, 46E).

Here the lower type of causation, transmitting motion or change from one body to another, is, in the same breath, declared to proceed ' of necessity ' and ' at random and without order '. This is the point rightly apprehended by Grote and emphasised by Professor Taylor in opposition to the identification of Necessity with natural law. But we could not follow Professor Taylor in his reduction of Necessity to a residuum of hitherto unexplained brute fact, which tends to vanish as our knowledge becomes more complete. That interpretation was inspired by the wish to make Plato's divine Reason an omnipotent ' God '. If it be accepted, then in the actual world, apart from any question of the point to which our knowledge has advanced, there will be no antagonist to confront the Demiurge, no intractable material restricting the effort of the craftsman to realise his design. ' Plato ', he writes, ' emphatically does not mean that some things are due to intelligence and others to mere mechanism.[1] "Mechanism" comes in only as the "subordinate" of intelligent purpose, which is the "principal" in all undertakings' (p. 300). With complete knowledge (if we could ever have it), Necessity, he holds, would 'vanish from our account of the world'. If so, then in the world as completely known by God it can have no place at all.

The question whether this view is consistent with the whole tenor of the *Timaeus* can only be decided by careful consideration of many passages, upon which the reader must judge for himself as he comes to them. It seems certain that the divine Craftsman is in some degree a mythical figure; taken literally, he has attributes inappropriate to the Reason which Plato believed to be operative in the world. The question at issue is now narrowed down to this: Are we to regard the given material on which the divine Craftsman works as mythical, in so far as it is represented as restricting his purposes and preventing him from producing a world that is perfect and not merely 'as good as possible'? Are there any forces now and always at work in Nature, that are not completely subdued by the persuasion of Reason? It is hard to think that Plato would have devoted a third part of the discourse to 'what comes about of Necessity' in contrast with 'the works of Reason', if he had meant that nothing comes about of Necessity save under the complete control of Reason. But the problem cannot be so easily settled; it must be left for discussion in detail. Here I can only indicate, without meeting possible objections, what I believe to be the true answers to the two remaining questions: (1) How is the lower type of causation subordinated to the higher? (2) What is the permanent and irreducible factor confronting Reason and never wholly subordinate?

If, for the moment, we remain on the surface and take Plato's analogy of the divine with the human craftsman at its face value, it is easy to illustrate the subordination of necessity to purpose. There is the necessity which Aristotle calls 'hypothetical' in

---

[1] It is hard for us to avoid the word 'mechanical', because, since Descartes claimed: *terram totumque hunc mundum instar machinae descripsi* and still more since the industrial revolution, scientific thought has been haunted by the analogy of the machine and we connect the 'laws of nature' with machine-like regularity. But the ancients did not use machines driven by their own power without human intervention; they used only tools guided by manual skill and intelligence. Such tools are means to the realisation of some designed order in the passive material. So the notion of order is not associated with the means, but with the designing intelligence and the end. It is characteristic that Plato regards the exact precision of the stars' movements as a proof of their intelligence (*Laws* 967B), not of their being subject to a mechanical necessity.

contrast to absolute necessity.[1] This is the necessity of the indispensable means to an end. Food is a necessary of life : we *must* have food, *if* we are to live ; but it is not necessary that we should live. If I wish to recover a debt, I may have to sail to Aegina to find my debtor ; but nothing compels me to sail. The necessity lies in the links connecting the purposing will at the beginning of the chain with the attainment of the purpose at the end ; we need not think of it as extending further in either direction. Reason and will are conditioned by this concatenation of indispensable means. So is it with the craftsman. If I wish to cut wood, I must make my saw of iron, not of wax. Iron has certain properties of its own, indispensable for my purpose. On the other hand, I can take advantage of this very fact to attain my end. I can make use of those properties to cut wood, though the iron in itself would just as soon cut my throat.

There is also the necessity residing in the properties themselves and governing their action. Fire has the characteristic power (δύναμις, as Plato and others call it) of burning heat. Fire can act only in one way ; it can heat other things, but not cool them. By virtue of this necessity of the fire's own nature, its action is so far regular. But just because it acts thus by constraint of its nature, Plato describes such causation as aimless or ' wandering '. The action is blind and undirected by purpose. If I strike a match to light a fire in my grate and warm myself, I am availing myself of the fire's power. The fire is indifferent to my purpose and has none of its own. If there is a wooden beam in my chimney, the fire may go on to burn down the house—a result neither foreseen nor desired. Once started by my voluntary action, the process of combustion will go on of itself. I did not ordain that process and it may get beyond my control. Yet, within certain limits I can direct its action into a channel leading to a foreseen and purposed end.

This notion of the hypothetical necessity of means to an end and of the partial subordination of the given means goes back to the *Phaedo*. Socrates complains that Anaxagoras, though he spoke of Intelligence ordering all things, did not carry this idea into the detailed account of the cosmos, or explain how every arrangement was planned ' for the best '. He fell back on the blind and aimless action of the elements. It was as if the presence of Socrates in the prison should be attributed to the action of his muscles in bringing him there, and not to his own purpose of abiding by the sentence of the court because he judged it better to do so. We ought to distinguish between the true reason or cause (αἴτιον)

---

[1] *Metaph.* Δ 5, where the various meanings of ' necessity ' are distinguished.

and ' that without which the cause would not be a cause '. It is the same contrast of the end with the indispensable means, the *conditio sine qua non* of the achievement of purpose. Socrates in the *Phaedo* says that this distinction ought to be applied to the explanation of the world as a whole, but that he himself had been unable to attempt that task. It is the task which, many years afterwards, Plato set himself to accomplish in the *Timaeus*. And here in fact we find him speaking of the Demiurge as making use of the lower kind of causes as auxiliaries (συναίτια) or subordinate instruments in his work of producing the best results possible (e.g. at 46c).

The question still remains, whether the analogy between the Demiurge and the human craftsman holds at this point or is to be explained away. The carpenter does not make the wood or ordain its natural properties and behaviour. Is the Demiurge in the same position of having to make the best he can of not wholly suitable materials, or did he himself endow the material he uses with all its properties and make them completely amenable to his own control ?

There is, indeed, one feature of the properties, once they exist, which makes them not wholly amenable. Physical qualities occur in groups of concomitants. The *Timaeus* contains an illustration of the disadvantage that may result. The function of bone is to protect from injury the seat of life, the brain and marrow. To that end bone must be hard. But its very hardness makes it too brittle and inflexible, and also liable to decay under excessive heat. Accordingly the skeleton needs to be wrapped about with soft and yielding flesh. The brittleness is a concomitant of the hardness, and it can be described both as necessary or inevitable and as ' accidental ' (συμβεβηκός). The ideas of necessity and chance are once more associated in the notion of the necessary accident.[1] In this instance brittleness *happens to be* an *inevitable* but undesirable concomitant of the useful quality, hardness. There is also the case in which two properties which would both be useful cannot be combined. We find, for example, that those parts of the body which are the seats of intelligence, above all the skull, have the thinnest covering of bone and flesh. ' The reason is that this frame, which is born and compacted of necessity (ἐξ ἀνάγκης), in no wise allows dense bone and much flesh to go together with keenly responsive sense-perception. For if these two characters had consented to coincide (εἴπερ ἅμα συμπίπτειν ἠθελησάτην), the structure of the head would have possessed them above all, and the human race, bearing a head fortified with flesh and

[1] Cf. 77A, συνέβαινεν ἐξ ἀνάγκης.

sinew, would have enjoyed a life twice or many times as long as now, healthier and more free from pain. But as it was, the artificers who brought us into being reckoned whether they should make a long-lived but inferior race or one with a shorter span but nobler'. Here the two desirable characters refuse to coincide as concomitants: they are incompatible. Necessity cannot be wholly persuaded by Reason to bring about the best result conceivable. Reason must be content to sacrifice the less important advantage and achieve the best result attainable. This last instance illustrates the truth of Galen's observation that the Demiurge is not strictly omnipotent. In arranging the world he could not group physical qualities in such a way as to secure all the ends he desired.

But we are still talking in metaphor. We have seen reason to regard the Demiurge, as such, as a mythical figure. Cosmos has always existed. It had no beginning in time and therefore no maker. The image of the craftsman is employed as the most simple and vivid means of making us realise that the world was not a chance product born of aimless natural powers but exhibits evidences of rational design, like a product of human art. There is a divine Reason at work, aiming at the best possible. It does not follow that this Reason stands to the world in precisely the same relation as the human craftsman to his materials and his product, though the craftsman may furnish the most convenient illustration. These considerations affect the status of the other factor, the craftsman's materials, or the chaos which confronts the Demiurge and which he is said to ' take over in a state of disorderly motion ' and reduce, so far as he can, to order. This chaos, again, is not to be taken literally. If the cosmos had no beginning in time, there never was a chaos before order was introduced. Chaos can only stand for some factor in the world as it exists at all times. The question then will be whether this factor is, now and always, in some measure chaotic and disorderly, or is, now and always, completely subordinate to the ends of Reason. It is here that I differ from Professor Taylor, who holds that the subordination is complete. The question cannot be argued till we come to the interpretation of the relevant passages in the text. I will only anticipate the conclusion so far as to say that, in my opinion, the body of the universe is not reduced by Plato to mere extension, but contains motions and active powers which are not instituted by the divine Reason and are perpetually producing undesirable effects. Further, since all physical motion has its ultimate source in a living soul, these bodily motions and powers can only be attributed to an irrational element in the World-Soul. It may be claimed that this theory preserves a sufficient and intelligible meaning for the statement that this world is a mixed product of the combination of Reason and Necessity—a Necessity that can also be called an Errant Cause. But we must not forestall the coming account of the Receptacle of Becoming and its chaotic contents.

48E–49A. *The Receptacle of Becoming*

For his fresh starting-point, Timaeus goes back here to the very beginning of his discourse : the distinction between the two orders of existence, the intelligible and unchanging model and the changing and visible copy. We now learn that the copy is not self-subsistent ; it needs the support of a medium, just as a reflection requires a mirror to hold it. Accordingly, a third factor has now to be added— a factor which had no place in the first part among the creations of Reason.

48E. Our new starting-point in describing the universe must, however, be a fuller classification than we made before. We then distinguished two things ; but now a third must be pointed out. For our earlier discourse the two were suffici- ent : one postulated as model, intelligible and always unchangingly real ; second, a copy of this model, which 49. becomes and is visible. A third we did not then distinguish, thinking that the two would suffice ; but now, it seems, the argument compels us to attempt to bring to light and describe a form difficult and obscure. What nature must we, then, conceive it to possess and what part does it play ? [1] This, more than anything else : that it is the Receptacle—as it were, the nurse—of all Becoming.

The third factor, not hitherto taken into account, is first presented as the Receptacle or nurse of Becoming. This Receptacle and its contents are to be analysed in a series of steps, which we shall do well not to anticipate. For some time yet Plato does not use the word ' Space ' ; it first occurs in the conclusion (52A), led up to by a series of images that are designed to elucidate gradually a nature more ' obscure and difficult ' than geometrical space.

We may note here, however, that the hitherto unrecognised third

---

[1] δύναμις, the active manifestation of the nature. Cf. δύναμις used of the ' force ' or significance of a word, and τὴν τῶν εἰκότων λόγων δύναμιν (48D), ' the *worth* of a probable account ', what it is good for ; also 64C, διὰ τὸ πυρὸς ἀέρος τε ἐν αὐτοῖς δύναμιν ἐνεῖναι μεγίστην ' because fire and air play the largest part in them ' (sight and hearing).

factor fills a gap in the scheme which Plato, in the *Republic*, had borrowed from Parmenides. He had there described the realm of objects of ' opinion ' as intermediate between the perfectly real and knowable and the wholly unreal and unknowable. But the *Sophist* has shown that the wholly unreal (τὸ παντελῶς μὴ ὄν) cannot even be named without self-contradiction. It is an absolute blank of nothingness. If the perfectly real Forms are to have the objects of opinion as images, there must be something, not totally unreal, to receive these images. The question that now confronts us is, what this Receptacle of *eidola* can be.

49A–50A. *Fire, Air, etc., are names of qualities, not of substances*

This question is first approached by a consideration of fire, air, etc., as the contents of the Receptacle. The point is that these are not permanent irreducible elements, not ' things ' with a constant nature. Plato rejects the old Milesian doctrine of a single fundamental form of matter, which was to serve both as the original state of 'things (ἀρχή) and as the permanent ground (φύσις) underlying change. He also rejects the belief of the pluralists who, in reply to Parmenides, had reduced all change to the re-arrangement in space of the four elements (Empedocles) or of ' seeds ' (Anaxagoras) or of atoms (Leucippus and Democritus). Plato's position was nearer to that of Heraclitus, who alone had rejected the notion of substance underlying change and had taught the complete transformation of every form of body into every other. We are now to think of qualities which are not also ' things ' or substances, but transient appearances in the Receptacle. The Receptacle itself alone has some sort of permanent being.

49A. True, however, as this statement is, it needs to be put in clearer language ; and that is hard, in particular because to

B. that end it is necessary to raise a previous difficulty [1] about fire and the things that rank with fire. It is hard to say, with respect to any one of these, which we ought to call really water rather than fire, or indeed which we should call by any given name rather than by all the names together or by each severally, so as to use language in a sound and trustworthy way. How, then, and in what terms are we to

---

[1] With προαπορηθῆναι and διαπορηθέντες (B, 7) compare Aristotle, *Met.* B, i, ' For those who wish to get clear of difficulties (εὐπορῆσαι) it is advantageous to state the difficulties (διαπορῆσαι) well ; for the subsequent free play of thought (εὐπορία) implies the solution of the previous difficulties.' Only to the man who has first faced the difficulties (τῷ προηπορηκότι) is it clear, what goal he is making for.

49B. speak of this matter, and what is the previous difficulty that may be reasonably stated ?

In the first place, take the thing we now call water. This, when it is compacted, we see (as we imagine) becoming earth and stones, and this same thing, when it is dissolved

c. and dispersed, becoming wind and air ; air becoming fire by being inflamed ; and, by a reverse process, fire, when condensed and extinguished, returning once more to the form of air, and air coming together again and condensing as mist and cloud ; and from these, as they are yet more closely compacted, flowing water ; and from water once more earth and stones : and thus, as it appears, they transmit in a cycle the process of passing into one another. Since, then, in this

D. way no one of these things ever makes its appearance as the *same* thing, which of them can we stedfastly affirm to be *this*—whatever it may be—and not something else, without blushing for ourselves ? It cannot be done ; but by far the safest course is to speak of them in the following terms. Whenever we observe a thing perpetually changing—fire, for example—in every case we should speak of fire,[1] not as ' this ', but as ' what is of such and such a quality ',[2] nor of water as ' this ', but always as ' what is of such and such a quality ' ; nor must we speak of anything else as having some permanence, among all the things we indicate by the

E. expressions ' this ' or ' that ', imagining we are pointing out some definite thing. For they slip away and do not wait to be described as ' that ' or ' this '[3] or by any phrase that exhibits them as having permanent being. We should not use these expressions of any of them, but ' that which is of a certain quality and has the same sort of quality as it perpetually recurs in the cycle '—that[4] is the description we should use in the case of each and all of them. In fact, we must give the name ' fire ' to that which is at all times[5] of

---

[1] πῦρ after προσαγορεύειν (D, 6) should perhaps be omitted.

[2] τὸ τοιοῦτον, a general expression for πυρώδης, ὑδατώδης, etc. Cf. Chalcid. *non est ignis sed igneum quiddam, nec aer sed aerium.*

[3] I omit καὶ τὴν τῷδε, as no convincing translation or correction of the words has yet been proposed. Tr.'s καὶ τὴν τοῦδε (' of this ' = relative to this) is perhaps the best ; but nothing in the context supports it.

[4] Taking οὕτω (before καλεῖν) as resuming the long phrase that precedes. τὸ τοιοῦτον ἀεὶ περιφερόμενον ὅμοιον is rightly explained by Tr. : ' the this-like which ever recurs as similar '. ἀεί can mean either ' from time to time ' or ' perpetually '.

[5] There is *at all times* (διὰ παντός) a certain amount of stuff that is fiery. This quality is sufficiently ' alike ' (ὅμοιον) to be recognised and named, though it is not an enduring substance, and is perpetually varying.

179

49E. such and such a quality ; and so with anything else that is in process of becoming. Only in speaking of that *in* which all of them are always coming to be, making their appearance and again vanishing out of it, may we use the words ' this '

50. or ' that ' ; we must not apply any of these words to that which is of some quality—hot or cold or any of the opposites— or to any combination of these opposites.[1]

The result so far is that fire and the rest are denied the status of elements or permanent things with an unchanging character. Their apparent[2] transformation in a cycle is described in terms borrowed from Anaximenes and Anaxagoras. Anaximenes had conceived that all things at all times really are air. Air is the permanent nature ; fire is air in a rarefied state ; when more closely packed, air becomes successively wind, cloud, water, earth, stone. Anaximenes thus took a step towards the doctrine clearly formulated after Parmenides, that qualitative change is reducible to the bringing together or separation in space of a number of unalterable elements. Anaxagoras, who explicitly identified all so-called ' becoming and perishing ' with the combination and separation of permanently real things, used similar language : ' From these things as they are separated off, earth is compacted. For out of clouds water is separated off, and from water earth, and from earth stones are compacted by the cold.' Empedocles also tried to abolish change of quality by reducing ' becoming and perishing ' to the mixture and interchange of his four unalterable things, fire, air, water, earth.

Plato rejects this interpretation, asserting the contrary view that there is change of quality without any underlying substance or permanent ground. The word ' quality ' (ποιότης) had been introduced for the first time at *Theaetetus* 182A, with an apology for its uncouthness. In pre-Socratic thought ' the hot ', ' the cold ', etc., had been treated as things (χρήματα) having each a characteristic power (δύναμις) in which its nature was manifested by action on other things. The coining of the word ' quality ' (ποιό-της, such-and-such-ness) as a general expression for hotness, coldness,

---

[1] ὅσα ἐκ τούτων. This may mean that fire (for instance) is a combination of sensible qualities, such as ' hot ', ' yellow ' (or orange or blue), etc., making up that ' fieriness ' (τὸ τοιοῦτον) which is sufficiently alike (ὅμοιον) for us to distinguish it from wateriness and other combinations of qualities. But the phrase might also cover compound bodies, mixtures of the four primary bodies.

[2] At 54B it will be remarked (as ὡς δοκοῦμεν, 49B, 8, and ὡς φαίνεται, C, 7, hint) that the transformation is not so complete as it appears. Earth cannot be transformed into the other three.

whiteness (θερμό-της, ψυχρό-της, λευκό-της), etc., marks the clear distinction of qualities from ' things ' or substances. Plato is now asserting that ' fire ' is properly only a name for a certain combination of qualities or ' powers ', which appear and disappear and are always varying. Such groups of qualities, though perpetually shifting, are sufficiently alike to be indicated by names ; but in referring to fire we ought not strictly to say ' this (thing) ', because the phrase suggests something which preserves a constant identity. We are to get rid of the notion of material *substance*.

In contrast with this stream of fluctuating qualities stands that *in which* [1] they make their transient appearances. The Receptacle is the only factor in the bodily that may be called ' this ', because it has permanent being and its nature does not change. What this Receptacle is, we do not yet know. Later on, when the Demiurge intervenes to introduce an element of rational order, he will form the primary bodies by fashioning for them geometrical shapes (53B). But here we are considering the bodily as it was ' before ' the Heaven was made. We must not imagine the qualities here described as existing in particles of any shape, regular or otherwise. There is nothing yet but a flux of shifting qualities, appearing and vanishing in a permanent Receptacle.

There is no justification for calling the Receptacle ' matter '—a term not used by Plato. The Receptacle is not that ' out of which ' (ἐξ οὗ) things are made ; it is that ' in which ' (ἐν ᾧ) qualities appear, as fleeting images are seen *in* a mirror. It is the qualities, not the Receptacle, that constitute ' the bodily ' (τὸ σωματοειδές). The term was used at 31B : ' That which comes to be must be bodily and so visible and tangible ; and nothing can be visible without fire or tangible without earth.' The contents of the Receptacle will presently be called ' bodies ' (σώματα 50B), but we must beware of taking this to mean ' particles ', as if the qualities had already received shapes.

50A–C. *The Receptacle compared to a mass of plastic material*

Turning now from the contents to the Receptacle, Plato begins to illustrate its nature by an image which, as he admits, is in some respects misleading. It is compared to a mass of plastic material, moulded and remoulded into various shapes. The nature of the material (gold) is permanent ; the shapes are formed only to be obliterated and give place to others.

---

[1] 49E, ἐν ᾧ ἐγγιγνόμενα φαντάζεται. This phrase ἐν ᾧ is consistently used in the following context to mean the Receptacle as a whole, not particular volumes in which events of a certain type take place '. This is one of Tr.'s importations from Whitehead (pp. 320–1).

50A. But I must do my best to explain this thing once more in still clearer terms.

Suppose a man had moulded figures of all sorts out of gold,[1] and were unceasingly to remould each into all the
B. rest : then, if you should point to one of them and ask what it was, much the safest answer in respect of truth would be to say ' gold ', and never to speak of a triangle or any of the other figures that were coming to be in it as things that have being,[2] since they are changing even while one is asserting their existence. Rather one should be content if they so much as consent to accept the description ' what is of such and such a quality ' with any certainty. Now the same thing must be said of that nature which receives all bodies. It must be called always the same ; for it never departs at all from its own character ; since it is always receiving all things, and never in any way whatsoever takes
C. on any character that is like any of the things that enter it : by nature it is there as a matrix for everything, changed and diversified by the things that enter it, and on their account it *appears* to have different qualities at different times ; while the things that pass in and out are to be called copies of the eternal things, impressions taken from them in a strange manner that is hard to express : we will follow it up on another occasion.[3]

Some critics have seen in this passage references to the later configuration of space by the geometrical shapes of the primary corpuscles.[4]

---

[1] ἐκ χρυσοῦ. The figures are made *out of* gold and *consist of* gold ; but the contents of the Receptacle are not made *out of* it. This is a point where the illustration is inadequate.

[2] μηδέποτε λέγειν ταῦτα ὡς ὄντα can also be construed : ' never to speak of a triangle, etc., as *these* (things), as though they had being ', and the contrast with τοιοῦτον following perhaps favours this.

[3] The reference may be to 52c (A-H.), or the promise may be unfulfilled (Tr.).

[4] Thus Baeumker (*Prob. d. Mat.* 131) identifies the ' things that pass in and out ' of the Receptacle with those shapes composed of elementary triangles. Tr. (rightly, I think) explains the transient ' characters ' as ' the characteristics of different sensible bodies, in fact the various sounds, colours, scents, etc., revealed to us in different regions ' (p. 326). But he adds that ' since Timaeus means at a later stage to account for all these qualities as consequences of the shapes of corpuscles, to all intents and purposes what he wants to insist on is that space itself has no specific " shape " of its own. He means, then, that space in all its regions is uniform or homogeneous. If it were not, its parts would not be indifferent to *all* configurations'. Tr. then strays into a discussion of modern non-uniform spaces—alternatives which Plato cannot have intended to exclude, because they could never have entered his mind.

But, since nothing has yet been said even about space, no one reading the *Timaeus* for the first time could associate the triangles and other figures moulded in the gold with the elementary triangles and solids later constructed by the Demiurge ; nor did Plato intend this. The figures mentioned belong solely to the illustration, the point of which is that the only thing we can call ' this ' and so treat as a thing with permanent properties of its own is the gold, not the shapes which are moulded, effaced, and remoulded. Similarly the Receptacle has a nature of its own, from which it never departs.

What corresponds to the figures of the illustration is ' the things that pass into and out of ' the Receptacle. What these things are we have been plainly told in the preceding paragraph ; they are those qualities—' any opposite or combination of opposites '— which ' are always coming to be in the Receptacle, making their appearance, and again vanishing out of it ' (49E). This was clear to some at least of the ancient commentators. A fragment of the lost part of Proclus' commentary [1] reads : ' Perhaps it is better to say that the term " things that pass in and out " is applied *not only to the qualities* (αἱ ποιότητες), but also to the forms immersed in matter (τὰ εἴδη τὰ ἔνυλα) ; for these, not the qualities, are likenesses (ὁμοιώματα) of the intelligible things ' (i.e. τῶν ὄντων ἀεὶ μιμήματα, 50C, 5). It is clear that Proclus had been discussing a current view that the qualities alone were meant. Proclus' further remarks show that by ' the forms immersed in matter ' (an Aristotelian phrase) he means copies, present in matter, of the eternal Forms of Fire, Air, Water, and Earth (not of any other Forms). He discusses whither these copies go, when they ' pass out '. Not into other matter ; ' for when fire is quenched and the matter becomes airy, we do not see other matter being kindled '. They must pass out simply into non-existence.[2] Proclus no doubt had in mind the

---

[1] Pr. iii, 357. This fragment and the other references to our passage in Proclus' commentary have been overlooked.

[2] Other passages in Proclus referring to this subject are : i, 233[14], 'Some forms (εἴδη) are inseparable from matter and are always coming into being from that which always is ; others come to be and pass away in time : thus corporeality (ἡ σωματότης) is always becoming and always in the region of matter ; but *the form (character, εἴδος) of fire or air enters matter and passes out, being separated and perishing owing to the victory of the opposite nature.'* i, 419[28], ' The eternally real was the model of unordered becoming, since it was from thence that *the unarticulated forms (characters, ἀδιάρθρωτα εἴδη) came to be present in the unordered before the Heaven came into being.'* These are the ' traces of the elements ' (τὰ ἴχνη τῶν στοιχείων—a reference to ἴχνη at 53B, 2). ii, 25[9] : In the case of fire there is (1) the *form* (εἴδος), an indivisible nature, the image of the cause of fire ; for there is a certain indivisible thing (the εἴδος ἔνυλον) even in divisible things ; from this results (2) an *extension*

theory of Forms as it is stated towards the end of the *Phaedo*. There the immutable and eternal Form is clearly distinguished from the character (μορφή, ἰδέα) present in things that are said to partake of the Form and bear the same name. Some such characters are grouped in pairs of opposites, tall and short, hot and cold. One member of such a pair will never admit its opposite : ' the hot *in us* ' can never become cold ; when we become cold, the hot character must either ' withdraw ' to make way for the cold, or it must ' perish '. Proclus decides for the latter alternative : what he calls the ' character immersed in matter ' must, he says, ' pass out ' into non-existence. His distinction between ' the form (character) immersed in matter ' and the ' quality ' is a piece of Neoplatonic subtlety. Plato speaks of the qualities as ' characters ' (μορφαί, ἰδέαι), as he had in the *Phaedo*, where μορφή and ἰδέα are used interchangeably and neither can mean ' shape '.[1] The things that pass into and out of the Receptacle are simply the opposite qualities or groups of qualities characteristic of the four primary bodies. They are called here ' copies of the eternal things ' ; and at 51B ' copies ' of Fire, Air, Water, and Earth, just before the passage which plainly asserts the existence of their originals, the intelligible Forms of just those four bodies. The Forms, ' in some strange manner that is hard to express ', impress their characteristic qualities on the Receptacle. But the Receptacle does not itself possess any of these characters or qualities, any more than gold in itself possesses triangular shape. The qualities do not belong to it ; they only pass in and out, like images crossing a mirror. They

of itself in the matter of the fire, and from this again (3) the *powers* (δυνάμεις) of fire, or *qualities* (ποιότητες) such as hotness, etc. (This is part of a misguided attempt to interpret the ἀριθμοί, ὄγκοι, and δυνάμεις of 30C, but it shows what Proclus meant by his distinction of the εἶδος ἔνυλον from the ποιότητες or δυνάμεις). The phrase ' unarticulated forms ' means the qualities as described at 52D ff., before the Demiurge endows them with ' geometrical shape and number '.

Simplicius, *Phys.* 539, 10, says that Plato in the *Timaeus* calls matter χώραν καὶ τόπον τῶν ἐνύλων εἰδῶν. It appears from 540, 13 ff., that this phrase ἔνυλα εἴδη was partly based on 53B, 4, διεσχηματίσατο εἴδεσί τε καὶ ἀριθμοῖς, which, in fact, refers to the geometrical shapes ; partly on 51A, 7, μεταλαμβάνον ἀπορώτατά πη τοῦ νοητοῦ, which Aristotle took as meaning that the Recipient partakes of the Forms (see p. 187).

[1] There is, for example, ' the character of three ' (ἡ τῶν τριῶν ἰδέα (104D)), the characters of evenness and oddness, and so on. The words are interchanged, e.g. at 104D, ἡ ἐναντία ἰδέα ἐκείνῃ τῇ μορφῇ. The term εἶδος is there reserved for the Form to which the character belongs, because the distinction is important to the argument (see especially 103E) ; but in the *Timaeus* Plato follows his usual practice of eschewing precise terminology, and uses εἶδος for character as well as μορφή and ἰδέα. A.-H. imports the word ' shape ' for μορφή (c, 1), and so does Tr.

are said to ' change and diversify ' [1] the Receptacle ; they form a constantly shifting pattern, ' presenting all diversities of aspect ' (D, 5), as some parts become fiery, others watery, and so on.

50C–51B. *The Receptacle has no qualities of its own*

The illustration of the man moulding all sorts of figures out of gold was sufficient for its purpose, to illustrate the contrast between the permanent nature of the Receptacle and the shifting qualities. Its defect is that gold is a stuff. that has sensible qualities of its own, persisting through all the variations of shape. Aristotle's objections to the illustration turn partly on this point.[2] But Plato himself proceeds to correct the defect. He has already said that the Receptacle does not in itself possess any of the characters that pass in and out, any more than gold as such possesses any of the shapes. It is now added that the Receptacle has no characters of its own ' before ' the qualities enter it, unlike the gold which has its own sensible properties.[3]

Before making this point, Plato introduces the image of the father, the mother, and the child, to illustrate the relations of the eternal Form, the Receptacle, and Becoming.

50C. Be that as it may, for the present we must conceive three
       things : that which becomes ; that in which it becomes ;
  D. and the model in whose likeness that which becomes is born.[4]
       Indeed we may fittingly compare the Recipient to a mother,
       the model to a father, and the nature that arises between
       them to their offspring. Further we must observe [5] that,
       if there is to be an impress presenting all diversities of aspect,
       the thing itself in which the impress comes to be situated,

---

[1] 50C, κινούμενόν τε καὶ διασχηματιζόμενον ὑπὸ τῶν εἰσιόντων. κίνησις is used as the general word for ' change ' (with its two species, locomotion and qualitative change) at *Parm.* 138B, *Theaet.* 181D. διασχηματίζεσθαι is used below (53B) of the pattern introduced by the creation of geometrical shapes ; but σχῆμα means appearance, manner, fashion, mode, etc., as well as shape, though no doubt the analogous figures (σχήματα) moulded in the gold suggested the word. Different qualities affecting different parts of a space must diversify it and form some kind of pattern, however vague in outline and irregular. Cf. the phrases ἰδεῖν ποικίλον πάσας ποικιλίας (50D, 5) and παντοδαπὴν ἰδεῖν φαίνεσθαι (52E, 1).

[2] They are summarised by Tr., p. 322.

[3] Cf. Baeumker (*Prob. d. Mat.* 132), whose analysis of the whole argument here is helpful, though I cannot accept all his conclusions.

[4] φύεται, ' born '. The next sentence takes up this metaphor as furnishing an appropriate image, which replaces that of the craftsman.

[5] νοῆσαι depends in thought rather on the χρή at c, 7, than on πρέπει, and perhaps also in grammar, the remark about the ' fittingness ' of the metaphor in φύεται being treated as parenthetical.

50D. cannot have been duly prepared unless it is free from all
E. those characters which it is to receive from elsewhere. For
if it were like any one of the things that come in upon it,
then, when things of contrary or entirely different nature
came, in receiving them it would reproduce them badly,
intruding its own features alongside. Hence that which is
to receive in itself all kinds [1] must be free from all characters ; just like the base which the makers of scented
ointments skilfully contrive to start with : they make the
liquids that are to receive the scents as odourless as possible.
Or again, anyone who sets about taking impressions of shapes
in some soft substance, allows no shape to show itself there
beforehand, but begins by making the surface as smooth and

51. level as he can. In the same way, that which is duly to
receive over its whole extent and many times over all the
likenesses of the intelligible [2] and eternal things ought in its
own nature to be free of all the characters. For this reason,
then, the mother and Receptacle of what has come to be
visible and otherwise sensible must not be called earth or
air or fire or water, nor any of their compounds or components [3] ; but we shall not be deceived if we call it a nature
invisible and characterless, all-receiving, partaking in some

B. very puzzling way of the intelligible and very hard to apprehend. So far as its nature can be arrived at from what has

[1] γένη is (as often) simply a synonym of ἰδέα, μορφή, εἶδος (character).
Plato varies the word, just as above (D, 7) he writes ἄμορφον ἀπασῶν τῶν ἰδεῶν
(= μορφῶν). None of the four words here means the eternal Form ; for
this is never ' received ' by the Receptacle. Note also that σχῆμα (' shape ')
is not used as a synonym for any of them, but confined to the shapes moulded
in gold or in some soft substance in the two illustrations (50A and 50E, 8).

[2] The conjecture νοητῶν (for πάντων) ἀεί τε ὄντων. can be supported by the
occurrence of the phrase at 37A, I. But πάντων or πάντα is required by the
sense. I suggest τῷ ⟨τὰ πάν⟩τα τῶν νοητῶν ἀεί τε ὄντων. The Receptacle is to
receive all the likenesses of the Forms concerned (the four primary bodies),
rather than likenesses of all the Forms there are. Cf. E, 5, τὸ τὰ πάντα
ἐκδεξόμενον γένη.

[3] ' Compounds ', i.e. complex bodies formed of more than one of the four
primary bodies. ' Components ', i.e. any qualities into which what we call
' fire ' or ' fieriness ' (etc.) might be analysed, e.g. the heat, yellowness, etc.,
of flame. Cf. 50A, 3, ὅσα ἐκ τούτων, where τούτων means the opposites (hot,
white, etc.), of which fire, etc., are composed. This statement formally
excludes the notion that the Receptacle is some subtler or more ultimate
kind of matter (such as ' the hot ', ' the cold ', etc.) beyond the four primary
bodies (cf. Fraccaroli, p. 89). At Sophist 243 the view that ' the hot and
the cold ' are the ultimately real things in nature is taken as typical of all
the early physicists. There is no reference to the triangles of which the
elementary figures are later to be composed, since these have not yet been
mentioned.

51B. already been said, the most correct account of it would be this : that part of it which has been made fiery appears at any time as fire ; the part that is liquefied as water ; and as earth or air such parts as receive likenesses of these.

The argument that the Receptacle must not possess in itself any quality like those which enter it, is preceded by the comparison of the eternal Form to the father and of the Receptacle to the mother. The connection of thought implies a current view of the part played by the mother in generation. In the *Eumenides* (660) Apollo argues that ' the mother of what is called her child is no parent (τοκεύς), but only the nurse (τροφός) of the new life sown in her. The parent is the begetter ; she is but a host (ξένη) harbouring the stranger plant '. Similarly, according to Diodorus (i. 80), the Egyptians regarded no child as a bastard, holding that the father is the sole cause of generation, while the mother furnishes only nourishment (τροφή) and room (χώρα) for the infant. The belief is mentioned several times by Aristotle, who debates whether the female contributes anything to generation or only provides the place (τόπος). He gives it as the opinion of Anaxagoras and other physicists that the seed comes from the male, the female only furnishing the place.[1] So here the Receptacle or ' nurse ' (τιθήνη, 49A) of Becoming is simply the place ' in which ' the qualities appear. If it had any qualities of its own, it would intrude its own features or visible appearance (ὄψις), as the mother's features might be expected to reappear in the child, if she contributes any part of its substance.

The Receptacle, then, has no visible appearance ; but is ' a nature invisible and characterless, all-receiving, partaking in some very puzzling way of the intelligible and very hard to apprehend '. ' Partaking of the intelligible ' is, unfortunately, an ambiguous phrase. Some have understood it as referring to ' the real informing of matter by the Ideas '[2] ; but Archer-Hind remarks that Plato's

---

[1] *de gen anim.* A19, init., B1, 763*b*, 30. The doctrine is still held by the natives of S.E. Australia : ' children emanate from the father alone and are merely nurtured by the mother ' (Frazer, *Totemism and Exogamy* i, 338. Contrast the Central Tribes who are ignorant that the father plays any part in begetting). In the *Life of Johnson* Boswell defends his ' partiality for heirs male ' by ' the opinion of some distinguished naturalists that our species is transmitted through males only, the female being all along no more than a *nidus* or nurse, as Mother Earth is to plants of every sort '. It follows that ' a man's grandson by a daughter has in reality no connection whatever with his blood '.

[2] So Zeller ; Baeumker (*Prob. d. Mat.* 133) ; Aristotle *Phys.* iv, 2, 209*b*, 12, Plato identified matter and space, τὸ γὰρ μεταληπτικὸν καὶ τὴν χώραν ἕν καὶ ταὐτόν (Simpl., *Phys.* 542 : He calls it τὸ μεταληπτικὸν in the *Timaeus*, μεταλαμβάνει γὰρ' ἀπορώτατά πῃ τοῦ νοητοῦ'). Tr. (p. 331) agrees with A.-H.

meaning is more fully expressed at 52B, where Space is said to be ' apprehended without the senses by a sort of bastard reasoning '. To ' partake of the intelligible ' will then mean ' to be an object of rational thought ', as opposed to being an object of the senses. Further discussion may be postponed to that later passage where Space has at last been mentioned.

In the present passage (where Space has not been mentioned) the words εἶδος, ἰδέα, μορφή, still bear the sense implied by the whole context : they mean sensible qualities, not ' shapes '. The last sentence speaks of part of the Receptacle being made fiery, part liquefied (made watery), and so on. The same language is used of the chaos described at 52D as existing *before* the Heaven was made or the Demiurge had designed the geometrical figures of the primary bodies. Plato's point is that the Receptacle has no inherent sensible qualities of its own, not that ' Space has no specific " shape " of its own ', or that ' we are not allowed to account for exceptional " appearances " in any region, as those who think of space as having a variable curvature would like to do, by suggesting that this region has a " different " geometry from others '.[1] It is a much more tenable position that, according to Plato, Space has a shape of its own, being coextensive with the spherical universe, outside which there is neither body nor void.[2]

51B–E.   *Ideal models of Fire, Air, Water, Earth*

Plato has just spoken of ' copies ' (μιμήματα) of Fire, Air, Water, and Earth being ' received ' by the Receptacle. This leads to the next question : Are there models to serve as originals for these copies ?

51B.   But in pressing our inquiry about them, there is a question that must rather be determined by argument.[3] Is there such a thing as ' Fire just in itself ' or any of the other things which we are always describing in such terms, as
c.   things that ' are just in themselves ' ? Or are the things we see or otherwise perceive by the bodily senses the only things that have such reality,[4] and has nothing else, over

---

[1] Tr., pp. 326, 328.

[2] See F. M. Cornford, *The Invention of Space*, Essays in honour of Gilbert Murray, Oxford, 1936.

[3] The emphasis falls, by position, on λόγῳ, ' by argument ', as opposed to ' what can be gathered from our earlier statements ' in the previous sentence.   Cf. the contrast of ὁ ὀρθὸς λόγος (λόγος in the true sense) and ὁ εἰκώς (56B, 4).

[4] τοιαύτην ἀλήθειαν, the independent and absolute reality, just mentioned, such as we ascribe to Forms.   So Stallbaum, A.-H.

51C. and above these, any sort of being at all ?   Are we talking
idly whenever we say that there is such a thing as an
intelligible Form of anything ?   Is this nothing more than
a word ?

Now it does not become us either to dismiss the present
question without trial or verdict, simply asseverating that
it is so, nor yet to insert a lengthy digression into a discourse
D. that is already long.   If we could see our way to draw a
distinction [1] of great importance in few words, that would
best suit the occasion.   My own verdict, then, is this.   If
intelligence and true belief are two different kinds, then these
things—Forms that we cannot perceive but only think of—
certainly exist in themselves ;   but if, as some hold, true
belief in no way differs from intelligence, then all the things
we perceive through the bodily senses must be taken as the
most certain reality.   Now we must affirm that they are
E. two different things, for they are distinct in origin and unlike
in nature.   The one is produced in us by instruction, the
other by persuasion ;   the one can always give a true account
of itself, the other can give none ;   the one cannot be shaken
by persuasion, whereas the other can be won over ; and true
belief, we must allow, is shared by all mankind, intelligence
only by the gods and a small number of men.

The alternative to be determined by argument is : whether those
combinations of qualities which we call bodies and which we see
or otherwise perceive through the bodily senses [2] have a fully
substantial existence in their own right, or are (as we have called
them) only copies of independently existing Forms.   The language
closely resembles *Parm.* 130D ff., where Parmenides questions
Socrates as to the extent of the world of Forms.   Socrates has no
doubt that there are separate Forms of terms such as Likeness,
Unity, Plurality, and also of moral terms, Just, Good, etc.   He is
doubtful about Forms such as Man ' separate from ourselves and
all other men ', and Fire, Water, etc.   This class corresponds to
the products of divine workmanship at *Sophist* 266B : ' ourselves
and all other living creatures and the elements of natural things—
fire, water, and their kindred '.   Living organisms and the four

---

[1] ὅρον ὁρίζειν, to draw a boundary-line (cf. *Gorg.* 470B) ; in this case the
boundary between the two orders of existence, corresponding to the two
kinds of apprehension next mentioned.

[2] The description shows that the ' copies ' are not the shapes of the corpuscles
of primary bodies, but the qualities which we perceive when we say ' Fire
is here '.   We do not perceive the corpuscles or their shapes.

primary bodies of which all other bodies are composed are the two classes of things in the physical world with the best claim to separate Forms. When it comes to hair, dirt, and other such undignified things, Socrates at first thinks it would be absurd to postulate Forms; these must be no more than ' just the things we see '.[1]

The present passage is concerned mainly with Forms of the primary bodies; and the reality of these Forms is affirmed on the same general grounds that make it necessary to believe in any Forms whatsoever. As in *Republic* v, the existence of two orders of objects—intelligible and sensible—is declared to follow from the indubitable distinction between rational understanding or knowledge and mere belief, which can be produced or shaken by persuasion. This characteristic of belief, even when true, was taken in the *Theaetetus* (201A) as fatal to the claim of true belief to rank as knowledge. Belief, moreover, can ' give no account of itself '. This characteristic is best illustrated by the *Meno*. The slave questioned by Socrates has produced true beliefs about the solution of a problem in geometry; but they will not become knowledge until he has been taken many times through the whole demonstration, grasped all the premisses, and seen how the conclusion must inevitably follow. His beliefs will then be unshakably secured ' by reflection on the reason ' (*Meno*, 85C ff., 97 E).

It is certain, then, that there are independently real Forms of Fire, Air, Water, and Earth. Fire ' just in itself ' is an eternal model, an object of intelligence, not of perception. We have been told that the name ' Fire ' is to be given to that which is of a certain quality, appearing in the Receptacle at any time in the cycle of change. This quality is the copy, bearing the same name as its model; the model itself is the meaning of the name ' Fire ', more or less clearly present to our thought whenever we use the word. Plato tells us nothing further as to its nature. It cannot be identified with the pyramid, the geometrical shape of the fire corpuscle. When we look at a fire, we do not see or think of pyramids; and when we say ' Here is fire ' we do not mean ' Here are pyramids '. What we perceive is a certain combination of shifting qualities in a certain place at a certain time—the yellowness we see, the hotness we feel. Such a combination, whenever and wherever it occurs, is sufficiently ' alike ' for us to name it ' fire ', and it is a fleeting copy or impress of an unchanging model. More than this Plato cannot tell us. We must not hope to get nearer

---

[1] *Parm.* 130D, ταῦτα μέν γε ἅπερ ὁρῶμεν, ταῦτα καὶ εἶναι. εἶδος δέ τι αὐτῶν οἰηθῆναι εἶναι μὴ λίαν ᾖ ἄτοπον. Cf. 51C, ἢ ταῦτα ἅπερ καὶ βλέπομεν . . . μόνα ἐστὶν τοιαύτην ἔχοντα ἀλήθειαν.

to his thought by translating his words into language that sounds to us scientific.[1]

There is no warrant for A.-H.'s remark that ' the list of ideas in the *Timaeus* includes, in addition to the ideas of living creatures, only the ideas of fire, air, water and earth ' (p. 180). In his introduction he goes further and suggests that Plato ought to have eliminated ideal types of the elements and would have eliminated them, ' had his attention been drawn to the subject ' (p. 35).[2] The unprejudiced reader may think that his attention was very clearly drawn to the subject in the passage before us. Nor will the Platonist easily believe that living creatures and the primary bodies alone have ideal Forms. How are mathematics and dialectic to be carried on, if the only unchanging objects of thought are the natural kinds of living creatures and the four primary bodies ? These are specially relevant to an account of the physical universe, and are therefore prominent in the *Timaeus*. We cannot infer that Plato no longer believed that there was such a thing as Justice ' just in itself ' or the Triangle ' just in itself '. The *Philebus* and the *Laws* would not bear out such a conclusion.

51E–52D. *Summary description of the three factors : Form, Copy, and Space as the Receptacle*

In the foregoing sections we started with the notion of a Receptacle of Becoming ; then passed to its contents, the sensible qualities

---

[1] Tr. (p. 334), for instance, says : ' The question is whether there is or is not a standard of scientific truth by which individuals can and *ought* to correct the deliverances of their senses.' ' Fire means the occurrence of events with some *definite* law or pattern in a region of the continuum, water the appearance of events of a different determinate pattern. It follows at once that only when this pattern is exactly realised do you have " real " or " pure " fire or water. If it is only imperfectly realised, you have not " pure " fire or water, just as we should say that " water " which proved on analysis not to be composed of hydrogen and oxygen in the proportions determined by the chemists is not " pure " water, but has " impurities ".' Plato's phrase ' Fire just in itself ' means, according to Tr., ' " fire which is just fire," " fire with no admixture of anything else ", exactly as we speak of " pure water ", " pure atmospheric air ", " pure gold ".' This account is in danger·of suggesting a confusion between an exact realisation of the pattern and the pattern itself. When we speak of ' pure water ' we mean something which, supposing it to exist, would be a perceptible thing which we could touch and drink.

Robin's account of the Form of Fire (*Phys. de Platon* 49) keeps nearer to Plato's own account, but involves theories about mathematical intermediates between Forms and sensibles and about Ideal Numbers which are too speculative for the scope of this book.

[2] In the Journal of Philol. xxiv, pp. 49 ff., Archer-Hind went the whole way and denied that the ontology of the *Timaeus* allows room for these ideas.

and their combinations, and finally to the ideal models. Next follows a summary description of these three factors, in the reverse order.

51E.
52.
•

This being so, we must agree that there is, first, the unchanging Form, ungenerated and indestructible, which neither receives anything else into itself from elsewhere nor itself enters into anything else anywhere, invisible and otherwise imperceptible ; that, in fact, which thinking has for its object.

Second is that which bears the same name and is like that Form ; is sensible ; is brought into existence ; is perpetually in motion, coming to be in a certain place and again vanishing out of it ; and is to be apprehended by belief involving perception.

B.

Third is Space, which is everlasting,[1] not admitting destruction ; providing a situation for all things that come into being, but itself apprehended without the senses by a sort of bastard reasoning, and hardly an object of belief.

This, indeed, is that which we look upon as in a dream [2] and say that anything that is must needs be in some place and occupy some room, and that what is not somewhere in earth or heaven is nothing.[3] Because of this dreaming

C.

state, we prove unable to rouse ourselves and to draw all these distinctions and others akin to them, even in the case of the waking and truly existing nature, and so to state the truth : namely that, whereas for an image, since not even the very principle on which it has come into being belongs to the image itself,[4] but it is the ever moving semblance of

---

[1] Taking ἀεί with ὄν (cf. A.-H.). The words are separated for the sake of euphony. Cf. 28A, 6, πρὸς τὸ κατὰ ταὐτὰ ἔχον βλέπων ἀεί, where ἀεί belongs to ἔχον.

[2] Taking πρὸς ὃ βλέποντες together (with A.-H.), an easy hyperbaton. Simplicius, *Phys.* 521, 31, paraphrases : ἀπὸ τῆς εἰς τὰ ἔνυλα ὀνειρατικῆς ἐμβλέψεως. Plato uses ἐγρηγορώς, not βλέπων, for a waking dream : *Soph.* 266c, ὄναρ ἀνθρώπινον ἐγρηγορόσιν. βλέπων normally means ' alive ', not ' awake '.

[3] Cf. Aristotle, *Phys.* iv, 1, 208a, 29. ' Everybody supposes that things which exist are *somewhere* ; the non-existent is " nowhere "—where is the goat-stag or the sphinx ? ' Simplic. *ad loc* describes this as a ' parody ' of our passage. Zeno (*Vors.* 19A, 24) assumed in one of his arguments that ' Everything that exists is somewhere ' or ' in some place '. Gorgias (quoted below) repeats this.

[4] This and other interpretations of the difficult clause ἐπείπερ οὐδ' αὐτὸ τοῦτο ἐφ' ᾧ γέγονεν ἑαυτῆς ἐστιν are discussed in the Appendix (p. 370). An image comes into being on the same principle or conditions as a reflection : there must be an original to cast it and a medium to contain it. Neither condition ' belongs to ' the image itself.

52C. something else, it is proper that it should come to be *in* something else, clinging in some sort to existence on pain · of being nothing at all, on the other hand that which has real being has the support of the exactly true account, which declares that, so long as the two things are different, neither can ever come to be in the other in such a way that the
D. two should become at once one and the same thing and two.

The three factors are here contrasted in three respects : (1) the sort of existence which they have ; (2) the manner in which they are known ; (3) the relations of the Form and of the copy to Space.

(1) Space, here named for the first time, is ' everlastingly existent and not admitting destruction '. The phrase differs only verbally from that applied to the Form, ' ungenerated and indestructible '. Here as elsewhere the Receptacle does not owe its existence to the Demiurge, but is represented as a given factor limiting his operations by necessary conditions. Space is thus essentially different from Time, which was ranked among the works of intelligence and had an archetype, eternal duration, of which it was an image. There is no archetype of Space,[1] which exists in its own right as surely as does the Form. By recognising Space as an independent and eternally existing factor necessary to the becoming of a world of sensible images, Plato has added to the old scheme borrowed in *Republic* v from Parmenides. The three things enumerated there were (1) the perfectly real and knowable, (2) the objects of opinion, (3) the absolutely unreal and unknowable. The third of these is not to be identified with Space, for Space is not unreal, and we can apprehend it. Plato's purpose is precisely to introduce Space, as an eternally real object, to fill the blank left by the totally non-existent in Parmenides' scheme, which consequently provided no support for any world of appearances.

(2) Space is apprehended, not by the senses, but ' by a sort of bastard reasoning ', and is ' hardly an object of belief '. It is not, like the Form, an object of genuine rational understanding ($\nu\acute{o}\eta\sigma\iota\varsigma$) ; nor is it, like the copy, apprehended by the senses and by judgment involving perception. Space is not sensible ; for it cannot be seen or touched. Is it, then, intelligible ? It is not a genuine intelligible object, because it has no status in the world of Forms ; these, as Plato goes on to say, are not in Space, nor are they extended, although we may imagine ' the Triangle ', for instance, as an extended figure. Space is rather a factor in the visible world ; and yet it is everlasting and imperishable, and can only be apprehended by thinking : so it ' partakes of the intelligible in a very puzzling

[1] Cf. Fraccaroli, p. 87.

193

way' (51B).   Plato may have in mind the process we call 'abstraction'—thinking away all the positive perceptible contents of Becoming, until nothing is left but the 'room' or place in which they occur.   'Hardly an object of belief' (μόγις πιστόν) seems intended to rule out 'belief' or 'opinion', in addition to perception and rational knowledge.   We normally 'believe' in the existence of a thing in the sense-world because we have perceived it,[1] but Space we cannot perceive.

(3) The Form is contrasted with Space in that the Form 'never receives anything else into itself from elsewhere', and with the copy in that 'it never itself enters into anything else anywhere'. The same thing was said of Beauty itself in the *Symposium* (211A) : this is 'never anywhere in anything else—in a living creature, for instance, or in earth or heaven—but is always in and by itself'. The contrast between the Form and the copy is dwelt upon in the rest of the paragraph in somewhat obscure terms.   The copy or image, not having the substantial existence of a perfectly real thing (ὄντως ὄν), but being 'the ever-moving semblance of something else', requires some medium 'in which' it may appear and disappear, like a mirror image.   Thanks to this medium, Space, it 'clings somehow or other to existence, on pain of being nothing at all'.   It is, in fact, an *eidolon*, defined in the *Sophist* (240B) as that which has some sort of existence (ὄν πως), but not real being (οὐκ ὄντως ὄν).   'The very principle (or condition) on which it comes to be' lies outside itself, partly in the medium which receives it, partly in the original which casts the image of itself on the medium.

The last part of the sentence contrasts with this fleeting semblance, the Form which has real being.   We should expect merely the statement that the Form, since it is self-subsisting, requires no medium and so is not in Space.   Plato complicates matters by adding the statement that neither is Space in the Form.   'That which has real being (the Form) has the support of the exactly true account' (no mere 'likely story') ; and this declares that 'so long as the two things are different, neither can ever come to be in the other in such a way that the two should become at once one and the same thing and two'.   The language is obscure, but the whole drift of the passage demands that the two things in question must be the Form and Space'.[2]   These must remain for ever distinct. The Form, we have been told, cannot receive anything into itself

---

[1] Cf. the use of πίστις for a state of mind intermediate between intellectual understanding and εἰκασία at *Rep.* 511E.

[2] Not the Form and the copy, as A.-H. and Fraccaroli suppose.   Above (c, 3) ἑτέρου τινός meant the Form, and ἐν ἑτέρῳ τινί (c, 4) meant Space. These expressions give τὸ μὲν ἄλλο and τὸ δὲ ἄλλο (c, 6) something to refer to.

from elsewhere. This applies to Space, which can never enter into the existence of Forms. Nor can the Form ever pass into anything else anywhere; it can never enter Space, and Space cannot receive anything more than the copy.

What remains obscure is the consequence stated in the last words : the result of the one coming to be in the other would be that ' the two would become at once one and the same thing and two '. Probably Plato is thinking of a somewhat archaic argument which figures in Gorgias' tract *On Not-being* (frag. 3) :

' If Being is unlimited, it is nowhere. For if it is somewhere, that in which it is is different from it ; and thus, being embraced by something which contains it, it will no longer be unlimited, since what embraces something must be larger than the thing it embraces, and nothing is larger than the unlimited. Hence the unlimited is not somewhere.[1] But again, neither is it contained in itself; for then *the same thing will be container* (τὸ ἐν ᾧ) *and content* (τὸ ἐν αὑτῷ), *and Being will become two things* : *place and body* ; *for the container is place and the contained, body.* But this is absurd. Therefore neither is Being contained in itself. Accordingly, if Being is everlasting, it is unlimited ; and if unlimited, it is nowhere ; and if it is nowhere, it does not exist.'

Gorgias is, so far as this argument goes, one of those who live in the dream which takes the sense-world as the sole reality,[2] and imagine that ' what is not somewhere in earth or heaven is nothing '. The argument is reproduced by Plato himself at *Parm.* 138A, in the criticism of the Parmenidean One, which excludes all plurality :

' Being such as we have described, it cannot be anywhere— neither in anything else, nor yet in itself.' If in anything else, it would be encompassed by that which contains it. If in itself, *the container as such must be one thing, the contained another ; the same thing cannot, as a whole, be both at once.* ' Hence the One would be no longer one, but two.' Therefore the One is not somewhere, not being either in itself or in anything else.[3]

In both passages the consequence that Being, or the One, ' will be no longer one thing but two '—container and content, follows from the supposition that it is contained ' in itself '. As applied to the Form and Space, it will come to this. If, like a god assuming visible shape, the Form were to enter Space, that would mean that

---

[1] This argument is echoed in the *Parmenides* 150E.

[2] A comparison which recurs frequently in the *Republic*, 476C, 520C, 534C.

[3] Aristotle, *Phys.* iv, 3, discusses at length the question whether, and in what sense, a thing can be ' in itself '.

it would become extended, and so Space would enter, as extension, into its existence. But in an extended thing, considered as self-contained, we can always distinguish the thing itself from the room or place it occupies. So Gorgias argues that 'Being will become two things, place and body'. In Plato's argument the two things will be the Form (which must retain its unity) and its extension, the space it has admitted ; and this last is the fundamental element of body. But Forms are essentially bodiless. So the Form cannot enter Space, nor can Space enter the Form as extension.

In this passage Plato comes nearer than anywhere else in the *Timaeus* to the problem of the *eidolon*. He contributes towards the solution an important factor which did not come into view in the *Sophist*. Space, as eternally self-existent, provides the copy with a 'room' or situation where it can 'somehow cling to existence' as ὄν πως and escape being nothing at all (παντελῶς μὴ ὄν). But the addition of this third factor does not, in itself, solve the difficulty of explaining how Becoming can ever occur. The two parents of Becoming—the Form and Space—are alike eternal and unchanging. How can an image cast by an unchanging object on an unchanging mirror be itself inconstant and fleeting ? Aristotle saw this objection to the theory of Forms, offered in the *Phaedo* as an explanation of becoming and perishing : 'If the Forms are causes, why is their generating activity intermittent instead of perpetual and continuous—since there always *are* participants as well as Forms ? '[1]

'There were others,' Aristotle adds, 'who thought that matter was adequate by itself to account for becoming ; matter originates the movement.' This account Aristotle considers more scientific than the theory of Forms : something which produces change of quality and transformation would be more capable of bringing things into being. But he rejects it on the ground that matter (in his own view) has only the passive power of being moved : water, for instance, has not the active power of producing a living creature without the co-operation of the 'form'. The powers (δυνάμεις) attributed by the theory he is criticising to the simple bodies are treated as 'instrumental' or auxiliary causes of generation ; the hot has the power to separate things, the cold to bring them together, and so on ; and the becoming and perishing of all other things are to result from these actions. But in the absence of the form, these powers cannot even be instrumental ; one might as well attribute the making of a table to the 'necessary' action of the saw or the plane.

[1] Ar., *de gen. et corr.* 335*b*, 18 (trans. Joachim).

This criticism recalls Plato's condemnation of the popular view that 'cooling and heating, compacting or rarefying, are not mere accessories, but the sole causes of all things' (46D). Plato himself, who does recognise the superior position of the Form, is entitled to treat the active powers of the primary bodies as accessory causes, amenable in some degree to the controlling direction of intelligence, though, left to themselves, they would produce random results by the blind necessity of their nature. They are things that can set other things in motion; but they require to be set in motion themselves. Neither the Form nor Space can act as the ultimate moving cause. Hence, although the Form has been compared to the father, Space to the mother, the Form cannot really supersede the Demiurge, or whatever he stands for, as the generator of Becoming. If, as we have concluded, the Demiurge is mythical, the moving cause can only be the World-Soul. It becomes more than ever difficult to resist the inference that the Demiurge is to be identified with the Reason in the World-Soul.[1]

52D–53C. *Description of Chaos*

So far we have been almost wholly concerned with the Receptacle of Becoming and the shifting qualities that appear in it and disappear, considered, so far as is possible, in abstraction from the element of rational design contributed by Reason. The Forms of the four primary bodies were only introduced towards the end, because a copy must have an original; but it has been emphasised that the Forms remain apart and cannot themselves enter the region of Becoming. Plato now sums up the three factors required for the production of a visible world, to which, as we have just seen, we must add the 'Demiurge' to produce it. He then passes to a description of the Receptacle and its contents, imagined as existing 'before' the ordered world came into being. We are now to hear what the Demiurge does when he 'takes over' this chaos.

52D. Let this, then, be given as the tale summed according to my judgment [2]: that there are Being, Space, Becoming—three distinct things—even before the Heaven came into being.

---

[1] The inference is drawn by W. Theiler (among others), who concludes that the Demiurge must be conceived ' *als Verdoppelung der Weltseele . . . als Hinausprojektion gleichsam ihrer künstlerisch wirkenden Seite* ' (Teleolog. Naturbetrachtung, 72). See above, pp. 34 ff.

[2] ψῆφοις λογίζεσθαι is to calculate with counters; but the singular ψῆφον seems to allude to τὴν ἐμὴν τίθεμαι ψῆφον, 51D. For τρία τριχῇ, A.-H. compares 89E, τρία τριχῇ ψυχῆς εἴδη, ' three distinct forms of soul '. Cf. also διχῇ (39A) for two ' distinct' motions in different planes.

52D. Now the nurse of Becoming, being made watery and fiery and receiving the characters of earth and air, and qualified by all the other affections that go with these,[1] had every

E. sort of diverse appearance to the sight ; but because it was filled with powers that were neither alike nor evenly balanced, there was no equipoise in any region of it ; but it was everywhere swayed unevenly and shaken by these things, and by its motion shook them in turn. And they, being thus moved, were perpetually being separated and carried in different directions ; just as when things are shaken and winnowed by means of winnowing-baskets and other instruments for

53. cleaning corn, the dense and heavy things go one way, while the rare and light are carried to another place and settle there. In the same way at that time the four kinds were shaken by the Recipient, which itself was in motion like an instrument for shaking, and it separated the most unlike kinds farthest apart from one another, and thrust the most alike closest together ; whereby the different kinds came to have different regions, even before the ordered whole consisting of them came to be. Before that, all these

B. kinds were without proportion or measure. Fire, water, earth, and air possessed indeed some vestiges of their own nature, but were altogether in such a condition as we should expect for anything when deity is absent from it. Such being their nature at the time when the ordering of the universe was taken in hand, the god then began by giving them a distinct configuration by means of shapes and numbers. That the god framed them with the greatest possible perfection, which they had not before, must be taken, above all, as a principle we constantly assert ; what I must now attempt to explain to you is the distinct formation

C. of each and their origin. The account will be unfamiliar ; but you are schooled in those branches of learning which my explanations require, and so will follow me.

The analysis of the bodily just concluded enables Plato to give a more detailed picture of that chaos 'taken over' at the outset by the Demiurge : 'all that was visible, not at rest, but in discordant and unordered motion' (30A). 'Visible' implies the

---

[1] 'Characters' ($\mu o \rho \phi \acute{a} s$) and 'affections' ($\pi \acute{a} \theta \eta$) both mean 'qualities'. The whole phrase is parallel to $\pi \hat{v} \rho$ $\mu \grave{\epsilon} \nu$ $\dot{\epsilon} \kappa \acute{a} \sigma \tau o \tau \epsilon$ $a \dot{v} \tau o \hat{v}$ $\tau \grave{o}$ $\pi \epsilon \pi v \rho \omega \mu \acute{\epsilon} \nu o \nu$ $\mu \acute{\epsilon} \rho o s$ $\phi a \acute{\iota} \nu \epsilon \sigma \theta a \iota$, $\tau \grave{o}$ $\delta \grave{\epsilon}$ $\dot{v} \gamma \rho a \nu \theta \grave{\epsilon} \nu$ $\ddot{v} \delta \omega \rho$, $\gamma \hat{\eta} \nu$ $\tau \epsilon$ $\kappa a \grave{\iota}$ $\acute{a} \acute{\epsilon} \rho a$ $\kappa a \theta'$ $\ddot{o} \sigma o \nu$ $\ddot{a} \nu$ $\mu \iota \mu \acute{\eta} \mu a \tau a$ $\tau o \acute{v} \tau \omega \nu$ $\delta \acute{\epsilon} \chi \eta \tau a \iota$ (51B), where $\mu \iota \mu \acute{\eta} \mu a \tau a$, like $\dot{a} \phi o \mu o \iota \acute{\omega} \mu a \tau a$ at 51A, 2, shows that the $\epsilon \dot{\iota} \sigma \iota \acute{o} \nu \tau a$ $\kappa a \grave{\iota}$ $\dot{\epsilon} \xi \iota \acute{o} \nu \tau a$ $\tau \hat{\omega} \nu$ $\ddot{o} \nu \tau \omega \nu$ $\dot{a} \epsilon \grave{\iota}$ $\mu \iota \mu \acute{\eta} \mu a \tau a$ of 50C are simply qualities.

presence of sensible qualities ; and there is also motion of a disorderly kind. So here the Receptacle, now identified ultimately
with Space, is qualified by the main characters (μορφαί) of fire, air,
water, and earth—hot, cold, moist, dry—and by other ' affections '
(qualities, πάθη) attendant upon these. The qualities are also
described as ' powers ' (δυνάμεις), with reference to their power of
acting on one another—the hot on the cold, the dry on the moist,
and so on—and on the senses of an observer, if we imagine someone
existing to observe them.[1] The powers are not evenly balanced,
because there is as yet no principle of measure and proportion [2] ;
and the result of their unregulated interaction is a swaying and
shaking of the Receptacle and all its contents. This is compared
to the shaking of a winnowing-basket, which has the effect of
separating the light chaff from the heavier grain. In the shaking
of the Receptacle the blind mechanical principle that like tends to
get together with like is operating ; and the effect would be that
the different ' kinds ' would drift or be thrust into different regions.
That is the nearest approach to a cosmic order that could result
from the purposeless interplay of dissimilar and unbalanced
qualities.[3] These qualities are called ' vestiges ' (ἴχνη) of fire, air,
water, and earth ; there was as yet nothing that fully deserved to
be called by those names (69B), because the Demiurge has not yet
fashioned the geometrical shapes of the primary bodies.

In this description, as in the whole foregoing account of the
Receptacle, there is not a single word implying that the contents

---

[1] Cf. 33A, ' hot things and cold and all things that have strong powers '
(δυνάμεις ἰσχυράς), and the commentary there (p. 53). The two words
μορφαί and δυνάμεις occur at the opening of the second part of Parmenides'
poem, where the opposites of sensible quality are added to the geometrical
Sphere of Being : frag. 8, 53, ' Mortals have decided to name two μορφαί '
(Fire and Night) ; frag. 9, ' But now that all things have been named Light
and Night, and the names corresponding to their several powers (τὰ κατὰ
σφετέρας δυνάμεις) have been assigned to these things and to those. . . .'
These are the names of what were later called ' qualities ', such as ' the hot ',
' the cold ' ; ' the light ', ' the heavy ' ; ' the rare,' ' the dense,' etc., arranged
in pairs of opposites (as in the Pythagorean Table of Opposites) under the
primary pair, Fire (Light) and Night. See F. M. Cornford, *Parmenides' Two
Ways*, Cl. Qu. xxvii (1933), p. 108.

[2] Contrast Empedocles' four elements, which are all equal in quantity and
evenly matched, so that they prevail in turn in the cycle of time (frag. 17).
The determination of the four primary bodies in a geometrical proportion
was a work of the Demiurge (31B), which has not yet taken place (cf. 69B).

[3] We are reminded of the materialists' doctrine in *Laws* x, 889B, that the
lifeless elements ' all move by the chances of their several powers (δυνάμεως),
and according as they clash and fit together with some sort of affinity—hot
with cold, dry with moist, soft with hard and in other mixtures that result,
by chance, of necessity, from the combination of opposites ' generate a world
' by nature and chance '.

of the Receptacle exist in the form of particles.[1] This notion, imported by some modern commentators, rests partly on the mistranslation of μορφή, ἰδέα, εἶδος already noticed, partly on the confusion of the winnowing-basket in Plato's simile with the sieve (κόσκινον), whose action in sorting grains of different kinds was used by Democritus (frag. 164) to illustrate the principle that like things tend to get together.   Then, since Democritus invoked that principle to account for atoms of like shape and size coming together in the void, this picture of particles in empty space is read into Plato's description.   Accordingly it is understood that the contents of the Receptacle are particles of irregular shape, and that the first task of the Demiurge was 'to reduce the particles to a number of definite shapes'.

There is really no warrant for attributing to Plato this atomistic picture of irregular particles moving at random in a void.   Atoms were completely determined particles of solid substance, separated by intervals of nothingness, which gave them room to move about. Plato's Space is not a void which remains completely distinct from particles moving in it ; it is a Recipient which affords a basis for images reflected in it, as in a mirror—a comparison that could not be applied to atoms and void.   Space is to him the 'room' (χώρα) or place where things are,[2] not intervals or stretches of vacancy where things are not ; and if he admits any void at all, it is only as the very smallest interstices (διάκενα) which the shapes of particles, when particles have been formed, do not allow them to fill (58A, B).

---

[1] This notion vitiates Tr.'s remarks on our passage (pp. 351 ff.). He first describes the contents of the Receptacle as 'all sorts of sense-data subject to no recognisable law and executing "random" movements'. But then he adds that 'the general effect would be like that of passing seeds of different sizes through a twirling sieve in which the meshes are of different sizes.   The tendency would be for particles of the same character to be sorted into the same heap.   Thus "before the universe was made" (53C, 7) there was already a tendency for like particles to be assembled together in distinct regions of space.' On p. 352 he even speaks (by an oversight) of 'the rain of the particles'—an Epicurean conception foreign to the earlier Atomists.

[2] τόπος obviously means the place where something is ; but χώρα also is τὸ ἐν ᾧ, the container of something, and has associations with χωρεῖν meaning to 'hold', 'have room for': 'the crater holds (χωρέει) 600 amphorae' (Hdt.). χώρα is used of the post, station, office, 'place' that a person holds : 'in the room of his father Herod'. χώρα is 'room' that is filled, not vacant space (κενόν). Simpl., Phys. 540, 33, τὸ δεχόμενόν τε καὶ χωροῦν αὐτὸ χώρα γίνεται τοῦ ἐγγιγνομένου, ἡ δὲ χώρα τόπος νενόμισται. 541, 26, τόπου καὶ χώρας καὶ ὑποδοχῆς καὶ ἕδρας τῶν εἰδῶν. 'Place' would, indeed, be a less misleading translation of χώρα than 'Space', because 'place' does not suggest an infinite extent of vacancy lying beyond the finite sphere of the universe.   It should be noted that 'unlimited' is not included among the attributes of Space at 52A, B, or anywhere else.

Apart from the mistranslation of μορφή as 'shape', the notion of irregular particles rests on the simile of the winnowing-basket. But the πλόκανον, as its name implies,[1] is 'woven' of basketwork, not a perforated sieve. It is the *liknon* or *vannus* described and figured by Jane Harrison,[2] a wide, shovel-shaped basket, high at one end and flattened out at the other, held by two handles projecting from the upper rim at the sides. 'The winnower takes as much of grain and chaff mixed as he can conveniently hold and

supports the basket against the knee. He then jerks and shakes the basket so as to propel the chaff towards the shallow open end and gradually drives it all out, leaving the grain quite clean. . . . The wind plays no part whatever in this process. . . . The chaff rises up and sprays over the shallow end.' Plato mentions the grains of corn (σῖτος) and the chaff (τὰ μανὰ καὶ κοῦφα), which, as he says, is 'carried away to another place and settles there'. There is no question of several kinds of grain of different sizes and shapes,

---

[1] πλόκανον is used again at 78c of the fish-trap (κύρτος, *nassa*), woven (πλέγμα . . . συνυφηνάμενος, 78B) of wicker-work or reeds.

[2] *Mystica Vannus Iacchi*, J. H. S. xxiii, 292 ff. ; *Prolegomena to the study of Gk. Relig.* 530. Daremberg and Saglio, s.v. *Vannus*. The sketch here given is based on a photograph of Mr. Wilson, Sir Francis Darwin's gardener, using a winnowing-basket of the pattern described. The *Nurses* (τιθῆναι) of Dionysus carry the infant god in a *liknon*. The Phrygian Earth *Mother*, Hipta, 'places a *liknon* on her head which she has wreathed with a snake and *receives* Dionysus' (ὑποδέχεται τὸν κραδιαῖον (?) Διόνυσον· τῷ γὰρ ἑαυτῆς θειοτάτῳ γίγνεται τῆς νοερᾶς οὐσίας ὑποδοχὴ καὶ δέχεται τὸν ἐγκόσμιον νοῦν Pr. i, 407 = Orph. *Frag.* 199, Kern). Cf. Nilsson, *Minoan and Mycenaean Religion*, pp. 493, 497. Did this image of the Mother or Nurse with the Child in the *liknon* suggest to Plato's mind the simile of the liknon-cradle in which the infant universe is rocked by the Nurse, Mother, Recipient of Becoming, as described again at 88c, D ?

as in Democritus' analogy of the sieve ; nor is there any ' twirling ' motion (δῖνος) in the case of the winnowing-basket. The contrast is between the density and heaviness of the corn and the lightness and fine texture of the chaff. It is to these qualities that the separation of like to like is due, not to differences of shape or size. ·In the application, it is things of like quality that come together. These things are the ' vestiges ' of fire and the rest, before any shape has been given to them. They come together on the principle, unanalysed (here as elsewhere) and assumed as obvious, that like things do come together.

With the Democritean sieve vanishes the last suggestion of discrete particles separated by vacant space. Plato's Recipient is not partly empty, but completely filled with the sensible qualities or ' powers ', tending to group themselves vaguely in indefinite masses, so that you say that one part is ' fiery ', another ' watery '. So much distinction must be presupposed because (as Parmenides had remarked, and as Plato repeats, 57E), if the whole contents were perfectly uniform, there could be no motion or disturbance of equilibrium. In accordance with tradition reaching back to Anaximander, the qualities are grouped in pairs of opposites ; if they are to exist at all, one part of the whole mass must be hotter, another colder ; one drier, another moister, and so on. The condition imagined resembles what Anaximander probably conceived in his ' unlimited ' mass, containing in indistinct confusion the opposites that came to be ' separated out ' of it. The ' separating out ' may have been partly due to the mutual repulsion of hostile opposites, which emerges as a distinct moving cause in the Strife of Empedocles. There is also the mutual attraction of like to like, invoked by most of the physicists and by Plato himself. Anaxagoras had another distinct moving cause, Mind ; but his description of the state of things after motion has been initiated resembles Plato's chaos. The first visible distinction arose as fire (*aether*) drew towards the outer part of the cosmic eddy, air towards the centre (frag. 2). At a somewhat later stage, before individual things were formed, ' all things being together, not even any colour was discernible ; that was prevented by the confusion of all things—of moist and dry, hot and cold, bright and dark ' (frag. 4). As the various ' things ' draw more completely apart, they become more and more distinct, and each part becomes more homogeneous, recognisable, and nameable.

· So Plato imagines his Receptacle filled and diversified with qualities, whose unlikeness and lack of equilibrium agitate the whole mass. The drift of like to like ' thrusts the most alike closest together and separates the most unlike farthest apart '. Thus there is a separating out of opposite qualities, illustrated by the action of the winnowing-basket making 'the dense and heavy things go one way, the rare and light move to another place and settle there '. The picture is left very vague. There is no mention of any whirling motion ; for winnowing-baskets are not ' twirled '. Circular motion is, to Plato, the rational motion, and he may have designedly excluded it. If any whirl or eddy did occur, it would be (as in the Atomists) only an accidental resultant of the six irrational motions—the rectilinear motions proper to the primary bodies and associated with the Errant Cause and blind Necessity. Granted that the qualities are active ' powers ', and that like necessarily tends to like, ' the bodily ', undirected by any intelligence, might be imagined as advancing so far, but no farther, towards a cosmic order.[1]

Such is the chaos taken over by the designing intelligence. How is it to be interpreted ? It is now generally agreed that this disorderly condition can never have existed by itself at a time before order was introduced. Bodily motion cannot exist without a soul to cause it. The World-Soul was a creation of the Demiurge, who put reason in soul, and soul in body. When soul was fitted to body, the world, as a living creature containing soul and reason, began its ' unceasing and intelligent life for all time '. Plato clearly means that there never was a time when the body existed without the soul, or the soul without the body. We must also, I think, rule out the notion, favoured by some ancient Platonists, that the soul of the world was at first irrational, having only the irrational motions, and then the Demiurge endowed it with reason and reduced it to order.[2]

It follows that chaos is, in some sense, an abstraction—a picture of some part of the cosmos, as it exists at all times, with the works of Reason left out, ' such a condition as we should expect for anything when deity is absent from it '. Now if you abstract Reason and its works from the universe what is left will be irrational Soul, a cause of wandering motions, and an unordered element of the bodily, itself moving without plan or measure. This bodily element is represented as consisting of qualities with active powers, moving within the Recipient which contains them, not as yet limited by the element of definite quantity, number, measure, shape. Now that we have ruled out the conception of discrete solid particles moving about and colliding in empty space, we need some other account of the way in which qualities can be supposed

[1] Cf. Robin, *Phys. de Platon* 42.

[2] Tr. (pp. 115 ff.) states and criticises this view as held by Plutarch and Atticus.

to be in motion. Here the *Timaeus* must be supplemented by evidence from the other late dialogues.

The *Theaetetus* states Plato's theory of the nature of sensible qualities.[1] The physical objects that yield our sensations and perceptions have no permanent qualities residing in them. They are described as actually being 'slow changes'. The only other thing we know about them is that they have the power (δύναμις) of acting on our organs and on one another. What we call a hot thing is a change that can make us 'feel hot' and can make another thing we call 'cold' hotter. This change, as opposed to locomotion, is a modification or qualitative change (ἀλλοίωσις) which is going on all the time, whether or not it gives rise to sensations and perceptions in any sentient organ. On the side of the percipient, the eye which sees or the flesh which feels is itself a physical object of the same kind, a qualitative change. Nothing that can properly be called an agent or a patient exists until an organ capable of sensation and an external object come within range of one another. When they do come within range, the 'powers of acting and being acted upon' come into play. In the case of sight, 'quick motions' pass between organ and object. The meeting and coalescence of the visual ray from the eye with the stream of light from the object (as described above, 45C) generate vision and colour. At that moment, but not before, the object can be said to become 'white' or 'black'. The whiteness exists only 'for' the percipient and so long as the perception lasts. Similarly the quality I call 'hot' has no independent existence in the object; the object becomes hot only when some sentient creature feels it hot. Before that, there exists only a change, which has the power of causing other changes in neighbouring objects. Plato speaks of qualities as changes, not as things which change, because that would suggest something permanent which undergoes change. The *Timaeus* has explained that the only permanent thing, which can be called 'this', is the Recipient.

In accordance with this theory of qualities, the Friends of Forms in the *Sophist* (246A ff.), among whom Plato must be included, confine real being to 'certain intelligible and bodiless Forms' and reduce the bodies which materialists regard as real to 'a moving process of becoming', with which we have intercourse by means of sensation and perception. They define this intercourse as 'the experiencing an effect or the production of one arising, as the result of some power, from things that encounter one another' (248B). Becoming has this 'power of acting and being acted upon'.

---

[1] *Theaet.* 155D ff. I have argued in *Plato's Theory of Knowledge*, pp. 48 ff., that, as Jackson held, this theory must be Plato's own.

All this is in agreement with the account of the perceptible likenesses of the Forms of Fire, Air, Water, and Earth, which are ' perpetually in motion, coming to be in a certain place and again vanishing out of it ' (52A). By virtue of their entrance, the Nurse of Becoming has ' every sort of diverse appearance to the sight ', in the sense that, if there were a spectator with eyes to see it, it would cause in him sensations of various colours. But in the absence of any spectator, there are, strictly speaking, no colours—only changes capable of causing such sensations. Space is accordingly described as filled with ' powers ' whose motions are in unordered and unbalanced agitation. Commenting on the statement (30A) that the Demiurge ' took over all that was visible ', Proclus (i, 387) remarks that the visible cannot be bodiless and without quality ; but must be something that ' already partakes of the Forms and possesses certain vestiges or appearances of them, in a condition of unordered and discordant motion ; for the image-like and unarticulated presence of the Forms produces in it various motions '. Elsewhere (iii, 275) he says : ' Nature, having implanted in the masses bodily powers immersed in matter, moves the masses by means of those powers—earth by its heaviness, fire by its lightness.'

Since no bodily changes can occur without the self-motion of soul, the other factor present in this chaos must be irrational motions of the World-Soul, considered in abstraction from the ordered revolutions of Reason. The disorderly moving mass must be conceived as animated by soul not yet reduced to order, but in a condition analogous in some ways to that of the infant soul described above (43A ff.). We must reject the view that Plato has reduced the bodily to mere empty space figured in the geometrical patterns which the Demiurge is now going to introduce. If that were so, the offspring of the two parents, the Forms and Space, would not be a moving process of becoming ; there would be no motion. The particles of the primary bodies, presently to be fashioned, cannot be, as it were, empty boxes—nothing but geometrical planes enclosing vacancy. They would then be inanimate and like Democritus' atoms, except that they would not be solidly packed with impenetrable substance. Indeed, the whole of Space would be empty—a void partitioned by geometrical planes. Plato's description throughout implies that the particles are filled with those changes or powers which are sensible qualities ; and that they are penetrated and animated by soul ; only so can the whole body of the cosmos be alive and moving and the primary bodies have characteristic motions of their own. The atoms of Democritus were not alive, and their movement was left unexplained,

According to Plato, in a world of Democritean atoms and void there could be no motion at all. The same would be true of his own particles, if they were not penetrated by soul. The activity of soul in every part of the physical universe is the only possible source of the active powers of bodies—of their motion in space and of their power of altering one another qualitatively and affecting our sense-organs.[1]

It may be added that all these motions are irrational. The movements in space characteristic of the primary bodies are recti-linear—those 'wandering motions' in all the six directions which have been repeatedly contrasted with the circular revolution of Reason. The qualitative alterations perpetually going on are inaccessible to any kind of scientific knowledge. They can cause sensations which, on the physical side, are themselves qualitative alterations of bodily organs and, on the mental side, yield percep-tions confined to the individual percipient, which can never rank as knowledge because subject and object are in a perpetual flux of change.

The abstract picture of the physical world without the guidance of Reason is illustrated by the myth in the *Statesman* (268D ff.). There are times when God himself helps to guide the revolution of the universe. Then, after an appointed period, he lets it go and the world is carried round in the reverse direction spontaneously (αὐτόματον) by the power of motion which it possesses as a conscious living creature. This reverse movement is implanted in it of necessity (ἐξ ἀνάγκης), because only the most divine things are always constant in the same state. The world, having a body, is subject to change ; but it keeps so far as possible to its own motion of rotation in one place. The least possible deviation is reversal of direction. The world cannot always turn itself ; that is possible only to the divine ruler of all things that are moved, and he cannot cause motion now in one direction, now in the opposite. Nor can there be two gods with opposite intentions to turn it. The only alternative is that at one time it is guided by divine causation, and acquires fresh life and renewed immortality from its maker ; at another, when it is let go, it turns itself in the reverse direction for many myriads of revolutions.

We are now living in one of the periods when the god's hand has

---

[1] In the *Laws* Soul is not merely called the source of motion (as at *Phaedrus* 245D), but more specifically ' the cause of the becoming and perishing of all things ' (891E); it ' controls all change and rearrangement ' (892A); it is the ' first becoming and change ' (896A); it originates all διάκρισις, αὔξη, γένεσις and their opposites (894B). See Theiler, *Teleolog. Naturbetrachtung* (1925), p. 70, who remarks on these passages that the World-Soul, as cause of becoming, leaves no room for any Demiurge beside itself.

been withdrawn from the helm. The reversals are marked by the greatest of all cosmic catastrophes; all but a remnant of life on earth is destroyed. The very current of life is brought to a standstill and set flowing in the contrary direction. When the steersman of the universe let go of the tiller and retired to his own conning-tower, the world began to turn the other way by fate and its own inborn impulse (εἱμαρμένη τε καὶ σύμφυτος ἐπιθυμία). The reversal caused earthquakes, which went near to destroying all life. As the disturbance began to settle down and calm followed the storm, the world began to be set in order and to move on its accustomed course, governing and caring for itself and all that it contained, and recalling, as well as might be, the teaching of its maker and father. But the memory grows dim and things begin to go worse, thanks to the admixture of the bodily element (τὸ σωματοειδές) inherent in the world's nature, which was full of confusion before it came into its present order. All that is good in the world comes from its maker; all the cruelty and injustice that it contains in itself and produces in living creatures come from its former chaotic condition. Hence in the former period when it was nurturing its living creatures under the god's guidance, it engendered great goods and few evils; but now that it is separated from him, as time goes on and forgetfulness grows, the old disorder threatens to prevail. Good things diminish, evils increase, and it comes in danger of utter destruction. Then at last the god, seeing its distress, and taking care that it shall not be shipwrecked in the storm of disorder and sink into ' the limitless ocean of Unlikeness ', will take the helm again. He will turn back the diseased and dissolving fabric to its former motion, order it and set it right, and save it from age and death.

As Proclus observes, the machinery of the reversal of the world's motion is a mythical device to represent as existing at separate times things which in fact are always coexistent in the cosmos.[1] The same is true of the description in the *Timaeus* of the condition of the world ' when divinity is absent from it ' as if it were a state of things that had existed ' before the Heaven was made '. If we discount these mythical devices, both myths present a picture of the universe as it would be if the works of Reason were abstracted, and the one may be used to illustrate the other. In the *Statesman* we find that when the god is absent, the world is still a living and

---

[1] Pr. iii, 273³¹. Simplicius, *Phys.* 1122, 3, too criticises Alexander for taking Plato's description of chaos (30A) as meaning that the cosmos had a beginning in time, preceded by a condition of disorder. He points out that the temporal separation of the two conditions is merely mythical in the *Timaeus*, as in the *Statesman*, where Plato imagines the Maker removed from the cosmos and contemplates its collapse into ' the ocean of Unlikeness '.

conscious creature capable of 'spontaneous' motion; and this motion is connected with Fate (εἱμαρμένη) and its own innate impulse (σύμφυτος ἐπιθυμία). The Fate of the *Statesman* is allied to the Necessity of the *Timaeus*; and the innate impulse may be connected with the blind attraction of like for like and those active powers which, as Proclus says, Nature (φύσις) implants in the bodily to produce motion. In both cases some semblance of order results by nature and chance without design. In the *Statesman* the world begins to settle down, after the convulsion, into its accustomed course. In the *Timaeus* the attraction of likes sorts out the powers or qualities into the regions to which their wandering motions carry them. And in both cases it is indicated that these tendencies, if left unchecked, would lead to a complete separation of the powers, comparable to the Reign of Strife in Empedocles' system.[1] The *Statesman* calls this 'the limitless ocean of Unlikeness'. The *Timaeus* later (58A) explains how the disintegration is in fact held in check by the interaction of the bodily particles, which have in the meantime received their definite geometrical shapes.

Following Proclus,[2] we may perhaps see a further analogy between the two opposite revolutions in the *Statesman* myth and the contrary movement of the circles of the Same and the Different in the World-Soul. The revolution of the Same towards the right symbolises the all-controlling movement of divine Reason. The revolution of the Different towards the left is the common factor in the proper motion of the planets and in particular of the Sun; and the movement of the Sun in the inclined circle is (as we have seen, p. 83) the cause of mutability, becoming and decay, in earthly things. As a motion of the World-Soul, contrary to that of Reason and yet overruled and controlled by it, the Different stands for a semi-rational element of innate impulse amenable to the persuasion of Reason. If the motion of the Same were arrested, as it was in the infant soul (43D), the uncontrolled revolution of the Different might be imagined as producing in the world an analogous disorder, in which the body of the living creature would 'move at hazard without order or method, having all the six motions, straying (πλανώμενα) every way in all the six directions' (43B). The source of this motion would be (in the condition supposed) the irrational element in the World-Soul.

These considerations may throw light upon the most obscure feature in the account of chaos—the statement that the Receptacle

---

[1] The comparison is made by Plutarch, *Fac. Lun.* 926E, F.

[2] Pr. i, 289, arguing against Severus, who had taken literally the reversal in the *Statesman* myth.

itself is shaken by the active powers it contains, and shakes them in return. This is immediately followed by the simile of the winnowing-basket. Plato cannot mean that Space really shakes or is shaken by the qualities. He is in a difficulty because he is attempting to describe the bodily, as if it could exist not only without Reason, but without soul; just as at the outset he spoke of the Demiurge taking over the visible in disorderly motion before the only source of all physical motion, the soul, was fashioned. Accordingly he has to help out an impossible situation by comparing the Receptacle to a winnowing-basket which both shakes and is shaken, leaving in obscurity the fact that, as surely as the basket needs someone to shake it, the bodily needs soul before any motion can occur.

Before we leave this account of the bodily, one question of great importance remains to be answered. The picture of chaos has represented the blind action of the powers producing ' by nature and chance ' results that are necessary and undesigned. These are ' causes ' of that lower order which can be made subservient to the purposes of Reason and persuaded to guide events ' for the most part ' to good ends. Are we to understand that, in the cosmos as actually organised by Reason, all these secondary causes are completely subdued ?[1] I do not think that this view can be defended ; but the answer, of course, depends on the interpretation of Necessity and the Errant Cause. I have maintained that Plato recognises in the working of the universe, a factor which confronts the divine Reason and is neither ordained nor completely controlled by it. This means that irrational and merely necessary motions and changes, with casual and undesigned results, actually occur in Nature at all times, as well as those which are subservient to rational ends. It is only ' for the most part ' that Reason can persuade Necessity. Were it otherwise, Plato's Demiurge would be represented as an omnipotent creator who had designed the whole contents of the universe, not as a craftsman who ' takes over ' materials in disorderly motion and does the best he can with them. And since, on Platonic principles, all physical motion must be due

---

[1] This appears to be Tr.'s view. He holds that ' if we could ever have complete knowledge, we should find that ἀνάγκη had vanished from our account of the world ' (p. 301) ; and on 46E (' causes which, being destitute of reason, produce their sundry effects at random and without order ') he remarks : ' It is not implied that there is really any such " random " agent working on its own account in the universe. . . . We have been told already that " mechanism " plays the part of an " understrapper " (ὑπηρέτης) to the " intelligent cause ". All that is meant is that *if* a mere mechanism were left undirected, to work of itself, the results would be " casual " (τὸ τυχόν) ' (p. 293).

to a living soul, I do not see how to escape the conclusion that the World-Soul is not completely rational. Besides the circular revolutions of the Reason it contains, there are the six irrational motions characteristic of the primary bodies. These bring about some desirable results, such as intelligence could purpose ; but the picture of their working below the level at which the Demiurge first takes a hand and introduces an element of rational design can hardly be accounted for unless we take it as representing an imperfectly subdued factor of blind necessity always at work in Nature.

53C–55C. *Construction of the figures of the four primary bodies*

From the abyss of bodily ' powers ' in complete abstraction from the works of Reason, we now ascend to the lowest level at which the element of order and design contributed by the Demiurge can be discerned in the turbulent welter of fire, air, water, and earth. The god endows the primary bodies with regular geometrical shapes. It is a reasonable conjecture that Plato's account of their structure is a deliberate correction of Democritus' atomism. Atoms were discrete particles of minimal size and definite shape ; but Democritus had given them any and every variety of shape. Plato will not leave this matter to chance.[1] He has adopted the Empedoclean doctrine that there are just four primary bodies ; and now he assigns to each the shape of one of the regular geometrical solids, so that there are just four shapes, chosen because they are the ' best '. Thus the operation of Reason is carried, so far as may be, into the dark domain of the irrational powers.

What follows here is entirely concerned with the construction of the regular solids which are taken as the figures of the four primary bodies. These figures alone are the work of the Demiurge ; the qualities previously considered are left out of account. The figures are not the actual shapes of existing particles, which can only be imperfect copies, but the perfect types, belonging to the intelligible world of mathematics. The theoretical construction of the regular solids had been completed by Theaetetus at the Academy. So far as we know, the assignment of these figures to the primary bodies is due to Plato and had not been anticipated by any earlier thinker.

The basic premiss of the following argument is not stated at the

---

[1] Cf. Ar., *de gen. et corr.* 325b, 24 : Both Plato and Leucippus postulate elementary constituents that are indivisible and distinctively characterised by figures ; but there is this great difference : the indivisibles of Leucippus are (1) solids, while those of Plato are planes, and (2) characterised by an infinite variety of figures, while Plato's figures are limited in number.

outset, but only after some preliminary remarks about solid figures in general. It is that four of the regular solids, namely the pyramid (tetrahedron), the octahedron, the icosahedron, and the cube, are the ' best ' figures that could be found for the purpose, for certain reasons that could be given in a fuller account than there is room for here. Of these, the first three have equilateral triangular faces, the cube has square faces. Since Plato intends to build his solids out of plane faces, we might expect him to take the equilateral triangle and the square as his elementary plane figures and proceed at once to construct the solids out of the proper numbers of such elements. It is by no means obvious why he does not take this simple course ; for the whole theory of the breaking down of solids into plane faces and the reformation of these into other solids, which is to explain the transformation of one primary body into another, could be worked out on this basis. For the present we shall note this as a point to be explained later, and follow the procedure which he actually adopts.

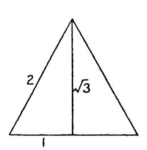 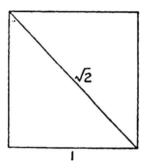

Having chosen as the best figures those regular solids which have the equilateral triangle or the square for their faces, Plato constructs these faces out of two triangles into which they can in fact be divided symmetrically. These are : (1) the half-equilateral, obtained by dropping a perpendicular from any angle of the equilateral to the opposite side. The sides of the half-equilateral have lengths corresponding to the numbers 1, 2, $\sqrt{3}$. It is accordingly described below (54B) as ' having the greater side (of the two containing the right angle) triple in square of the lesser '. (2) The other elementary triangle is the half-square, whose sides correspond to the numbers 1, 1, $\sqrt{2}$. This right-angled isosceles triangle and the half-equilateral (which is one type out of the infinite variety of right-angled scalene triangles) are thus taken as the two irreducible ' elements ' for the construction of all the four solids. These results are led up to by some general remarks about the construction of solids out of planes and of planes out of triangles, which somewhat disguise the fact that the choice of the regular solids is really the premiss determining the whole procedure.

53C.   In the first place, then, it is of course obvious to anyone that fire, earth, water, and air are bodies; and all body has depth. Depth, moreover, must be bounded by surface[1]; and every surface that is rectilinear[2] is composed of triangles. Now all triangles are derived from two, each having one

D.   right angle and the other angles acute. Of these triangles, one has on either side the half of a right angle, the division of which is determined by equal sides (the right-angled isosceles); the other has unequal parts of a right angle allotted to unequal sides (the right-angled scalene). This we assume as the first beginning of fire and of the other bodies, following the account which combines likelihood with necessity[3]; the principles yet more remote than these are known to Heaven and to such men as Heaven favours.

These introductory remarks lead up to the last statement that Plato intends to construct his solids out of triangles without pursuing the analysis of the triangles themselves into simpler principles. Any rectilinear plane face is 'composed' of triangles, in the sense that it can be divided up into triangles; and the triangle, as the surface contained by the minimum number of straight lines, is 'assumed' as the irreducible 'element' of all such figures. Plato indicates that there is something arbitrary in starting from this assumption. If planes can be constructed of triangles, triangles themselves can be constructed of lines, and lines can be expressed as numbers. We have already had a hint of this in the phrase 'giving them a distinct configuration by means of (geometrical) shapes and *numbers*'. This suggests that the 'remoter principles', known to mathematicians, are lines and numbers.[4]

---

[1] 'Surface', not 'plane', since some solids have curved surfaces.

[2] ἡ ὀρθή (sc. φύσις). Cf. 54A, 2, the isosceles has only one φύσις (kind, or type). βάσις is the 'face' of a solid.

[3] The account is 'likely' because we cannot know that the whole theory which assigns the figures of regular solids to the primary bodies is the truth. It is combined with 'necessity' in that the necessarily given properties of Space and the logical necessity of geometrical construction are involved in the consequences of the initial assumption.

[4] Miss A. T. Nicol in an interesting paper on Indivisible Lines (C.Q. xxx, p. 125) writes: 'The complete process is described at *Laws* 894A in answer to the question 'How does γένεσις take place?' δῆλον ὡς ὁπόταν ἀρχὴ λαβοῦσα αὔξην εἰς τὴν δευτέραν ἔλθῃ μετάβασιν καὶ ἀπὸ ταύτης εἰς τὴν πλησίον, καὶ μέχρι τριῶν ἐλθοῦσα αἴσθησιν σχῇ τοῖς αἰσθανομένοις. The ἀρχή is the indivisible line, the second stage the indivisible surface, the next the indivisible solid, and the last is the solid perceived by the senses. We see now why there is no mention

Plato's reason for stopping short at triangles was perhaps the need to keep his exposition within reasonable bounds. His business at the moment is to explain the transformation of some of the primary bodies into one another ; and for this physical purpose all he needs is triangles which can be reformed into solids of a different pattern. Possibly he also means that he intends to build up his solids by the rough-and-ready means of putting together triangular surfaces, as if they were pieces of a child's puzzle. This is, of course, a very different procedure from the construction of the figures by the methods of geometry. Philosophic mathematicians, ' men favoured by Heaven ', will know that a very long chain of propositions reaching back to the ultimate premisses of the science is needed before you can solve problems so difficult as the construction of the regular solids, as set out in Euclid xiii.

The result, so far, is that any triangle can be divided into, and so composed out of, two right-angled triangles, obtained by dropping a perpendicular from any angle to the opposite side. The resulting triangles may be both isosceles, or both scalene, or one isosceles and one scalene. It is now pointed out that the right-angled isosceles is a unique type (since the figure is completely determined by the right angle and the two equal sides), but there are infinitely many types of right-angled scalene (since any side may be of any length). Out of this infinite variety we have to choose the best ; and this is asserted to be the half-equilateral. Two such triangles compose the equilateral ; and the equilateral is, in fact, the face of three of the regular solids ; and these three solids are chosen because their regularity is itself a perfection and because, all having the same type of face, any one of them can be taken to pieces and those pieces can be regrouped to compose one or more of the other solids. We shall thus be able to explain how three of the primary bodies can be transformed into one another. Such is the real train of thought, disguised by Plato's design not to mention the regular solids he has selected until the end of the section.

53D. Now, the question to be determined is this : What are the
E. most perfect bodies that can be constructed, four in number, unlike one another, but such that some can be generated out

---

of indivisible lines in the *Timaeus*. The *Timaeus* is a myth of the physical world, and therefore has no need to go further back than the surface, the stage where in descending from the ἀρχή the third dimension becomes possible ; for without the third dimension there is no sensation '. Cf. Alexander Polyhistor's summary of Pythagorean doctrine (D.L. viii, 25) : From the One and the Indeterminate Dyad are derived successively numbers, points, lines, surfaces, *solid figures*, *sensible bodies*, whose elements are fire, water, air, earth.

53E. of one another by resolution ?   If we can hit upon the answer to this, we have the truth concerning the generation of earth and fire and of the bodies which stand as proportionals between them.   For we shall concede to no one that there are visible bodies more perfect than these, each corresponding to a single type.[1]   We must do our best, then, to construct the four types of body that are most perfect and declare that we have grasped the constitution of these things sufficiently for our purpose.[2]

54.   Now, of the two triangles, the isosceles is of one type only ; the scalene, of an endless number.   Of this unlimited multitude we must choose the best, if we are to make a beginning on our own principles.   Accordingly, if anyone can tell us of a better kind that he has chosen for the construction of these bodies, his will be the victory, not of an enemy, but of a friend.   For ourselves, however, we postulate as the best of these many triangles one kind, passing over all the rest ; that, namely, a pair of which compose the equilateral triangle.   The reason is too long a story ; but if anyone

B. should put the matter to the test and discover that it is not so,[3] the prize is his with all good will.   So much, then, for the choice of the two triangles, of which the bodies of fire [4] and of the rest have been wrought : the one isosceles (the half-square), the other having the greater side triple in square [5] of the lesser (the half-equilateral).

Plato once more indicates that his choice of the right-angled isosceles (the half-square) and the particular right-angled scalene which is the half-equilateral, is not so arbitrary as it appears.   He has in reserve an explanation too long to be given here.   I shall attempt to supply this explanation later (pp. 231 ff.), when we have accumulated all the points that remain obscure without it.   One such point we have already noted : why does Plato subdivide the

---

[1] ' Type ' (γένος) here seems to mean ' type of solid figure ', as in the next sentence.

[2] ἱκανῶς : ' sufficiently ' in order to explain the physical transformation of the primary bodies (' these things '), whereas the full geometrical construction would be a much longer business.

[3] Reading μή. With δή, the sense is : ' he who tests this and proves it to have this property '.   But this is the wrong sense : Plato could prove it himself, if the reason were not too long a story.

[4] τὸ τοῦ πυρὸς, sc. σῶμα, supplied from the following σώματα.   Not ' substance ' (A.-H.), but the geometrical solid figure.

[5] κατὰ δύναμιν.   Cf. Eucl. x, Def. 2, ' Straight lines are commensurable in square (δυνάμει σύμμετροι) when the squares on them are measured by the same area ' (though they may be incommensurable in length).

equilateral and the square which form the faces of his solids ? Another point arises here : the choice of the smaller elements into which he subdivides them is not determined by the figures they compose. There are other ways of dividing an equilateral or a square symmetrically into smaller triangles. Why are the half-square and the half-equilateral better than any possible alternatives ? A third question, about which nothing is said here, is the size of the elementary triangles. It will appear later (57C) that each of the solids built up out of them must be of several sizes. Does this involve that the elementary triangles must be of more than one size ? I shall argue (p. 233) that, as the present passage might seem tacitly to imply, one size of each elementary triangle is sufficient, thanks precisely to one of the merits of these particular triangles which Plato refuses to dwell upon here.

The choice of the two elementary triangles has now been made. Before going on to construct out of them the equilateral and square faces of his solids, Plato recalls once more the physical process of transformation which the structure of the solids is to account for. Where this transformation was first mentioned (49B, c), it was said that we see, ' as we *imagine* ', all four primary bodies being transformed into one another in a cycle. We are now told that this cannot, in fact, occur, because the transformation is to be effected by breaking down the solids into their elementary triangles and regrouping these into other solids. That this is what happens is simply assumed as part of the theory. But one of the chosen solids, the cube, is built of a different triangle from the other three. It follows that earth, to which the cube will be assigned, cannot be transformed into fire or air or water. The transformation must be confined to these three.

54B.   We must now be more precise upon a point that was not clearly enough stated earlier.[1] It appeared as though all the four kinds could pass through one another into one another ; but this appearance is delusive ; for the triangles

c.   we selected give rise to four types, and whereas three are constructed out of the triangle with unequal sides, the fourth alone is constructed out of the isosceles. Hence it is not possible for all of them to pass into one another by resolution, many of the small forming a few of the greater and *vice versa*. But three of them can do this ; for these are all composed of one triangle, and when the larger bodies are broken up,

---

[1] At 53E it was said that *some* (but not all) primary bodies could pass into one another by resolution. The next sentence refers to 49B, c.

54C. several small ones will be formed of the same triangles, taking on their proper figures ; and again when several of the smaller bodies are dispersed into their triangles, the total

D. number made up by them will produce a single new figure of larger size, belonging to a single body. So much for their passing into one another.

The exclusion of earth from the cycle of transformation is simply a consequence of the decision to assign the cube to earth. Other physicists (including Aristotle) felt no objection to earth being transformed ; and the exclusion is certainly not dictated by any facts of observation, to which, indeed, Plato makes no appeal.[1] Plato now proceeds to build the four regular solids. He begins with the construction of the equilateral triangular face which is common to the pyramid, the octahedron, and the icosahedron. The ' element ' is the half-equilateral, ' whose hypotenuse is double of the shorter side in length '. The equilateral is formed by putting together six (not, as we should expect, two) of these elements in the following figure :

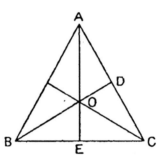

54D. The next thing to explain is, what sort of figure each body has, and the numbers [2] that combine to compose it.

First will come the construction of the simplest and smallest figure (the pyramid). Its element is the triangle whose hypotenuse is double of the shorter side in length. If a pair

[1] Aristotle, *de caelo* 306a, 2, bluntly says that the exclusion of earth is unreasonable, contrary to observed facts, and dictated by *a priori* principles. I cannot see the force of Tr.'s defence : ' The obvious reply is that as to the evidence of our senses, Aristotle's statement rests on a misinterpretation, and as to ' plausibility ', if experience really *does* provide examples of the transformation of the other ' roots ' but none of the transformation of earth, it is a ' plausible ' view that a particle of earth has a geometrical structure radically different from those of the other three bodies ' (p. 404).

[2] ' Numbers ' may mean ' numbers of units ', i.e. elementary triangles ; but there are also the numbers of faces, angles, etc., in the solid.

54D. of such triangles are put together by the diagonal,[1] and this
E. is done three times, the diagonals and the shorter sides
    resting on the same point as a centre, in this way a single
    equilateral triangle is formed of triangles six in number.

Here is another seemingly arbitrary feature, which has never
been satisfactorily explained.[2] Plato has himself remarked that
the equilateral can be formed of two elementary triangles (54A).
Why does he actually take six ? A reason will be suggested later
(p. 234). Having constructed the equilateral face, Plato next
builds the pyramid (with four such faces), the octahedron, and the
icosahedron.

54E. If four equilateral triangles are put together, their plane
     angles meeting in groups of three make a single solid angle,
55. namely the one (180°) that comes next after the most obtuse
     of plane angles. When four such angles are produced, the
     simplest solid figure is formed, whose property is to divide
     the whole circumference into equal and similar parts.[3]

     A second body (the octahedron) is composed of the same
     (elementary) triangles when they are combined in a set of
     eight equilateral triangles, and yield a solid angle formed by
     four plane angles. With the production of six such solid
     angles the second body is complete.

     The third body (the icosahedron) is composed of one
B.   hundred and twenty of the elementary triangles fitted
     together, and of twelve solid angles, each contained by five

---

[1] Probably, the diagonal of the resulting figure, the trapezium *CDOE*, viz.
the hypotenuse *CO*. Cf. κατὰ διάμετρον at 36C. Since there is no question
of proper geometrical methods of construction, but only of fitting pieces
together as in a puzzle, there is no objection to building an equilateral out
of trapezia. Not using diagrams, Plato simply describes the figure as briefly
as he can.

[2] A.-H.'s suggestion will not bear examination. Cook-Wilson's (Tr., p. 374),
that the division is symmetrical with respect to A, B, C, is true, but why
is this important ? Taylor's reference to the ' centre of gravity ' is hardly
relevant, even ' if the triangle be supposed to have weight (being thought
of as a very thin uniform lamina) ', which is itself a questionable supposition.
Heath (*Euclid* ii, 98) adduces the theorem, attributed to the Pythagoreans,
that six equilateral triangles, or three hexagons, or four squares, placed
contiguously with one angular point of each at a common point, will just
fill up the four right angles round that point.

[3] Martin rightly understands περιφερές as the circumference of the sphere
in which the pyramid is supposed to be inscribed (so Tr., p. 376). In *Euclid*
xiii, 13, ' To construct a pyramid, to comprehend it in a given sphere ', etc.,
is stated as a single problem. The last words mean that the pyramid is
regular. It is the ' simplest ' figure, because 4 is the smallest number of
faces that can contain a solid.

55B. equilateral triangular planes [1] ; and it has twenty faces which are equilateral triangles.

The second elementary triangle, the half-square, is now used to construct the square face of the cube. Here again Plato uses more elements than are necessary—four instead of two. This is another point which we will hold in reserve.

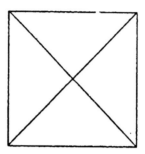

55B. Here one of the two elements, having generated these bodies, had done its part. But the isosceles triangle went on to generate the fourth body, being put together in sets of four, with their right angles meeting at the centre, thus forming a single equilateral quadrangle.

C. Six such quadrangles, joined together, produced eight solid angles, each composed by a set of three plane right angles. The shape of the resulting body was cubical, having six quadrangular equilateral planes as its faces.

Plato has only four primary bodies, which are now provided for. But there remains the fifth regular solid, the dodecahedron, for which some use must be found.

55C. There still remained one construction, the fifth ; and the god used it for the whole, making a pattern of animal figures thereon.

The dodecahedron is not constructed. Plato knew that its pentagonal faces cannot be formed out of either of his two elementary triangles : it was in fact constructed by means of an isosceles triangle having each of its base angles double of the vertical angle.[2]

---

[1] Tr. (p. 376) understands γωνιῶν with ἐπιπέδων (as in the previous sentence), τριγώνων being dependent genitive : ' five plane angles which belong to equilateral triangles '. It is true that a solid angle is defined (*Euclid* xi, Def. 11) as ' contained ' (περιεχομένη) by more than two plane angles ; but Plato speaks above of the solid angle as consisting of (ἐκ) plane angles, and the construction of the genitives is difficult. Also at 55c, ἐξ ἐπιπέδους τετραγώνους ἰσοπλεύρους means ' six equilateral quadrangular *planes* '.

[2] See Heath, *Euclid* ii, p. 98.

Not requiring a dodecahedron with plane faces for any primary body, the Demiurge ' uses it for the whole ', i.e. for the sphere, to which this figure approaches most nearly in volume, as Timaeus Locrus remarks. The meaning is explained in Wyttenbach's note on *Phaedo* 110B, where Socrates says that the spherical Earth, seen from above, would resemble ' one of those balls made of twelve pieces of leather ' marked out in a pattern of various colours. ' To make a ball, we take twelve pieces of leather, each of which is a regular pentagon. If the material were not flexible, we should have a regular dodecahedron ; as it is flexible, we get a ball.' [1] So here Plato imagines a flexible dodecahedron expanding into spherical shape. The word διαζωγραφῶν is ambiguous. It might mean ' giving it a pattern of various colours ' ; but this seems hardly appropriate to the sky. On the other hand, the whole sky is covered with ' animals '—not only the twelve signs of the Zodiac, but all the other constellations. [2]

55C–D. *Might there be five worlds?*

At this point Plato interrupts his argument to reopen the question, whether there is more than one ' cosmos '.

55C.   Now if anyone, taking all these things into account, should raise the pertinent [3] question, whether the number of worlds

D.   should be called indefinite or limited, he would judge that to call them indefinite is the opinion of one who is indeed indefinite about matters on which he ought to be definitely informed. [4] But whether it is proper to speak of them as being really one or five, he might, if he stopped short there, more reasonably feel a doubt. Our own verdict, indeed, declares the world to be by nature a single god, according to the probable account ; but another, looking to other

---

[1] Burnet, *ad loc.*, referring to Wyttenbach, who cites Plut., *Qu. Plat.* 1003C : the dodecahedron, because of the number of its elements and the bluntness of its angles, εὐκαμπές ἐστι καὶ τῇ περιτάσει καθάπερ αἱ δωδεκάσκυτοι σφαῖραι κυκλοτερὲς γίνεται καὶ περιληπτικόν. Pr. iii, 141, also connects our passage with the *Phaedo.*

[2] Burnet's suggestion that ' the real allusion is to the mapping out of the whole apparently spherical heavens into twelve pentagonal regions for the purpose of charting the constellations ' is quoted with approval by Tr. (p. 377), but not supported by any evidence that the heavens were so mapped by astronomers. Burnet took the suggestion from Newbold, *Arch. G. Phil.*, N.F. xii (1906), 203, to which he refers.

[3] ἐμμελῶς. The question is not πλημμελές, ' out of tune ' with the subject in hand.

[4] The pun on the two senses of ἄπειρος occurs again at *Philebus* 17E and possibly at *Theaet.* 183B.

55D. considerations, will judge differently. He, however, may be dismissed.

This passage is extremely puzzling. There is nothing in the previous context to suggest a plurality of worlds, except a certain resemblance between Plato's solids and the atoms of Leucippus and Democritus. The Atomists believed in innumerable coexistent worlds, not because bodies are composed of atoms, but because atoms are unlimited in number and void is infinite in extent. Plato would deny both these premisses, and he has nothing but disapproval for the Atomist's philosophy.

In the earlier passage he refers to (31A, B), it was argued ' according to the probable account ' that the world would be the better for possessing the uniqueness of its model. Why is it now suggested that there might be something to be said for five worlds ? Five is not the number of the primary bodies, but only of the regular solids that have been mentioned. Why should there be as many worlds as there are regular solids ? There is no evidence that anyone had ever believed in five worlds [1] ; and since the association of the regular solids with the primary bodies is, so far as we know, a novel theory of Plato's own, there was no reason why anyone should. Even if any philosopher had believed in five elements, that would be no ground for thinking that there might be five worlds, each consisting of one element.

Plutarch twice refers to our passage. In the tract *On the E at Delphi* (xi) he understands Plato to assign ' the five perfect figures in Nature ' to five primary bodies, earth, water, air, fire, and ' heaven ' or, as some call it, ' light ' or ' aether ' or ' the fifth substance, the only body to which circular motion is natural and not constrained or otherwise accidental '. On this view, the five ' *cosmoi* ' mean the five *regions* proper to these bodies, within the unique universe. The suggestion is repeated by Heracleon in the treatise *On the Cessation of Oracles* (422F), in reply to Demetrius' protest that Plato had not supported his suggestion of five worlds by any appeal to reason or probability. Heracleon cites the critics who referred the doctrine to Homer on the ground that he had partitioned the universe into five regions (*cosmoi*) : heaven, water, air, earth, Olympus. Earth and Olympus were left ' common '; the three intermediate regions were divided among the three gods (*Il.* xv, 189). Similarly Plato, assigning the perfect figures to the ' differences ' (διαφοραί) of the Whole, ' spoke of five *cosmoi*, of

[1] I have argued (C.Q. xviii, 1934, p. 14) that there are no grounds for regarding Petron, with his eccentric doctrine of 183 worlds arranged in a triangle, as earlier than Plato.

earth, water, air, fire, and last the one which embraces these, the *cosmos* of the dodecahedron, widely spread and with many turnings, to which he assigned the figure suitable to the revolutions and movements of the soul ' (i.e. the sphere). Further on, the somewhat confused explanations quoted from Theodorus of Soli and Ammonius' criticism of them may suggest a closer parallel between Homer's Earth and Olympus, which are left out of the Division, and Plato's cube (earth) and dodecahedron (Heaven), which, being based on elementary triangles of a different type from one another and from the rest, take no part in the transformation of the primary bodies. Homer's ' heaven ' (fire), air, and water, will correspond to the three intermediate bodies which are transformed into one another, being all based on the same triangle.[1]

This, so far as I know, is all the light contributed by the ancients, who seem to have found the passage as obscure as it is to us. Their suggestions are irrelevant, unless they were right in connecting the dodecahedron with the fifth form of body, aether, which first appears in the *Epinomis* (981c). Zeller held that this ' deviation from his earlier doctrine ' must be ascribed to Plato himself, because Xenocrates in his *Life of Plato* spoke of his carrying the Division of animals to the point where he reached ' the elements of all animals, which he called five figures or bodies, namely aether, fire, water, earth, air '.[2] It is conjectured that Xenocrates' statement must be based on Plato's oral teaching towards the end of his life.

Either our passage must be given up as inexplicable, or we must see in it a veiled allusion to the possibility of a fifth form of body, which, in any case, would be thought of as of a higher order than the four we are here concerned with, and would have no part in the physical processes of transformation which it is Plato's object to explain. It still remains a puzzle, why Plato should speak of the notion that there are five *cosmoi = regions* in one world as an alternative to a single cosmos = *world* or an indefinite number of worlds.[3]

---

[1] The scheme attributed to Pherekydes by Damascius, *de princ.* 124[b] (after Eudemus, frag. 117), is curiously parallel. The three first principles are Zas (*heaven*), Chthonie (*earth*) and Time ; then Time makes out of his own seed *fire, air, water*. These are connected with Pherekydes' word πεντέμυχος, which, the writer says, may mean πεντέκοσμος.

[2] Zeller ii[4], 951. The passage from Xenocrates is quoted by Simplicius, *Phys.* 1165, 33 (and twice elsewhere), who adds ὥστε ὁ αἰθὴρ πέμπτον ἄλλο τι σῶμα ἁπλοῦν ἐστι καὶ αὐτῷ (τῷ Πλάτωνι) παρὰ τὰ τέτταρα στοιχεῖα. See Harward's note on *Epinomis* 981B.

[3] Miss A. T. Nicol writes to me : ' Plato may be suggesting that by substituting the dodecahedron and leaving out the other regular solids in turn, as he has just left out the dodecahedron in the description of the elements, five different *cosmoi* could be obtained.'

55D-56C. Assignment of the regular figures to the four primary bodies

55D.   Let us next distribute the figures whose formation we have now described, among fire, earth, water and air.

To earth let us assign the cubical figure ; for of the four
E.   kinds earth is the most immobile and the most plastic of bodies.[1] The figure whose bases are the most stable must best answer that description ; and as a base,[2] if we take the triangles we assumed at the outset, the face of the triangle with equal sides is by nature more stable than that of the triangle whose sides are unequal ; and further, of the two equilateral surfaces respectively composed of the two triangles, the square is necessarily a more stable base than the triangle, both in its parts and as a whole. Accordingly
56.   we shall preserve the probability of our account, if we assign this figure to earth ; and of the remainder the least mobile to water, the most mobile to fire, and the intermediate figure to air. Again, we shall assign the smallest [3] body to fire, the largest to water, and the intermediate to air ; and again the body with the sharpest angles to fire, the next to air, the third to water.

Now, taking all these figures, the one with the fewest faces (pyramid) must be the most mobile, since it has the sharpest cutting edges and the sharpest points in every
B.   direction, and moreover the lightest, as being composed of the smallest number of similar parts [4] ; the second

---

[1] ' Plastic ', as retaining any shape into which it is moulded. ' Immobile ' is equivalent to ' hard to move ' (δυσκίνητος below ; ' unyielding ', ' sluggish ', A.-H.), not ' stable ', for the icosahedron (water) is said to be the hardest to move of the other three bodies, whereas it is the least stable of them (a fact noted at 58D). In a mixed mass of solids of all the types the cubes would be the hardest to shift, the pyramids the easiest because their edges and points are sharp, so that a slighter thrust would push them between the rest. The next sentence argues that earth is hardest to shift and most plastic because it is also the most stable.

[2] βάσις here takes its meaning from the βάσεις preceding : the face which serves as base for a solid to stand on ; accordingly, as applied to the triangles, it means their faces considered as possible bases for solids (though none of the solids actually has an elementary triangle as its face), not, as the ' base ' of a plane figure would ordinarily mean, one of the lines containing it. Cf. Tr., p. 380.

[3] As Tr. remarks (p. 381), we shall hear later that there are several grades of size for each primary body, but that point is left out of account until it is actually mentioned at 57C. It is here assumed that all three bodies have equilateral faces of the same size.

[4] ἐλαφρότατον must mean ' lightest ' (not ' nimblest ', A.H., which = εὐκινητότατον), since the reason is the small number of parts ; so Aristotle, de caelo 299B, 31. Cf. 58E, Water composed of large particles is harder to

56B. (octahedron) must stand second in these respects, the third (icosahedron), third. Hence, in accordance with genuine reasoning as well as probability, among the solid figures we have constructed, we may take the pyramid as the element or seed of fire [1] ; the second in order of generation (octahedron) as that of air ; the third (icosahedron) as that of water.

    Now we must think of all these bodies [2] as so small that
C. a single body of any one of these kinds is invisible to us because of its smallness ; though when a number are aggregated the masses of them can be seen. And with regard to their numbers, their motions, and their powers in general, we must suppose that the god adjusted them in due proportion, when he had brought them in every detail to the most exact perfection permitted by Necessity willingly complying with persuasion.

The last paragraph adds to the determination of the shapes the adjustment in due proportion of the numbers, motions, and powers of the primary bodies. ' Numbers ' or ' quantities' ($\pi\lambda\eta\theta\eta$) probably means the total quantities of the several kinds, as at 57C the same word means the main mass of each kind. We conjectured that these quantities were the terms in the geometrical proportion fixed by the Demiurge at 31B ff. The motions and powers are those varying and active qualities which figured in the description of the bodily as without plan or measure before the geometrical shapes were fashioned (53A). The mention of them is immediately followed by a reference to Necessity ; they belong to the Errant Cause, but the Demiurge introduces as much order and proportion as Necessity allows. As we have remarked, the particles of the primary bodies are not simply portions of empty space partitioned

---

move and *heavy*; 59C, Bronze is *lighter* than gold because it has larger interstices. According to the analysis of lightness and heaviness at 62C ff., a larger quantity of any primary body is heavier than a smaller one, and this only means that it offers a greater resistance to the attempt to force it away from its proper region. Since it will be forced into the region of another element and have to make its way through that, the ' nimblest ' body will also be the ' lightest '. It will be easier to force a fire pyramid in among the octahedra of air, than to force an octahedron in among the pyramids.

[1] ' Element ' (' unit ', Tr.) because, when the pyramid is broken up into the elements proper (the triangles), fire ceases to exist as such, with the ' motions and powers ' characteristic of it (though on this point Plato is not quite clear later). ' Seed ' (a term applied to the microscopic bodies in Anaxagoras' system) is added to show that ' element ' has this sense here.

[2] All the four, to which these concluding remarks apply (not, as Tr. says, p. 382, the three last named only).

off by geometrical planes, but animated and filled with motions (changes) and powers, which are vigorously exercised in the warfare described in the next paragraph. As in the *Theaetetus* (above, p. 204), these changes and powers are conceived as existing in the primary bodies, apart from any effects they may produce on sentient beings.

56c–57c. *Transformation of the primary bodies*

Now that the regular solids have been assigned to the four kinds of body, the transformation of some of them into others can be described in terms conditioned by the assumptions of that theory. These are: (1) that particles can be broken down into their triangular or square faces, and these faces again into the elementary triangles out of which they were ' put together '; (2) that these elementary triangles can continue to exist, drift about in space, and recombine into the same or different figures ; with the limitation (3) that earth triangles, being of a different pattern, can only recombine as earth particles. The description of the various processes is extremely condensed, and has consequently been mis-understood.

The first section is concerned with the upward transformation, in the direction from earth through the intermediate bodies to fire. Three cases of resolution are described, the principal agent being fire, the most active, mobile, and penetrating of the four solids. We are told how it acts on earth, water, air, breaking down the less mobile figures, so that their elementary triangles are set free to recombine.

56c. Now, from all that we have said in the foregoing account concerning the kinds,[1] the following would be the most
D. probable description of the facts.

Earth, when it meets with fire and is dissolved by its sharpness, would drift about—whether, when dissolved, it be enveloped in fire itself or in a mass of air or of water—until its own parts somewhere encounter one another, are fitted together, and again become earth ; for they can never pass into any other kind.

---

[1] Reading ὧν περί A²FWY : ὧνπερ A. Burnet, Rivaud and Tr. prefer ὧνπερ, but this involves translating ἐκ πάντων by ' on the whole account ', which is not normal Greek for ἐκ πάντων τούτων, and ὧνπερ τὰ γένη προειρήκαμεν by ' the things whose kinds we have named ', whereas the things in question *are* the ' kinds '. If a copyist happens to omit a stroke which has to be added by a corrector and the result is a reading which can only be construed by forcing the language, it is not really scientific to defend the mistake as a ' *lectio difficilior* '.

Fire breaks down earth. This first case is the action of heat on solid bodies ('earth'), whether these are actually burnt in a flame or 'enveloped' in a surrounding mass of air or water. The faces of the earth-cubes, or their constituent triangles, can only recombine as cubes. There can be no transformation into water or air or fire. This, as we have seen, is simply a consequence of assigning the cubical figure to earth. In the other two cases transformation does occur.

56D.   But (1) when water is divided into parts by fire, or again by air, it is possible for one particle of fire and two of air to arise
E.   by combination ; and (2) the fragments of air, from a single particle [1] that is dissolved, can become two particles of fire.

(1) Fire (or air) breaks down water. When water is boiled, it passes into hot steam and vapour, which disappears into the air ; or again water is evaporated by air heated by the sun. The neighbouring air will become warmer. The theory explains this by supposing that the 20 faces of each icosahedron (water) are regrouped as 2 octahedra (air) and 1 pyramid (fire). The presence of the fire-pyramid will account for the air being warmed. The numbers of the faces thus make a complete transformation of water (partly into air, partly into fire) '*possible*'. It is not said that the transformation must always be complete ; but it is Plato's purpose to show that, if you start with any number of icosahedra of water, their constituents *can* be completely transformed into fire pyramids, after passing through the intermediate stage here described, in which some air particles and some fire particles are produced. The intermediate stage seems to be regarded as at least normal, if not necessary ; for, so far as the figures go, there is nothing to prevent the 20 faces of the icosahedron from reforming immediately as 5 pyramids. The result, then, at this intermediate stage is 1 fire-pyramid and 2 octahedra of air. (2) Fire continues its dissolving action and breaks down the newly formed octahedra, each of which can yield two pyramids by recombination. The final result of heating air is that it is wholly converted into fire. The upward transformation of water, through air and fire, into nothing but fire is thus theoretically possible.

We now pass to the downward transformation of (3) fire into air, (4) air into water.

56E.   And conversely, (3) when a little fire, enveloped in a large quantity of air or water or (it may be) earth, is kept in motion

---

[1] μέρους must here mean 'particle' (cf. μέρη ὕδατος, 60E, 61B), though μέρη at D4 means the elementary triangles.

56E. within these masses which are moving in place, and makes a fight, and then is overcome and shattered into fragments, two particles of fire combine to make a single figure of air. And (4) when air is overpowered and broken small, from two and a half complete figures, a single complete figure of water will be compacted.

In the downward transformation (3) the most compact, mobile, and active body has to be overpowered by more unwieldy antagonists, which must consequently outnumber it. Hence we are told that there must be a large quantity of the other bodies enveloping only a small amount of fire. And the fire is represented as putting up a fight ; like the rest, it is, in fact, forcing its way through the others owing to the attraction of like to like, and, being the sharpest, it will inflict more damage than it receives. But, if sufficiently outnumbered, it will be overpowered and shattered. The faces of two fire particles will then reform as a single air-particle, and the transformation will again be complete. As before, it is not contemplated as likely that 5 fire pyramids should combine immediately as 1 icosahedron of water. (4) The process continues in the reduction of the resulting air particles to water. In Nature this occurs, presumably, in the formation of mist, cloud, and rain ; but Plato speaks as if the air-particles required to be ' overpowered ' by the clumsier water particles outnumbering them. The result may be complete transformation into water, since $2\frac{1}{2}$ octahedra will yield the 20 faces required for a water particle.

The complete transformation, upwards from water to fire and downwards from fire to water, has now been accounted for. Two points remain to be added : (α) that no further change can occur, beyond these limits, in either direction ; (β) that the process will not stop until the limits are reached. Both points are made with reference (a) to the upward, (b) to the downward, process.

56E. Let us reconsider this account once more as follows.
(a) When one of the other kinds is enveloped in fire and
57. cut up by the sharpness of its angles and edges, then (α), if it is recombined into the shape of fire, there is an end to the cutting up ; for no kind which is homogeneous and identical can effect any change in, or suffer any change from, that which is in the same condition as itself. But (β) so long as, passing into some other kind, a weaker body is contending with a stronger, the resolution does not come to an end.

Plato here lays down the principle that the ' active power ' ($\delta\acute{\upsilon}\nu\alpha\mu\iota\varsigma$ $\tau o\tilde{\upsilon}$ $\pi o\iota\epsilon\tilde{\iota}\nu$) of a primary body can act only on what is unlike it : the hot can only modify the cold, and so on. Hence when the upward transformation is complete, nothing more can happen.

In the second sentence (which has been misunderstood), ' passing into some other kind ' is contrasted with ' if it is recombined into the shape of fire ' above. The case in question is that of water (not air) under the action of fire. Water does not turn straight into fire, but passes through the intermediate stage, in which the water particle is regrouped as two particles of air and one of fire. Here it is ' passing into some other kind ' than fire, namely air ; and in that form the weaker body (air) is still contending with the stronger (fire), which originally assailed it. If the fire is strong enough it will not stop until it has broken down the remaining air particles and turned the whole into fire.

The same two principles are now applied to the downward transformation.

57A. And, on the other hand, (b) when a few smaller particles
B.   are enveloped in a large number of bigger ones and are being shattered and quenched,[1] then, (α) if they consent to combine into the figure of the prevailing kind, the quenching process comes to an end : from fire comes air, from air, water. But (β) if they (the smaller particles) are on their way to these [2] (air or water), and one of the other kinds meets them and comes into conflict, the process of their resolution does not stop until either they are wholly dissolved by the thrusting and escape to their kindred, or they are overcome and a number of them form a single body uniform with the victorious body and take up their abode with it.

In this downward transformation, smaller bodies are being broken down by larger ones : fire by air, air by water. (α) The faces of the fire particles can recombine completely as air, those of the air particles as water, and the process must then stop, for

---

[1] ' Quenched ' shows that Plato is thinking in particular of fire enveloped in larger particles (as at 56E, 2) ; but the statement applies also to air passing straight into water, as the last words of the sentence show.

[2] Reading ἐὰν δ' εἰς ταῦτα (Y) or αὐτά (A) ἴῃ. As Tr. says, εἰς αὐτά (or ταῦτα) ἴῃ cannot mean ' assail the others ' (A.-H.), ' setzen sie aber den kampf fort ' (Apelt) ; this should be ἐπ' αὐτά, and the meaning is irrelevant. Tr.'s own suggestion that εἰς ταὐτά ἰέναι means ' come to terms ' is not supported by Apol. 36c (which he quotes), where εἰς ταῦτα ἰόντα means ' by engaging in these pursuits ' ; and in any case ἀλλήλοις would be required. εἰς ταῦτα ἰέναι ' to be passing into these kinds ' is like εἰς ἄλλο εἶδος ἐλθεῖν (56D, 5), εἰς ἄλλο τι γιγνόμενον (57A, 5). An exact parallel is τὸ ἐξ ἀέρος εἰς ὕδωρ ἰὸν ὁμίχλη, ' mist is that which is on the way from air to water ' (66E).

the same reason as before: that like cannot act on like. (β) The second statement, like its parallel above, has been misinterpreted. It explains what may happen to fire particles when passing through the air stage 'on the way' to water. In that phase they may be assailed by bodies of another kind. The breaking-down process will then continue, and not be arrested at the point where all the fire has become air. One of two results may follow: either the loose triangles into which the newly formed air particles are broken down again may escape to reform elsewhere as fire-particles, their original 'kindred'[1]; or they may finally recombine as water ('the victorious body'). In the latter case they take up their abode in the region of water, and the process can go no farther.

The next paragraph explains how the perpetual transformation of particles involves perpetual changes of direction. When a set of air particles, for example, is converted into fire, they will cease to be attracted towards the main mass of air, and begin to move towards the main mass of fire.

57C. Moreover, in the course of suffering this treatment, they are all interchanging their regions. For while the main masses of the several kinds are stationed apart, each in its own place, owing to the motion of the Recipient, the portions which at any time are becoming unlike themselves and like other kinds are borne by the shaking towards the place of those others to which they become like.

The principle of motion here invoked is the attraction of like to like, which already operated in the chaos described at 52D ff. (p. 198). The shaking of the Recipient, acting like the winnowing-basket, 'separated the most unlike kinds farthest apart from one another, and thrust the most alike closest together; whereby the different kinds came to have different regions, even before' the ordered whole consisting of them came to be'. If this motion were unchecked, it would sort out the primary bodies into four distinct regions. But the process of transformation is constantly modifying that tendency, as particles reformed into different kinds change the region towards which they drift. The motion, in so far as it is explained at all, is attributed to the qualities or powers of the primary bodies—those 'vestiges of their own nature' which they possess in abstraction from their geometrical shapes (53B). As we have seen (p. 205), it must be finally due to blind irrational impulse

---

[1] It will be suggested later (on 58A–C) that this alternative may be due to the expansion in volume, consequent on the change of figure, leaving no room for some of the loose elements to reform in any regular shape.

in the soul that animates the whole body of the world. Reason has now contrived to establish some check upon the tendency by endowing the bodies with shapes which can be transformed into one another. The motions are still blind and mechanical, but Reason has subordinated these ' secondary causes ' to its purpose of keeping an ordered world in being.

If we attempt to go behind Plato's description and ask after the ' real ' nature of the process of dissolution and recombination, it is doubtful whether we can expect any certain answer. One thing is obvious : the transformation is based on surfaces and the numbers of elementary triangles they contain, not on volumes. Thus the eight faces of two fire pyramids make up an octahedron of air ; but the volume of the octahedron will be more than twice that of the pyramid. Also the reason why earth cannot pass into any other figure lies in the different shapes of the elementary triangles ; it is not that the volumes will not fit. But what is to be made of the picture of plane surfaces being broken up and the fragments drifting about till they find others to combine with ? It cannot be taken literally. It was not necessary for Aristotle to point out that geometrical planes cannot behave in this way ; and there is no warrant for Martin's suggestion that the surfaces are thin plates of corporeal matter, forming boxes with a hollow interior. What is the ' matter ' of which these plates are composed ? And is the interior hollow in the sense of an absolute vacancy ? Plato does not say so, but speaks of the contents of the figures as qualities or ' motions and powers '.[1] The whole description of the warfare of the primary bodies in the process of transformation implies that these powers are actively operating. Without them the geometrical figures could not move at all or break one another down. The qualities are evidently conceived as existing in the primary bodies quite independently of the sensations and perceptions of any possible observer.

Plato's theory is capable of explaining why Nature contains a number of definite recognisable substances, e.g. gold, silver, copper, etc., and the other ' homœomerous ' substances which had figured in the systems of Empedocles and Anaxagoras. These stand out as distinct things with a more or less constant group of characteristic qualities ; there is not an infinite gradation of intermediate sub-

---

[1] Cf. Rivaud, p. 80 : *La théorie des figures élémentaires est destinée à expliquer comment l'ordre s'introduit dans le chaos mouvant des qualités. Par leurs propriétés définies et invariables, ces figures mettent une certaine fixité dans le devenir. Mais, elles n'en forment pas la substance, qui reste constituée par les qualités changeantes.* Tr.'s discussion of this question is vitiated by his use of Whitehead's terminology (p. 409).

stances, through which gold, for example, shades off into copper. Modern chemistry has explained the abrupt transitions which do occur by the shifting of elementary factors from one definite pattern to another. Plato's elementary factors are pictured as triangular surfaces, whose regrouping provides for a sudden change from water or air to fire.   Such an account seemed to him 'likely', as covering phenomena which Democritean atomism could not satisfactorily explain ; but he did not mean it to be taken as a literal statement of what 'really' happens.

The most questionable feature is the suggestion that triangular surfaces which are not at the moment the surfaces of any solid body can drift about by themselves. It will be suggested later (p. 274) that this feature disappears when further details of the theory are disclosed ; the account so far given is (as we shall soon find) by no means complete. The description of the particles which cause sensations of smell (66D) seems to imply that these 'loose' triangles may be combined in *irregular* solids of a size intermediate between the sizes of the regular solids.

57c–d. *Each primary body exists in various grades of size*

So far nothing has been said about the size of particles, except that they are all microscopic. We now hear for the first time that there are specific varieties of fire and the rest, due to the different sizes of the triangles and consequently of the particles they compose.

57c. In this way, then, the formation of all the uncompounded and primary bodies is accounted for. The reason [1] why there are several varieties within their kinds lies in the construction of each of the two elements : the construction in

d. each case originally produced its triangle not of one size only, but some smaller, some larger, the number of these differences being the same as that of the varieties in the kinds. Hence, when they are mixed with themselves or with one another, there is an endless diversity, which must be studied by one who is to put forward a probable account of Nature.

These differences of size in the primary bodies introduce considerable complications, which have not been sufficiently studied.

[1] τὸ (τοῦ, FY) δὲ ἐν τοῖς εἴδεσιν αὐτῶν ἕτερα ἐμπεφυκέναι γένη τὴν ἑκατέρου τῶν στοιχείων αἰτιατέον σύστασιν, μὴ μόνον ἐν ἑκατέραν (sc. σύστασιν) μέγεθος ἔχον τὸ τρίγωνον φυτεῦσαι κατ' ἀρχάς, ἀλλ' ἐλάττω τε καὶ μείζω . . .
The subject of φυτεῦσαι is ἑκατέραν (σύστασιν), not the Demiurge. That σύστασιν has the active sense, ' putting together ', will be argued below.

How will they affect the process of transformation, which has been based on the assumption that the various bodies concerned are composed of triangles of the same size, so that they can be regrouped in the several figures ? This difficulty must now be considered in conjunction with other obscure points that have been noted by the way. In order to keep this topic within reasonable bounds, Plato has stated only the main features of his theory, hinting at the outset that he had in reserve a ' fuller account ' (πλείων λόγος, 54B) of the peculiar merits of the two triangles chosen as elements. If we can reconstruct this, we shall be able to clear up a number of difficulties that have never been explained.

The present passage tells us that there is a limited number of varieties of each primary body, corresponding to different sizes of pyramid, octahedron, etc., and that there are indefinitely numerous combinations of all these varieties in composite bodies. At 58c three varieties of fire will be mentioned. The list may not be complete ; but the number is certainly limited, and probably small, for all the sizes must be microscopic. What is not yet clear is how there come to be different sizes of pyramid, etc. The reason, we are told, lies in ' the construction of each of the two elements (the half-equilateral and the half-square) : the construction ' in each case originally produced its triangle not of one size only, but some smaller, some larger '. The editors take ' construction ' (σύστασιν) in the passive sense, ' structure ', which the word often bears elsewhere. They understand the whole statement to mean no more than that each of the two elementary triangles is of more than one size. If that were all, Plato would have wrapped this simple statement in a singularly clumsy form. But the sentence cannot bear that meaning : it is impossible to say that the *structure* of each elementary triangle *produced* the triangle which is the structure in question.[1] The word σύστασις must have its active sense, ' putting together '.[2] We may paraphrase as follows : ' The reason why there are several varieties within their kinds lies in the way in which the elementary triangles of each of the two sorts are put together : in each case the elementary triangles were put together in such a way as to produce the corresponding triangle

---

[1] Translators evade this absurdity by resorting more or less to paraphrase, or by ignoring the fact that ἑκατέραν (σύστασιν) is the subject of φυτεῦσαι.

[2] Corresponding to the active συνιστάναι. Cf. 89c, τὰ τρίγωνα . . . συνίσταται, ' the triangles are put together '; 54E, ' if 4 equilateral triangles are put together (συνιστάμενα) so that ', etc. 55A, ' a second body is composed of the same triangles when they are put together (συστάντων) in a set of eight ', etc. At 89A, σύστασις means the ' bracing ', ' pulling together ' of the body by gymnastic exercise. Tr. there compares *Laws* 782A πόλεων συστάσεις καὶ φθοράς.

(the half-equilateral or the half-square [1]) not of one size only, but some smaller, some larger '. We have to discover what can be meant by putting together elementary half-equilaterals or half-squares in such a way as to form larger or smaller half-equilaterals or half-squares. But first we may state the difficulties that arise from the interpretation commonly assumed. The consequences are easily deduced.

Let us suppose that there are three sizes of the right-angled scalene triangle which serves as element in the construction of pyramid, octahedron, and icosahedron, and that each of the three is to be an irreducible element. We shall then have three corresponding grades of solids, each with an element somewhat larger than the one before:

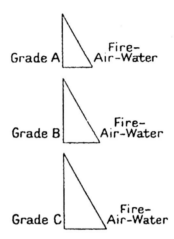

The solids within each grade will be capable of transformation into one another. There will thus be three parallel processes of transformation; but there can be no transformation from one grade to another. Nothing in the previous account of the process has led us to expect this startling and unsatisfactory result: that one variety of fire can be transformed only into one variety of air or water. Nor is there later any hint of such barriers between grades of solids. Further there is an element of vagueness and casualness in the supposition of three unspecified sizes with no definite ratio between them, which ill accords with the emphatic declaration that the Demiurge fixed the proportions with the greatest possible accuracy down to the last detail (56c). In the summary at 69B we shall hear again that the god, finding his materials in a dis-

---

[1] ' The triangle ' cannot mean the equilateral triangular face of the three solids, because the corresponding face of the fourth solid, the cube, is not a triangle, but a square.

orderly condition, introduced proportion and symmetry into the internal relations of each one and their mutual relations, in every way that was possible. If this is true, he must have established some definite ratio between the sizes ; and, if he could, he would surely fix upon some ratio that would permit of transformation occurring freely between the grades.

Now there is one way in which this desirable result can be attained. Transformation between grades will be possible, if the larger scalene triangles are some definite multiple of the smaller, and if one larger triangle (say of Grade *B*) can be composed of two or more smaller triangles (of Grade *A*). Consider the consequences of this supposition. The smallest scalene (Grade *A*) will now be the highest common measure of the others (*B* and *C*), which will be multiples of it ; and that smallest scalene will be the irreducible element— the στοιχεῖον proper. The larger scalenes will be obtained by putting together two or more of the smallest. Thus (as everything said hitherto has led us to expect) there will be, strictly speaking, only one irreducible element of each of the two types, scalene and isosceles, not an unspecified number. The grades will now be related by definite ratios, and transformation between grades will be possible.

The next question is, whether larger scalene or isosceles triangles can be built up out of smaller ones of the same type in the way proposed. They can, because of a certain property possessed by both elementary triangles, which we can now see to be one of the reasons why they are the ' best ' that could be selected. It is a property that had been regarded as characteristic of an ' element ' :

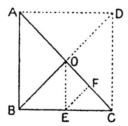

either of the two triangles can be subdivided without limit into parts of the same type as itself. The half-square *ABC* is divisible into two smaller half-squares *ABO, BOC* ; and Plato does in fact so divide it, when he constructs the whole square face, *ABCD*, of the cube. *BOC*, again, can be subdivided by dropping the perpendicular *OE* ; *OEC* by the perpendicular *EF*, and so on *ad infinitum*. In the same way the half-equilateral, *ABC*, can be subdivided into smaller half-equilaterals by bisecting the angle at

C and dropping the perpendicular *OD*. It is actually so subdivided into three elementary scalenes in Plato's figure (p. 216) ; and the

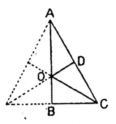

subdivision can be carried on *ad infinitum*.[1] Plato, however, does not continue the process indefinitely. He stops at a minimum triangle (*OBC*) of each type, which is taken to be atomic. He then builds the square out of 4 half-squares, the equilateral out of 6 half-equilaterals.

Here we encounter one of the points that we noted as never having been satisfactorily explained. Why use 4 half-squares to construct a square when 2 would suffice ? Why 6 half-equilaterals, when 2 would suffice—a fact which Plato himself mentioned where the scalene element was originally described as ' such that a pair of them compose the equilateral triangle ' (54A) ? Evidently he was aware that there were at least two ways of composing an equilateral out of this element. The seemingly arbitrary procedure can be explained by supposing that, in the earlier construction of the four solids, Plato intended to describe solids of an intermediate size— Grade *B*—not of the smallest possible grade (*A*). He deliberately used more elementary triangles than would have been required, if he had had only one grade of solid in mind. Solids of the smallest grade could be produced quite simply by forming the equilateral and the square each out of two elements ; and he could have built his pyramids, cubes, and the rest just as well. Moreover, if there were only one grade of solids, all the transformations described could occur between them. He chose to describe solids of a larger grade because he wanted to suggest that there are in fact several grades, and that when these larger solids are broken down into elements, those elements can be recombined in several ways. Thus the 6 scalenes in the equilateral face of a pyramid can recombine, in pairs, to make three equilateral faces for pyramids or octahedra or icosahedra of the lower grade. Or, as in the cases actually described, the transformation may be into solids of the same grade.

[1] Since the triangles, not the solids, are Plato's ' elements ', this meets Aristotle's objection that not every part of a pyramid or cube is a pyramid or cube. ' Homœomereity ' was first clearly defined by Anaxagoras, but Empedocles had no doubt already assumed that every part of fire was fire.

Or again, the elements might go towards the formation of solids of a still higher grade (*C*), and of as many more higher grades as are required to account for the actual varieties of the primary bodies in question. An advantage of this scheme is that it would make it possible for there to be more—perhaps many more—varieties of (say) water [1] than there are of fire ; and yet transformation could occur between them all.

The results may be illustrated by taking some of the possible transformations of Plato's solids of the intermediate grade (*B*) into solids of the smallest grade (*A*) :

	GRADE B			GRADE A	
Solid	*Equilateral faces composed of 6 elements*	*Elements*	*Elements*	*Equilateral faces composed of 2 elements*	*Solid*
1 pyramid	= 4	= 24	= 12	=	3 pyramids or 1 octahedron + 1 pyramid.
1 octahedron	= 8	= 48	= 24	=	6 pyramids, or 3 octahedra, or 1 icosahedron + 1 pyramid.
1 icosahedron	= 20	= 120	= 60	=	15 pyramids, or 6 octahedra + 3 pyramids, or 3 icosahedra.

Further, we can see what would suggest to Plato's mind the building of his solids out of these elements, namely, the proper geometrical construction of the figures as it appears in Euclid. The problem : ' To construct a pyramid, and to comprehend it in a given sphere, etc.,' is solved in *Euclid* xiii, 13. The solution either was due to Theaetetus or had been discovered earlier ; it was certainly known to Plato. We have seen an allusion to the circumscribing sphere at 55A. Now the demonstration involves a theorem proved in the preceding proposition (xiii, 12) : ' If an equilateral triangle be inscribed in a circle, the square on the side of the triangle is triple of the square on the radius of the circle.' The figure, as it appears in Euclid, is given by the unbroken lines in the accompanying diagram. *BD* = the radius *OD*. It is also the side of a hexagon inscribed in the circle (as the demonstration mentions). The dotted lines inside the circle complete the hexagon, as in the figure at iv, 15 (' In a given circle to inscribe an equilateral and equiangular hexagon '). The diagram so completed exhibits Plato's equilateral triangle *ABC* divided into the 6 elementary triangles. The hexagon contains 12 of these elementary triangles, all equal to one another. In this proposition (xiii, 12), Euclid

---

[1] ' Water ' is a wide term, covering the fusible metals (58D ff.).

proves that the square on *AB* is triple the square on the radius or on *BD*, which = the radius. Thus the gist of the proposition

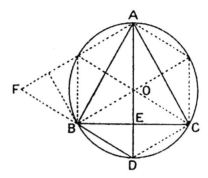

is that the larger triangle *ABD* is *of the same type* as the half-equilateral *ABE* :

In *ABE*, *BE* = 1, *AB* = 2, *AE* = $\sqrt{3}$

In *ABD*, *BD* = 1, *AD* = 2, *AB* = $\sqrt{3}$.

In other words, if you add one elementary triangle *BED* to the half-equilateral *ABE*, you obtain a larger half-equilateral *ABD*.

The figure suggests some further consequences : (1) All the smallest triangles are of Plato's elementary type. (2) Two of them form the equilateral *OBD*, the base of the solids of the smallest possible grade.   (3) Six of them form the equilateral *ABC*, actually constructed by Plato, as the base of his Grade *B* solids. (4) This large equilateral, *ABC*, can be regarded as composed of two half-equilaterals, *AEB* and *AEC*, each of the same type as the three elements which compose it. (5) There is also (as *Eucl.* xiii, 12, demonstrates) a still larger half-equilateral *ABD* of the same pattern, composed of 4 elements.   Two of these make up a larger equilateral *ADF*, containing 8 elements, and also divisible into 4 of the smallest equilaterals (*OBD*, etc.).   *ADF* would serve as base for solids of the next largest type (Grade *C*).   (*b*) In the same way, you can go on to still larger bases and solids, as shown, for example, in the following diagram, in which the five equilateral triangles, *ODC*, *ABC*, *ADE*, *AFG*, *AHJ* are respectively composed of 2, 6, 8, 18 and 24 elements.

It will now be clear why Plato chose the right-angled scalene for his element, instead of the equilateral.   This was one of the obscurities we noted.   The pyramid, octahedron, and icosahedron all have the equilateral triangle as their base.   Why did not Plato simply take the equilateral triangle as the element of these solids, and the square as the element of the cube ?   These figures have the strongest claim to simplicity and ' beauty ' or perfection : all their

sides and all their angles are equal.[1]   The simplest way to make a
pyramid or octahedron or icosahedron is to put together 4, or 8,

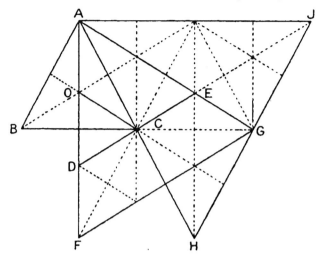

or 20 equilateral triangles.   The transformation of the elements (if
it is confined to one grade of solids) can be effected just as well by
breaking down the solids into their equilateral faces and regrouping
these, without breaking down the faces themselves into smaller
elements.   Why, then, did Plato analyse his equilateral face into
6 elements, which have less *prima facie* claim to perfection ?   And
why did he analyse the square face of the cube into 4 half-squares ?

The objection to the simpler procedure, which takes the equilateral
and the square as faces, is this.   If we take the equilateral as

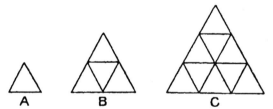

elementary, the smallest pyramid (Grade *A*) will have one of these
elements as its face.   The face of the next largest pyramid (on our
supposition that there is a definite ratio between the grades, such
that transformation from grade to grade is possible) can only be
formed by putting together 4 elementary equilaterals.   But then
the Grade *B* pyramid will have a face 4 times the size of the Grade *A*
face ;   and its volume will be more than 4 times as great.   In
order to obtain the next largest grade (*C*), we shall need 9 elements,

---

[1] So Speusippus (frag. 4, quoted by Tr., p. 370) ranks the equilateral first
among triangles, second the half-square, third the half-equilateral.

and so on. The result would be that the intervals in size between the grades of solids would be very large. The three varieties of fire, for instance, would be too far apart ; and if (as seems to be the case) there are considerably more than three varieties of water, it would be hard to suppose that the icosahedra could all be microscopic.

The merit of Plato's elementary triangles is that they can yield a series of equilateral or square faces which are much closer together in size. In the light of Euclid's diagram and Plato's own hints we obtain the following series of equilateral faces :

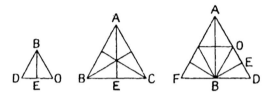

Grade *A* consists of 2 elements, as indicated at 54A. Grade *B* is Plato's figure, with 6 elements. Grade *C* has 8 elements, or 2 half-equilaterals, *ABD* (as in Euclid's figure) and *ABF*.

The same argument applies to the square face of the cube. If the square be taken as the element, the Grade *A* cube will have 1 element as its face ; the Grade *B* face will require 4, and the

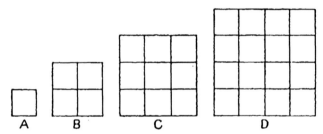

volume will be 8 times as great ; the Grade *C* face will take 9 elements and the volume be 27 times as great, and so on. The volumes of the varieties of earth-cube will soon exceed microscopic proportions. On our hypothesis the square bases will be as follows, with 2, 4, 8, 16, etc., elements :

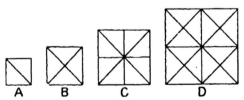

As before, Grade *B* is Plato's figure. The resulting cubes will be much closer together in size than on the alternative plan.

We have now found a satisfactory meaning for Plato's statement that the existence of several varieties (grades) of each primary body is due to the way in which the elementary triangles of either pattern are put together to form larger or smaller triangles of the same pattern. Our hypothesis not only removes all the difficulties which have perplexed the commentators and satisfies Plato's declaration that symmetry and proportion were introduced down to the smallest detail, but finally restores to the text before us its only possible sense. It is not surprising that Plato has left to the reader the task of reconstructing from a few hints the ' longer account ' which he held in reserve. Aiming at extreme compression, he was content to state only the main outline of this theory, as of his astronomy. The rest would be accessible to men ' favoured by heaven ' with a knowledge of geometry, and Timaeus has remarked (53C) that his friends are in this respect well equipped. They had, in fact, only to look at the figure which now illustrates Euclid xiii, 12, and observe how that proposition is used in the construction of the pyramid.

57D–58C. *Motion and Rest*

The next paragraph is concerned with motion and rest. It has still to be explained why the four primary bodies are not sorted out by the attraction of like to like into four separate homogeneous masses. They would then settle down into a permanent state of rest, since that attraction is the only mechanical force active in the chaos described earlier, and, when it had completed its work, nothing more could happen. The answer given here is not that the world is animated by a self-moving soul, which can and must constantly keep motion going. In this second part of the discourse mechanical explanations are to be given, so far as they will go ; we are concerned with secondary causes and what happens of Necessity. The mover here is of that lower order which is itself moved and transmits motion to other things. We are now being told ' in what manner it is of the nature of the Errant Cause to produce motion ' (48A).

The first point is that motion of the mechanical sort can exist and be kept going only in a condition of heterogeneity ; and the theory of the solids has provided for this.

57D. Now concerning motion and rest, if we do not agree in what manner and in what conditions they arise, many difficulties will stand in the way of our subsequent reasoning. Some-

E. thing has already been said about them, but there is this to be added : motion will never exist in a state of homo-

57E.   geneity. For it is difficult, or rather impossible, that what
        is to be moved should exist without that which is to move
        it, or what is to cause motion without that which is to be
        moved by it. In the absence of either, motion cannot exist ;
        and they cannot possibly be homogeneous. Accordingly,
        we must always presume rest in a state of homogeneity, and
        attribute motion to a condition that is heterogeneous.
58.   Further, inequality is a cause of heterogeneity ; and the
        origin of inequality we have already described.[1]

That motion presupposes heterogeneity was a main point in the
reasoning of Parmenides, who declared that if Being is one in
every sense, it must be homogeneous (ὅμοιον) throughout, a continuous plenum, and motionless. Plato himself has asserted that
like, though it can attract, cannot alter or otherwise act upon like,
and when once like things have come together, with no admixture
of the unlike, no further change can occur (57A). It follows that,
if there is to be motion, body can never have existed in the form
of a single substance, as the earliest Monists had supposed. There
must be at least two kinds of body with different powers, so that
the one may act on the other. Following Empedocles, Plato has
provided four, with opposite powers. The mechanical motion of
chaos was imagined as going on before the primary bodies were
given their regular shapes ; it was not a consequence of their
having those shapes or any other shape. The sorting out of like
to like in different regions was attributed to the mutual attraction
of like qualities (' motions and powers '). Now, however, the
bodies have received their shapes, and these shapes are unequal
in size in two ways : each kind of body has a particle of a different
size from the rest, and there are several grades of each kind.
This may be what is meant by inequality as the cause of heterogeneity ; so the ' origin of inequality ' has been stated in the
preceding paragraph.[2] The different sizes of particles are dwelt
upon in the explanation that follows. This is a new factor, governing the behaviour of the primary bodies as they actually exist in
particles of unequal sizes.

It is probable, however, that ' inequality ' refers also to the
' powers that were neither alike nor evenly balanced ' in the original
chaos (52E). ' The unequal ' (τὸ ἄνισον) is an alternative phrase

---

[1] Obviously the mover here cannot be the soul, which belongs to a higher
order of existence. It could not be spoken of as either heterogeneous and
unequal, or homogeneous and equal, with the moved.

[2] At 58D, ' heterogeneity ' (ἀνωμαλότης) is used for the non-uniformity due
to the unequal sizes of water particles, and so frequently.

for the 'Great-and-Small' or the 'More-and-Less' in quality. In the *Philebus* qualities are represented as unlimited ranges (ἄπειρα), since there is always a hotter in one direction and a less hot in the other. 'The equal' $\left(\frac{1}{1}\right)$ is a limit of quantity introduced anywhere in this range, with 'the unequal' extending indefinitely on both sides. Accordingly the powers called hot and cold, moist and dry, etc., were imagined as swaying in unbalanced conflict before the Demiurge introduced limit, measure, numerical proportion.

One effect of the introduction of geometrical shapes has already been mentioned (57C). The reformation of the elements of disintegrated particles into particles of another kind is constantly shifting the direction of the drift of like to like. We thus have a somewhat more definite picture of the disorderly motion as a mechanical process. But, if that were all, the result would still be chaotic, shifting, rectilinear motions, not unlike the Democritean chaos of atoms moving casually in all directions, colliding with one another and so changing the directions of one another's motions. If there were an unlimited void (as in Atomism), the result would be that like bodies would tend to get together in homogeneous masses, casually, at different places. And, if the atoms were unlimited in number, as the Atomists assumed, vortices might arise, here and there, and similar bodies might be sorted out by the eddying motion into concentric layers. In Plato's scheme the primary bodies are limited in number, and the result imagined is the formation of four homogeneous masses, which, when completely separated, would cease to move through one another or to change in place. So the whole would come to permanent rest. Empedocles had pictured such a condition as arising in the Realm of Strife and persisting until Love began to pour in from outside and break it up.

The next paragraph introduces a further factor distinguishing Plato's scheme from Atomism. The primary bodies, limited in quantity, are all confined within the spherical boundary of the world's body, as described earlier (33B). Outside the sphere it is probable that there is (as in Parmenides and Aristotle) not even empty space. But even supposing that there were a void outside, we are now given a mechanical reason why the primary bodies all hold together in the spherical shape and are packed inside it as closely as possible; and why they perpetually move through one another instead of sorting themselves out into separate regions, like to like.

58A. But we have not explained how it is that the several bodies have not been completely separated apart in their kinds and so ceased to pass through one another [1] and to change their place. We must, then, resume our explanation as follows. The circuit of the whole, when once it has comprehended the (four) kinds, being round and naturally tending to come together upon itself, constricts them all and allows (or tends to allow) no room to be left empty. Hence

B. fire has, more than all the rest, penetrated in among all the others [2] ; and, in the second degree, air, as being second in the fineness of its particles ; and so on with the rest. For the kinds that are composed of the largest particles leave the largest gaps in their texture, while the smallest bodies leave the least.[3] So the coming-together involved in the condensing process [4] thrusts the small bodies together into the interstices between the large ones. Accordingly, when the small are set alongside the large, and the lesser disintegrate [5] the larger, while the larger cause the lesser to combine, all are changing the direction of their movement, this way and that, towards their own regions ; for each, in

C. changing its size, changes also the situation of its region. In this way, then, and by these means there is a perpetual safeguard for the occurrence of that heterogeneity which provides that the perpetual motion of these bodies is and shall be without cessation.

The difficulty here lies chiefly in the opening sentence of the explanation :

'The circuit (περίοδος) of the whole, when once it has comprehended the (four) kinds, being round (κυκλοτερής) and naturally

---

[1] τῆς δι' ἀλλήλων κινήσεως (as distinct from φορᾶς, locomotion) may mean transformation. Cf. 54B, δι' ἀλλήλων εἰς ἄλληλα γένεσιν ἔχειν, and 49C.

[2] Cf. 78A : ' Of all the kinds fire has the smallest particles and consequently passes through (διαχωρεῖ) water, earth, and air and all bodies composed of these, and nothing is impervious to it.'

[3] The icosahedra composing a mass of water, however closely packed, must, owing to their shape, leave larger gaps between them than those left between the octahedra (of the same grade) in a mass of air.

[4] ἡ τῆς πλήσεως σύνοδος. For the form of this phrase cf. Phaedo 97A, ἡ σύνοδος τοῦ πλησίον ἀλλήλων τεθῆναι, and Tim. 76C, τῇ πλήσει τῆς ψύξεως.

[5] διακρινόντων. If the particles were atoms, διακρίνειν could only mean ' separate ' and συγκρίνειν ' bring together ' ; but since particles can be broken up into elementary triangles, the breaking down and recombining of these elements may be meant ; and the reference to change of figure seems to imply this. As we have seen (56D), disintegration is chiefly caused by the smallest body, fire. At 56E we learnt how the larger bodies (air and water) cause fragments of the lesser to recombine in the larger figures.

tending to come together upon itself (πρὸς αὑτὴν συνιέναι), constricts (σφίγγει) them all, and allows (or tends to allow, ἐᾷ) no room to be left empty.'

There are several verbal ambiguities : περίοδος can mean either 'circumference' or 'revolution'; σφίγγει can mean 'constricts' (compresses) or simply 'embraces'. 'Circuit' covers both 'circumference' and 'revolution' precisely because the word περίοδος ('journey round'), like *circuitus* and 'periphery' ('carrying round'), properly means both. The sphere is defined in Euclid xi, Def. 14, as the figure traced 'when, the diameter of a semicircle remaining fixed, the semicircle is *carried round* (περιενεχθέν) and restored again to the same position from which it began to be moved'. Hence the word circum-ference (περιφορά), the 'carrying round' of the semicircle whose sweep leaves this trace. The two notions of rotatory movement and of the spherical figure traced by it are inextricably associated. It is probable that both notions are present here, for one of the works of Reason was to endow the body of the world with spherical shape and with rotation (33B ff.), though the word κυκλοτερής (*teres atque rotundus*) is more appropriate to shape than to movement, and is applied to the shape at 33B. That earlier passage is here recalled. We were told, moreover, that no part of the bodily was left outside the sphere, just as we are told here that the circuit of the whole completely comprehends the primary bodies. Spherical shape, then, including all the bodily, is a new factor here introduced, as a work of Reason, which imposes conditions on the movement of the primary bodies, as hitherto described.

We know also that the spherical mass is rotating, and that its motion at the circumference is the swiftest of all motions. But Taylor is right in objecting to the view of Archer-Hind and others that the rest of the sentence ascribes to spherical rotation a constricting force. Archer-Hind translated: 'The *revolution* of the whole, when it had embraced the four kinds, being circular, with a natural tendency to *return upon itself*, *compresses* everything and suffers no vacant space to be left.' He speaks of the whole globe being 'subject to a mighty constricting centripetal force' or 'inward pressure', which he compares to the force 'exerted in winding a hank of string into a round ball'. 'This', he says, 'is the second of Plato's two great dynamic powers.'[1] But a spherical mass, rotating as a whole, does not set up any constricting force.

[1] Mondolfo, *L'infinito nel pensiero dei Greci*, p. 315, also speaks of the pressure exercised by rotatory motion contracting the universe into a smaller space, and infers that there must be an infinite void outside.

Anyone who has been driven or swung round a sharp curve will be aware that the motion tends to throw him outwards and would do so if it were not counteracted by another force acting towards the centre.[1] Taylor points out this defect and denies any reference to forces of pressure. He understands περίοδος to mean ' circumference ', and σφίγγει as meaning ' encompasses round about '. Timaeus, he thinks, says ' only that " the round of the all " encompasses the particles of the " roots ", so that no space is left for them to " drift " off to infinity in. It is the round shape and finitude of the all, not an imaginary centripetal force set up by the " rotation " which " leaves no space vacant ".' ' The οὐρανός being finite and round ; the particles cannot get too far away from one another ' (p. 398). That is all.

I agree that ' circuit ' means (at least primarily) the ' circumference ', which comprehends the whole of the primary bodies. This is, indeed, rendered almost certain by the parallel passage (81A) where the substances in the blood are said to be ' comprehended by the living creature (the microcosm) framed like a heaven (οὐρανοῦ) ' to include them, and to be ' constrained to reproduce the movement of the All ', as here described. On this view the meaning will be : (1) that Reason has confined all the particles within a spherical figure, so that similar particles cannot congregate just anywhere in an unlimited space ; and (2) that the rotatory movement natural to a sphere, coming round as it does in a closed circle and not expanding outwards (as, say, a spiral would do),[2] allows no vacant intervals to be formed between the congregating masses inside. The particles are always packed together as closely as possible within a rigid boundary. Hence the attraction of like to like has to operate inside the closed sphere. Consequently fire, instead of being free to fly off to a distance from air or water, must force its way through the coarser particles in order to reach its like ; air must force its way through water, and so on.[3] The result will be (as Plato says) that, so far as possible, no vacant

---

[1] Anaxagoras (*Vors.* A71) invoked this tendency to account for rocks being torn off the earth τῇ εὐτονίᾳ τῆς περιδινήσεως and flung outwards to form the heavenly bodies (Tr., p. 397).

[2] Contrast the world-forming vortex in Anaxagoras, which ' began to revolve first from a small beginning, but is now spreading further, and will spread further still ', as it takes in more and more of the surrounding mass (frag. 12).

[3] Cf. Albinus' paraphrase : διότι τῇ τοῦ κόσμου περιφορᾷ σφιγγόμενα συνωθεῖται καὶ συνελαυνόμενα πρὸς ἄλληλα φέρεται τὰ λεπτομερέστατα εἰς τὰς τῶν ἀδρομερεστέρων χώρας. διὰ τοῦτο δὲ μηδὲν κενὸν ὑπολείπεται σώματος ἔρημον (*Didasc.* xiii). *Tim. Locr.* 98E, ἅπαντα δ' ὦν πλήρη ἐντί, οὐδὲν κενὸν ἀπολείποντα. συνάγεται δὲ τᾷ περιφορᾷ τῶ παντός, καὶ ἠρεισμένα τρίβεται μὲν ἀμοιβαδόν ...

interstices will be left within the sphere. A heterogeneous mass of particles with all the shapes and sizes described cannot be packed so as to leave no interstices at all. But ' the coming-together involved in the condensing process' (ἡ τῆς πιλήσεως σύνοδος) is always driving the smaller particles into the spaces left between the larger. Fire, as the smallest, sharpest, and most easily moved, penetrates, above all, in among the rest and exercises its disintegrating power by breaking up the clumsier particles of air and water. All this has been described already. Particles of all three kinds are shattered, and their elements reform as other kinds and change the direction of their drift towards their new kindred.[1] The result finally stated is that this transformation perpetually maintains that heterogeneity which is the condition of perpetual motion. We can now see, in fact, why the four kinds have not permanently come to rest, in separate regions, each as a homogeneous mass in which no change could occur.

It is probable that Plato is allowing for another factor that would certainly be involved, though nothing very clear is said about it. The ' thrusting together ' of the particles may be partly produced by the changes of volume resulting from transformation. When two fire pyramids are reformed as one air octahedron, the volume is increased.[2] This would be obvious to anyone who had ever followed the geometrical construction of the regular solids or seen models of them. Such increases of volume must occur in the *downward* transformation, from fire through air to water, as described at 56E and 57B. This is the ' condensing process ', as appears from the traditional account of it at 49C. The expansion of volume will set up thrusts and cause further disturbances : intercepted particles will be crushed by their expanding neighbours ' coming together ', and their debris will fill the original interstices. At 57B one of the things that may happen to the smaller particles in the downward transformation is that ' they are wholly dissolved by the thrusting and escape to their kindred '. When the elements of some of the smaller particles recombine as water-icosahedra, they leave no room for others to reform as regular bodies of any size. If this factor of expanding volume is allowed for, we must suppose

---

[1] It is not explained where earth comes into the scheme. There is nothing to show what sizes the earth cubes have, as compared with the other bodies. Cubes can be packed so as to leave no interstices ; yet at 60E we hear that interstices between earth cubes are so large that fire or air particles can slip into them without disturbance.

[2] I cannot understand why Tr. thinks that Timaeus is supposed not to know this, though Plato knew it (p. 395). A moment's reflection would show that a double pyramid with its 6 faces must have less volume than an octahedron with 8 faces.

that on the whole a balance is maintained : the expansion involved in the downward condensing process in one place must be compensated by the contraction involved in the upward process towards fire somewhere else. This compensation is in full accord with the Heraclitean conception of the 'way up and down ',[1] along which transformation is always taking place in both directions.

It is not explicitly stated, here or elsewhere, that the main masses of the primary bodies form four concentric spherical layers, with fire on the outside (in the stars) and earth at the centre. This is no doubt assumed as an obvious fact, recognised in other cosmologies. The order of the layers could be explained as due to the rotatory movement (a work of Reason), sifting the more mobile particles towards the circumference, the less mobile towards the centre, on the familiar analogy of an eddy in water collecting the heavier floating objects at its centre. Aristotle says that this analogy was invoked by ' all those who try to generate the heavens, to explain why the earth came together at the centre ' (*de caelo* 295a, 13).

58c–61c. *Varieties and compounds of the primary bodies*

We have learnt that each of the regular solids exists in a limited number of different grades of size. There are, accordingly, (1) a corresponding number of grades of (say) water, each composed of particles of uniform size. There are, besides, (2) non-uniform varieties of water (etc.), consisting of particles of water only, but of more than one grade ; and (3) compounds of more than one primary body, e.g. earth and water. Since all grades of all the primary bodies can enter into such compounds, there will be a very large variety of possible combinations (57D). These three classes embrace the recognisable substances which occur in Nature, many of which have received names, while others are nameless. They form the subject of the following section.

The three classes are not kept entirely distinct ; but the account starts from the simplest case, namely (1) varieties of fire, or air, or water, which differ in the grade of the particles severally composing them, and the transformation of any one of these varieties into another variety of the same primary body, by change of some or all of the particles to a higher or lower grade.

58c. Next we must observe that there are several varieties of fire : flame ; that effluence from flame which does not burn but gives light to the eyes ; and what is left of fire in glowing

D. embers when flame is quenched. And so with air : there is

---

[1] ὁδὸς ἄνω κάτω, cf. 58B, 9, πάντ' ἄνω κάτω μεταφέρεται.

58D. the brightest and clearest kind called ' aether ', and the most turbid called ' murk ' and ' gloom ',[1] and other nameless kinds, whose formation is accounted for by the inequality of the triangles.

Light, which Plato regards as a body given off by flame, has already been described at 45B. It is similar to the visual current of ' pure fire ' which is so fine that it alone can filter through the close texture of the eyeball. We may infer that it consists of particles of smaller grades than flame or glowing heat. It has the quality or ' power ' of brightness, but not that of heat, possessed by the other two varieties. We do not feel light as hot, presumably because of the extreme fineness of the pyramids ; the pricking of their points would not disturb the coarser fabric of flesh. In the later account of colour (67D ff.), at least three grades of fire are invoked, corresponding to differences of colour.

It is not actually stated whether each of the varieties of fire and air here mentioned consists of pyramids or octahedra all of the same grade. The last words, ' whose formation is accounted for by the inequality of the triangles ', are ambiguous. They might mean that two varieties of air differ in that each is composed of uniform octahedra of a different grade from the other, or that some varieties are composed of octahedra of different grades. Plato may have intended to leave this question open, being aware that he had no means of deciding it one way or the other. In either case, however, if our hypothesis was correct, the transformation of one variety into another variety of the same primary body is possible ; and such transformation between grades is in fact described in the following paragraph about water. We may understand, accordingly, that ' murk ', for example, can be transformed into aether by being broken down into its elementary triangles, which then reform into a larger number of smaller-grade octahedra.

Water, liquid and fusible : melting and cooling of the fusible.—The formations consisting of water particles (icosahedra) present an additional complication. They include a large number of varieties, which will be classified under two main types, according as they

---

[1] ὀμίχλη and σκότος here are varieties of air, not mixtures of air and water ; so the word ' mist ' is better avoided, though ' fog ' might serve. They consist of octahedra of a larger grade than those of ' aether '. Possibly also the particles are more closely packed together, like those of the ' violently compressed air ' at 61A (so A.-H.). Later (66E) we shall hear of mist (ὀμίχλη) and vapour (καπνός) as composed, apparently, of irregular particles intermediate in size between icosahedra of water and octahedra of air and formed as air is passing into water or water into air.

are normally found in the liquid or in the solid state.   The liquid
type will include what we commonly call fluids : ordinary water,
oil, wine, etc.   The fusible type will contain the metals, gold,
copper, etc.   But fluids can pass from the liquid to the solid state,
as when water freezes into ice ; and the metals are ' fusible ' solids,
i.e. they can pass into the liquid state.   Accordingly, before the
varieties of fluids and metals can be described and illustrated, it
is necessary to define the characteristics of liquid formations
(τὸ ὑγρὸν γένος) and solid formations (τὸ πεπηγὸς γένος), as such, and
explain the processes of heating and cooling, whereby a fusible
substance can pass from one condition to the other, without changing
its nature ; for molten copper is still copper.   It is not suggested
that the melting process can turn copper into any other metal, still
less into a fluid, such as ordinary water or oil.

58D.   Of water, the primary division is into two types : (1) the
liquid, and (2) the fusible.

    (1) The liquid, because it contains portions of the small
grades of water, unequal in size, is in itself mobile and can
be readily set in motion by something else, owing to its
non-uniformity and the shape of its figure.

E.   (2) The other (fusible) type composed of large and uniform
particles is harder to move than the former and heavy,
being set hard by its uniformity.   But under the action of
fire, making its way in and breaking it down, it loses its
uniformity, and consequently becomes more mobile ; and
when it has become quite easy to move, under the thrust
of the neighbouring air it is spread over the ground.   Each
of these two processes has received a name : ' melting ' for
the reduction in bulk of the particles, ' flowing ' for the
spreading over the ground.

59.   When, on the contrary, the fire is being expelled from it
again, since the fire does not pass out into vacancy, the
neighbouring air receives a thrust and itself thrusts together
the liquid mass, while it is still quite easily moved, into the
places left by the fire and makes it a homogeneous combination.   The liquid, being so thrust together and regaining its
uniformity, as the fire which created the lack of uniformity
departs, settles into its original state.   The departure of the
fire is called ' cooling ' ; the contraction that follows on its
withdrawal is referred to as ' being in a solid state '.

The characteristic of the liquid state is mobility.   A liquid is
mobile ' in itself ', that is, by virtue of its structure, and hence
easily set in motion by an impulse from outside.   A formation of

water particles will be mobile when it consists of particles that are
(1) comparatively small, (2) of different grades, not 'uniform'.[1]
When it is added that mobility is also due 'to the shape of the
figure', Plato seems to be thinking of the contrast between (normally
liquid) water and (normally solid) earth, the icosahedron being a
less stable figure than the cube (55E). Similarly at 59D ordinary
water is said to be soft because its faces are less stable than those
of earth cubes. This is strictly irrelevant here, since all water
formations consist of icosahedra, and this contrast has nothing to
do with the difference between the liquid and fusible types of water.

The solid state is characterised by immobility, due to (1) comparative largeness and (2) comparative uniformity of the particles.
The process by which a fluid, such as water, is solidified into ice
will be described later (59D). Here Plato considers only the
melting and cooling of the fusible type—the metals. It follows
from the above definitions of the liquid and the solid, that the
reduction of a metal to the liquid state must involve (1) reduction
in size of some of the water-particles, and (2) the consequent
diminution of its original uniformity.

The description of melting and cooling is interesting because it
involves that transformation between grades of the same kind of
body which our hypothesis has shown to be possible. The earlier
account of the action of fire on water (and air) was concerned only
with the transformation of one primary body into another, as a
process occurring within a single grade; for at that stage no
differences of grade had been mentioned. In melting and cooling
there is no such transformation: the 'water' remains water
throughout; only its structure is changed.

Metal in the solid state consists, then, of icosahedra which are
comparatively large in size and uniform in grade: absolute uniformity is not intended, for gold is said to have the 'most uniform'
particles, and copper has particles of several grades (59B). Metal
is also described as 'heavy', apparently in the same sense that
fire was called the lightest of the three bodies (fire, air, water), 'as
being composed of the smallest number of similar parts', while
water was the heaviest (56B). Owing to the size and uniformity
of its particles metal is set hard, or solid (πεπηγός). Melting begins
when fire particles from outside penetrate into the interstices
between the icosahedra. The resulting change is less violent than
the process described earlier, where the conflict of fire and water
led to the transformation of the vanquished body into the victorious

[1] 'Uniform' will be used for ὁμαλός, meaning 'of the same grade in size';
'homogeneous' will mean 'consisting of only one kind of primary body',
not a compound of several kinds.

one.   In melting there is no conversion of fire into water or of water into fire.   Apparently the fire particles remain uninjured within the mass of metal, which consequently becomes hotter.   But as more and more fire particles force their way in they begin to break down some of the icosahedra into their triangles.   These dislocated fragments are able to reform themselves into icosahedra of a smaller grade.   The metal thus passes into the liquid condition, since it now contains an increasing proportion of smaller particles, which may be further reduced to yet smaller grades.   In this way melting means ' the reduction of the bulks ' (of particles) [1] to a lower grade in size.   Apparently this reduction does not suffice to make room for all the fire particles.   The metal consequently expands, and the expansion sets up a thrust from the surrounding air which is itself, perhaps, supposed to be expanded by fire invading it from the source of heat ; for the air near melting metal is heated.   The thrust spreads out the now fluid metal and sets it flowing over the ground.

In the reverse process of cooling, the fire particles are expelled into the neighbouring air, which they heat and expand.   The metal is at first in the liquid state, consisting of particles of unequal sizes.   The thrust communicated to the air by the fire which is expanding it causes the metal to pass back from the liquid state to the solid.   This means that the small icosahedra reform themselves as large icosahedra of their original grade, filling once more the spaces left vacant by the fire as it passes out.   The result is described by three phrases :   the water ' regains its uniformity ' (in respect of the size of icosahedra) ;   it settles ' into its original state ' ; and the air ' makes it a homogeneous combination ' [2] (of one primary body only), by thrusting the water together into the spaces vacated by fire.   Some details of the two processes are not completely explained.   Plato would not have vouched for them all ; he only means that something of this sort must happen, and the mechanical explanation can at any rate be carried thus far.

Some varieties of the fusible type (metals) : gold, adamant, copper.—From this general description of the liquid and fusible types we

---

[1] τὴν τῶν ὄγκων καθαίρεσιν has not been understood by the commentators, who have not worked out the possibilities of transformation between grades of the same body.   They have supposed that the elementary triangles are dilated by fire (Martin) or that particles are simply reduced to loose triangles, ' vagrant plane surfaces ' (Tr.).

[2] αὐτὸν αὑτῷ συμμείγνυσιν must mean ' makes it a combination of water with water ' (*not* ' with air '), instead of a combination of water and fire. Cf. συμμειγνυμένα αὐτὰ πρὸς αὐτά (57D, 4) of the combination of a primary body with itself (in that case, with other grades of itself).   αὐτῷ cannot mean ' with air ', because there is no air in the normal metal.

now pass to an account of a few varieties of the metals, which are species of the fusible class, found normally in the solid state.

59.   Of all these fusible varieties of water, as we have called
B.   them, one that is very dense, being formed of very fine and uniform particles, unique in its kind, tinged with shining [1] and yellow hue, is gold, the treasure most highly prized, which has been filtered through rock and there compacted.

The ' scion [2] of gold ', which is very hard because of its density and is darkly coloured, is called adamant.

Another has particles nearly like those of gold, but of more than one grade ; in point of density in one way it surpasses gold [3] and it is harder because it contains a small portion of fine earth ; but it is lighter by reason of containing
C.   large interstices.   This formation is copper, one of the bright and solid kinds of water.   The portion of earth mixed with it appears by itself on the surface when the two substances begin to be separated again by the action of time ; it is called verdigris.

It would be no intricate task to enumerate the other substances of this kind, following the method of a probable account.   When a man, for the sake of recreation, lays aside discourse about eternal things and gains an innocent pleasure from the consideration of such plausible accounts
D.   of becoming, he will add to his life a sober and sensible pastime.   So now we will give it rein and go on to set forth the probabilities that come next in this subject as follows.

Gold is ' unique in its kind '.   This apparently means that it is the only metal formed, solely or almost solely, of icosahedra of the finest grade.   Hence it will be very dense in texture, since the smallest icosahedra will leave the smallest interstices between them.

' Adamant ' seems to have been originally a poetic term for steel or tempered iron.[4]   In Hesiod's *Theogony* 161, it is applied to the metal created by Earth for the sickle used by Cronos to emasculate Ouranos.   This metal was grey (πολιός) and has no connection

---

[1] At 68A, ' shining ' (στίλβον) and ' bright ' (λαμπρόν) are treated as names of colours which can be mixed with yellow, etc.

[2] The English ' scion ' (like ὄζος) means the eye, bud, or sprout, used in grafting, and then ' offshoot ', ' offspring ' (in poetry) ; *auri nodus* (Pliny).

[3] πυκνότητι δὲ, τῇ μὲν χρυσοῦ. The δὲ τῇ of A, omitted by other MSS., has been corrected by some editors to δ' ἔτι. As the text stands, the μὲν is answered by δὲ after τῷ at B, 8.

[4] See P.-W., *Enclyc. svv. Diamant, Stahl.*

251

with gold. Adamant is the material of the shaft of the Spindle of Necessity in the myth of Er (*Rep.* 616c), symbolising, as Proclus says (*ad loc.*), 'the unalterable and unsubduable' (ἀδάμαστον). Later the name was transferred to the diamond, because of its extreme hardness. In the *Statesman* (303E) a simile drawn from the process of refining gold mentions the removal of earth and stone, leaving 'mixed with the gold its precious kindred that can only be removed by fire : copper and silver, and sometimes adamant'. Archer-Hind records the opinion of Professor W. J. Lewis that Plato's adamant was probably haematite. But it is also conjectured that the 'dark colour' (μελανθέν) may be due to some misapprehension, and that adamant here is the diamond, the *auri nodus* described by Pliny as a rare gem which was found in gold mines and had been supposed to be formed only in gold. It may have been imagined that the purest and most precious part of gold was condensed in the diamond. Hence 'adamant' was called 'the flower of gold' (ἀδάμας τοῦ χρυσοῦ τὸ ἄνθος, Pollux). It is quite likely that Plato had never seen a diamond, and, from the little he had heard of it, imagined it to be a metal.

The structure of copper is somewhat obscure. How can copper be denser than gold, and yet contain larger interstices and so be lighter ? The meaning may be that copper is denser 'in one way' (τῇ μέν), namely in so far as icosahedra of different sizes can be more closely packed, the small ones helping to fill the interstices between the larger. If this were the only difference between copper and gold, copper would be nearer to the liquid condition and so the softer, and it would also be the heavier. But it contains (apparently as a foreign body, not part of its constitution) an admixture of earth cubes. The cube being the body which is hardest to dislodge, the presence of earth makes copper actually harder than gold. Also there will be large interstices between the cubes and the icosahedra, which cannot be packed very closely ; hence copper is somewhat lighter than gold.

From this account of the metals, it does not appear that there is any bar to the transmutation of any metal into any other. The earth in copper is not, apparently, a constituent in its specific structure, since it can work its way to the surface without the copper ceasing to be copper. Thus all the metals consist solely of water icosahedra, and free transformation between grades of icosahedra should make any one convertible into any other. It would be interesting to know whether the alchemists were encouraged by this theory to attempt the transmutation of metals.

Solidification of fluids : Water, hail, ice, snow, hoar-frost.—We now turn to the liquid type, embracing the fluids. The next paragraph is parallel to the earlier account of the melting and cooling of metals : it describes the solidification, or freezing, of fluids, which are normally in the liquid state. Ordinary water is taken as the obvious example. It is described as containing an admixture of fire, not as part of its constitution but as a foreign body which maintains the fluid in a liquid condition.

59D. Water that is mixed with fire and is fine and liquid (it is called ' liquid ' because of its motion and its rolling course over the ground [1]), and also soft because its bases give way, being less stable than those of earth—water of this sort, when it is separated off from fire and air and left by itself, becomes more uniform, and at the same time is thrust

E. together upon itself by the action of the particles that are passing out of it.[2] Water so compacted, when it suffers this change to the extreme degree above the earth, is hail ; when on the ground, ice. Water that is less affected and still only half congealed is called ' snow ' above the earth, and when congealed from dew on the ground is known as ' hoarfrost '.

Water, in its normal condition as a liquid, consists of comparatively small icosahedra of several different grades, this non-uniformity making it mobile (58D). In the solid state, as ice, the icosahedra are more uniform. The change to the solid state is due to passing out of the fire particles, whose presence maintained it in the liquid state. It is also said to be left free of air. Nothing has been said of the presence of air in ordinary water. Possibly there is a reference to what happens to a small quantity of fire when surrounded and outnumbered by the larger particles of water. We were told that the fire pyramids might be shattered and reformed, first as air octahedra, and finally as icosahedra of water, or they might escape to their kindred (57B). It may be meant here that some of the fire particles are thus reformed as air before they finally escape.[3] However this may be, the water particles become more uniform, on the same principle as in the cooling of metals (59A), where the small icosahedra reformed as larger ones to fill the spaces left by the withdrawal of fire, under pressure from the thrust of the expanding air in the neighbourhood. The increase in

---

[1] Understanding (with Lindau) that ὑγρόν is connected with ὑπὲρ γῆς (or γῆν, Tr.) ῥέον. But the text is suspicious and barely defensible.

[2] The particles do not pass out into vacancy but (as before, 59A) expand the surrounding air, which thus exerts pressure on the liquid.

[3] Tr.'s tentative suggestion (p. 419) that some of the *water* particles might be broken up and reconstituted as air by the action of heat, thus forming steam, would fit a heating but not a cooling process.

size and uniformity results in the immobility characteristic of the solid state.[1]

Some varieties of the liquid type (juices).—Some illustrative examples of the liquid type are now given. These are the juices of plants, which are compounds of water particles of a large number of grades (as contrasted with the small grades which alone occur in ordinary water).

59E.   Mixtures composed of most of the grades of water are given the general name of juices, being filtered through the plants that grow out of the earth ; while their several differences

60.   are due to the variety of combinations.[2] A great number of the varieties they present are nameless ; but the four kinds which contain fire are specially conspicuous [3] and have received names. One is wine, which heats soul and body together ; next, the oily kind, which is smooth and divides the visual current and therefore appears bright and shining to the view and glistens : [4] resin, castor oil, olive oil itself, and all the rest that have the same property ; third,

B.   the kind that relaxes the contracted pores in the region of the mouth to their normal condition,[5] producing sweetness by this property, has received the general name of honey [6] ; last, a kind that dissolves the flesh by burning, a frothy substance distinct (?) from all the other juices, which is called ' acrid juice '.[7]

---

[1] 58E. Cf. 62B, where moisture is rendered immobile by compression which causes the particles of different grades to reform themselves in more uniform sizes.

[2] Taking δέ (after διά) as answering μέν (after σύμπαν). The two clauses σύμπαν μὲν . . . λεγόμενοι and διὰ δὲ . . . σχόντες are parallel, and the main sentence is resumed with τὰ μὲν ἄλλα κτλ. The participles λεγόμενοι and σχόντες are attracted to the gender of χυμοί.

[3] That διαφανής means ' conspicuous ' rather than ' transparent ' seems probable from the parallel at 67A : the varieties of smell are so indistinct as to be *nameless*; the only *clearly discernible* (διαφανῆ) distinction is between pleasant and unpleasant.

[4] Cf. 67E.

[5] The correct explanation is due to A.-H. So Tr. The property (δύναμις) resides in the object as its power of acting on the sense-organ. Sweetness is a quality of the honey produced by commerce with the sentient organ and existing only while sensation is taking place (*Theaet.* 159D). Sweetness as a quality of the sensation experienced is described at 66c, and attributed to the same causes.

[6] The ' honey ' found in flowers, rather than the stuff made by bees. The word also covers sweet gums exuded by certain trees, e.g. ἐλαιόμελι.

[7] ἐκ πάντων ἀφορισθὲν τῶν χυμῶν, ὀπὸς ἐπωνομάσθη. The meaning of ὀπός here is doubtful. Galen says that there are a very great number of ὀποί, since the word means the thick and sticky stuff that flows from an incision in any root

Varieties and compounds of earth: stone and earthenware; soda and salt; glass and wax.—The varieties of earth are illustrated by several types of formation. (*a*) Stone and earthenware come from earth which has lost the moisture it originally contained; and, when formed, they are insoluble by water. (*b*) There are less solid formations, like soda and salt, which remain soluble by water. (*c*) Some compounds of earth and water can be dissolved only by fire, not by water. Such are glass and wax.

60B.   Of the varieties of earth, that which has been strained through water becomes a stony substance in the following way. When the water mixed with it is broken up in the mixing, it changes into the form of air [1]; and when it has

C.   become air, it rushes up towards its own region. But there was no empty space surrounding it [2]; accordingly it gives a thrust to the neighbouring air. This air, being heavy, [3] when it is thrust and poured round the mass of earth, squeezes it hard and thrusts it together into the places from which the new-made air has been rising. Earth thrust together by air so as not to be soluble by water [4] forms stone, the finer being the transparent kind consisting of equal and homogeneous particles, the baser of the opposite sort.

   The kind that has been robbed of all moisture by the

D.   rapid action of fire, a formation more brittle than the other,

---

or stalk; but it is more specifically used of silphium juice (ὀπὸς κυρηναικός). Theophrastus uses ὀπός for plant-juice, and specially for silphium and for fig juice, used in curdling milk. In our passage it obviously means bitter juice, and is probably (like μέλι) the general name for a whole class.

ἀφορισθέν is difficult. (1) It seems manifestly untrue that a juice bitter enough to burn the tongue is *secreted* from all the other juices—honey, for instance. (2) It is not clear why this juice should be said to be *marked off* or *distinguished from among* (ἐκ) the whole number of juices. It may, however, be remembered that χυμός means both 'juice' and 'taste' or 'flavour' (65C). Does the phrase mean that these juices have a distinctive flavour, markedly different from any other?

[1] Here the transformation of one body into another is involved, as we suggested might be the case in the solidification of water (59D).

[2] Reading οὐ περιεῖχεν αὐτόν with A², A.-H. In the alternative ὑπερεῖχεν αὐτῶν, 'there was no empty space above them', the plural αὐτῶν is hard to explain.

[3] 'Heavy' in the popular sense, since it exerts a downward pressure (as well as in other directions) which reacts on the earth, as at 59A. This has no connection with the natural drift of every body to its proper region.

[4] ἀλύτως ὕδατι, 'insolubly by water'; Ar., *Meteor.* 383b, 10, τὰ μὲν ἄλυτα, τὰ δὲ λυτὰ ὑγρῷ; cf. ὕδατι οὐ λυτά, 60E. So A.-H., comparing λυτῶ πάλιν ὑφ' ὕδατος (60D). Precious stones and crystals, consisting of 'equal and homogeneous' (ὁμαλός here apparently means this, since 'equal' must mean 'of the same grade', 'uniform') particles, can hardly contain icosahedra of water as well as earth cubes.

60D. is what we have named 'earthenware';[1] but sometimes, when some moisture is left and the result is earth that is fusible by fire, the dark-coloured stuff produced when it cools is lava (?).[2]

There are, again, two kinds which are left in the same way when a great amount of water has departed from the mixture; but the particles of earth in their composition are finer and they have a saline taste. These become only half-solid and are soluble again by water. The one, which cleanses from grease and dirt, is soda[3]; the other product, which blends

E. agreeably in the combinations of flavour, is salt, a substance which, according to human convention, is pleasing to heaven.[4]

In contrast to substances like salt and soda, which are soluble in water, there are others, such as glass and wax, which will not dissolve in water, but can be melted into a liquid state by fire only. These are compounds of earth and water, and the reason (given in the third paragraph) why only fire can dissolve them is that they are so closely compacted that water particles from outside cannot gain admittance, whereas the smaller fire particles can. This explanation is preceded by an account of the conditions under which homogeneous masses consisting only of (1) earth, or (2) water, or (3) air, can or cannot be dissolved.

60E. The compounds of both (earth and water) which are soluble by fire, but not by water, are compacted in that manner for the following reason.

(1) Masses of earth are not dissolved by fire or air, because

---

[1] The γῆ κεραμίς of *Critias* 111D is the natural potter's clay, impervious to water. But 'brittle' seems to show that Plato here means earthenware or pottery. Ar., *Meteor.* 383a, 20, gives κέραμος ὀπτώμενος as an example of a soft, but not liquid, substance, that does not thicken but solidifies, when the moisture leaves it.

[2] *Meteor.* 383b, 9, mentions a group of substances solidified by dry heat, namely (1) κέραμος and 'some kinds of stone (λίθων) that are formed out of earth burnt up by fire, such as millstones' (οἷον οἱ μυλίαι); these are insoluble by liquid; (2) natron (νίτρον) and salt, which are soluble by liquid. If νίτρον (FY), the variant for λίτρον (A), is read in the next paragraph below (D, 8), this group corresponds to Plato's. This suggests that we should read λίθος ⟨μυλίας⟩, which is mentioned as a specially hard kind of stone, *Hipp. Maj.* 392D. In Aristotle 'millstone' apparently means lava. He says it can be melted (by heat) and become fluid, and that when the 'fluid mass begins to solidify (by cooling) it is black, and it comes to be like lime (τίτανος)'. Another possibility is λίθος ⟨λιπαραῖος⟩; see A.-H. on our passage.

[3] λίτρον A, νίτρον FY.

[4] Reading κατ' ἀνθρώπων νόμον with Bernadakis, on Plut., *Qu. Conv.* 984F, Πλάτωνος δὲ τῶν ἁλῶν σῶμα κατὰ νόμον ἀνθρώπων θεοφιλέστατον εἶναι φάσκοντος. (Cf. Tr., p. 426.)

60E.   their particles are smaller than the interstices in the texture of earth, so that, having plenty of room to pass through without forcing their way, they do not loosen the earth but leave it undissolved ; whereas the particles of water, being larger, make their passage through by force and so loosen the earth and dissolve it. Earth, then, is dissolved in

61.   this way by water only, when it is not forcibly compressed ; but if it is so compressed only fire can dissolve it, for no entrance is left for anything but fire. (2) Water, again, when most forcibly compressed is dispersed by fire only ; though when the consistency is weaker, both fire and air disperse it, air by its interstices, fire actually breaking it down into its triangles ; while (3) air forcibly compressed cannot be resolved by anything save into its elements, and when not so compressed, is dissolved only by fire.

The upshot of this parenthesis on the dissolving of earth, water, and air, in so far as it concerns the class of substances under consideration (such as glass and wax), which are compounds of earth and water, is as follows. It has to be explained why these compounds can be dissolved by fire, but not by water. Both the earth and the water contained in them are highly compressed, and we have seen that in that state of compression earth and water alike yield no entrance to the larger bodies ; only fire is small enough to find a way in between the particles. Hence water from outside can only flow harmlessly over the surface of glass or wax, while fire can enter the interstices between the water particles in the compound and melt it, in the same way that water can dissolve earth that is not too closely packed.

61.   So in these bodies compounded of earth and water, so long

B.   as water occupies the interstices of the earth in such a body, though these may be forcibly compressed, the particles of water assailing it from outside can find no entrance, so that, flowing round the whole mass, they leave it undissolved ; whereas the particles of fire make their way into the interstices between the water particles and, acting upon them—fire upon water—in the same way that water acts upon earth,[1] are

---

[1] Reading τοῦτο πῦρ ὕδωρ with Cook-Wilson and Tr., as symmetry demands. ἀέρα can be explained as a dittography of the ἀπερ- which follows, displacing ὕδωρ. Rivaud retains ἀέρα at the cost of mistranslation : ' elles agissent sur l'eau comme l'eau sur la terre ou le feu sur l'air.' ἀπεργάζεσθαι with double accus. meaning ' to do something to something ' occurs at Charm. 173A. Water acts on loosely compacted earth by forcing its way through and so driving apart the earth cubes (60E, 7).

61B.   the only agents that can cause the compound body to be
dissolved and set flowing. Of these compounds, some
contain less water than earth, namely all kinds of glass
   c.   and any varieties of stone that are called fusible; others
contain more water, namely all the substances with a consistency like that of wax or incense.

61C–64A.   *Tactile qualities, as they appear to sensation and perception*

Starting from space, the ultimate factor reached in the analysis
of the bodily, Plato has now built up the primary bodies, illustrated
the possibilities of transformation within or between their various
grades, and finally reached the recognisable and nameable compound
substances found in Nature. A few of these have been described
in some detail, so as to illustrate the sort of way in which the
theory could explain their observed properties, although their
structure is beyond the range of observation and the whole amounts
to no more than a likely account.

Thus, from the fresh starting-point of the second Part, we have
returned, as it were from below, to the point reached from above
at the end of the first Part—the senses and their external objects.
So far these objects have been considered as they are supposed to
exist in their own right, independently of the effects which they
produce in commerce with sentient organs. But, as Plato now
remarks, since our knowledge of their existence is due entirely to
sensation and perception, their properties could not be mentioned
save in terms implying our perception and so anticipating the
account, which has not yet been given, of the organs of sense and
the sentient part of the soul. The next section deals with certain
qualities common to all external bodies and perceived by the sense
of touch, which is diffused over all the fleshy parts of the living
organism: hot and cold, hard and soft, heavy and light, smooth
and rough. Some attempt is made to connect the character of
our sensations with the supposed structure of the external objects.

61C.   We have now, perhaps, sufficiently illustrated the varieties
due to diversity of shapes, combinations, and transformations
of one body into another. Next we must try to make clear
how it is that they come to have their qualities.
First, then, our account at every point must assume the
existence of sensation; but we have not yet described the
formation of flesh and all that belongs to flesh, or the mortal
part of soul.[1] Yet no adequate account can be given of

----

[1] It has been stated (42A) that the implanting of the immortal part in a
body entails, of necessity, the faculty of ' sensation (αἴσθησις) arising from
violent affections ', pleasure and pain, desire and passion. But the only

61D. these apart from all those qualities that are connected with sensation, nor yet of the latter apart from the former; and to treat of both together is hardly possible. We must, then, first assume one side, and afterwards turn back to examine what we have assumed. So, in order that our account may proceed from the kinds of body to their qualities, let us take for granted what is involved in the existence of body and soul.[1]

'Qualities (affections) connected with sensation' (τὰ παθήματα ὅσα αἰσθητικά) are distinguished from those properties which bodies are supposed to possess in the absence of any sentient being, such as the shapes of the microscopic particles, which are never perceived. We know from the *Theaetetus* that external objects cannot properly be said to be white, or hot, or sweet in themselves. Before they come within range of some sentient organ they possess only 'the power of acting or being acted upon', and the same is true of the organ itself. Only when organ and object are in commerce with one another is the organ affected in such a way that (if the change penetrates through the body and reaches the soul) we have the sensation we call 'seeing white' or 'feeling hot'. And it is only while this situation endures that the object *becomes* white or hot *for* the percipient. Thus the 'affection' of the organ and the 'affection' of the object occur (to use Plato's phrase) as 'a pair of twins', born of the marriage between the two. It is with such affections that the present section is concerned, and, among them, only with those belonging to the generally diffused sense of touch. It may be noted that here, where we are approaching this subject as it were from below, we proceed from the lowest of the senses, touch, through taste and smell, to the highest, hearing and sight.

61D. First, then, let us see how it is that we call fire 'hot'. We may study this question by observing the rending and cutting effect of fire upon our bodies. We are all aware that the

E. sensation is a piercing one; and we may infer the fineness of the edges, the sharpness of the angles, the smallness of the particles, and the swiftness of the movement, all of which properties make fire energetic and trenchant, cleaving and piercing whatever it encounters. When we recall the forma-

62. tion of its figure we see that this substance, more than any

___

senses so far dealt with in any detail were sight and hearing (45B ff.), not touch, and nothing has yet been said about the differences of quality as perceived in colours and sounds.

[1] The mortal parts of the soul and the main bodily organs are reserved for the third part of the discourse, from 69A onwards.

62. other, penetrates the body and divides it minutely and naturally gives the affection we call ' hot ' its quality and its name.[1]

The opposite quality is obvious enough, but it shall not go without an explanation. The particles of fluids in the neighbourhood of the body, when they enter it, thrust out the particles smaller than themselves,[2] and not being able to insert themselves into their places they compress the moisture in us and solidify it by reducing what was not

B. uniform and was therefore in motion to immobility, resulting from uniformity and compression. But a thing that is unnaturally contracted struggles, pushing itself apart again into its normal state. This struggling and shaking is called trembling and shivering ; and the name ' cold ' is given to this affection as a whole and to the agent producing it.

' Hard ' is applied to anything to which our flesh yields, ' soft ', to anything that yields to flesh ; and hard and soft things are also so called with reference to one another. A thing is yielding when it has a small base ; the figure com-

C. posed of square faces, having a firm standing, is most stubborn ; so too is anything that is specially resistant because it is contracted to the greatest density.

It is probable that Plato is here improving on Democritus, whose account of sensations presents parallel attempts to connect the character of a sensation with the shapes of the atoms in the body which yields it.[3] This line of explanation was pursued further by the later atomists of ancient and modern times. Meyerson [4] quotes the following from Lémery (*Cours de Chymie*, 1675) :

' Comme on ne peut pas mieux expliquer la nature d'une chose aussi cachée que l'est celle d'un sel, qu'en attribuant aux parties qui le composent des figures qui correspondent à tous les effets qu'il produit, je dirai que l'acidité d'une liqueur consiste dans des parties de sel pointues, lesquelles sont en agitation ; et je ne

---

[1] As if θερμός were κερμός. But cf. *Crat.* 412D ff. δίκαιον = δια-ιον (τάχιστον, λεπτότατον) = πῦρ or τὸ θερμόν.

[2] I take these smaller particles to be fire and air (not a smaller grade of water-particles). Cf. the descriptions of cooling by loss of fire-particles (59A), and of freezing by isolation of water-particles from fire and air and consequent increase of uniformity (59D, E). There seems to be no reason why larger water-particles from outside should expel smaller water-particles already present in the body ; and we do not in fact sweat when we are getting cold. The moisture in us is not expelled but compressed.

[3] Theophr., *de sens.* 60 ff., compares and contrasts Democritus' treatment with Plato's.

[4] *De l'Explication dans la science* i, 285.

crois pas que l'on me conteste que l'acide n'ait des pointes, puisque toutes les expériences le montrent ;. il ne faut que le goûter pour tomber dans ce sentiment : car il fait des picotements sur la langue semblables ou fort approchants de ceux que l'on recevrait de quelque matière taillée en pointes très fines ', etc.

Theophrastus, however, emphasises, as a point of difference between Democritus and Plato, that Democritus reduced all *sensa* (αἰσθητά) to ' affections of the sense which undergoes alteration ', whereas Plato did not deprive them of their independent reality (φύσις, cf. 61, καθ' αὑτὰ ποιῶν ταῖς οὐσίαις). Plato does, in fact, say in the *Theaetetus* that the physical object which causes my sensation of whiteness (hotness, etc.), although it is not white when no one is perceiving it, is ' saturated with whiteness ' when it is perceived, and ' becomes a white thing ' (156E). He thus speaks as if the object acquired, for so long as it is perceived, a quality which it does not possess at other times. The particles of fire have sharp points and fine edges at all times. These characteristics, consequently, cannot constitute its hotness ; and it is hard to see how the hotness can have ' independent reality ', unless this means that it is an object of which I am aware in perception rather than a change that occurs in the sense organ or a quality of my sensation.[1] As we have seen, however, Plato differs from Demo-

---

[1] Tr. (p. 430 ff.) holds that Plato is speaking of perceptible qualities that are not dependent on the mind of a percipient. Neither our bodies nor our minds play any part in making them. The objective fact of which we are directly aware in the sensation itself is ' that our flesh is being lacerated or pierced by the " hot " body—as wholly " objective " a process as the cutting of a loaf by a knife '. ' The πάθημα which we call θερμόν is thus as strictly a πάθος of the particles of fire as the penetrating of a loaf is a πάθος of a bread-knife. In both cases, to explain fully the πάθημα of the knife or of the fire, we have to take into account what it does to a second body. The second body is not necessarily my own. The fire divides a log of wood which it sets on fire exactly in the same way and for the same reason that it divides my flesh. Only it is due to the fact that I have a body and that my sensations are aroused in connexion with it that heat is a directly " sensed " object. If I had no body, or none which the tetrahedra of fire could penetrate, fire would still be θερμόν, but its heat would no more be directly revealed to me by sense than the distinctive character of a magnetized iron bar is. It would be in the coals, but it would be something inferred, not felt.'

This account seems to be incompatible with the *Theaetetus*, which asserts that the sense-organ does play a part in making the hotness we perceive, and that fire is not hot save when perceived. What resides at other times in the coals is not hotness, but the power of producing sensation and hotness in conjunction with the organ. When the fire pierces and lacerates the insensitive log there is, strictly, no hotness in the fire or in the wood.

I am assuming that the theory of sensation in *Theaetetus* 155D ff. is Plato's own (see *Plato's Theory of Knowledge*, 49) ; and Tr. admits that ' from Plato's

critus in that his particles are not impenetrably solid lumps, but contain 'motions (changes) and powers'. These may be called independently existing qualities, which can cause in our souls sensations such as feeling hot.

Hard and soft seem to have less claim than hot and cold to be qualities distinct from the characteristics possessed by bodies at all times, whether perceived or not. The next pair, heavy and light, have as little.[1] These two pairs of opposites are the two which Democritus connected with the inherent properties of his atoms and did not treat as mere 'affections of the sense'.[2] Plato himself remarks that bodies are called hard or soft 'with reference to one another', apart from any sentient organ.

62C.   'Heavy' and 'light' may be most clearly explained by examining them together with the expressions 'above' and 'below'. It is entirely wrong to suppose that there are by nature two opposite regions dividing the universe between them, one 'below', towards which all things sink that have bodily bulk, the other 'above', towards which everything is reluctant to rise. For since the whole heaven is spherical

D.   in shape, all the points which are extreme in virtue of being equally distant from the centre, must be extremities in just the same manner; while the centre, being distant by the same measure from all the extremes, must be regarded as at the point 'opposite' to them all. Such being the nature of the ordered world, which of the points mentioned could one call either 'above' or 'below' without being justly censured for using a quite unsuitable term? The central region in it does not deserve to be described as being, in its nature, either above or below, but simply at the centre; while the circumference is not, of course, central, nor is there any difference, distinguishing one part of it from another with reference to the centre, which does not belong equally to some part on the opposite side.[3]

own point of view, the theory would be perfectly acceptable as an account of "pure" sensation' (*Plato, the Man and His Work*, 1926, p. 330). But, on Tr.'s peculiar theory, the doctrine here belongs to Timaeus and may not be acceptable to Plato.

[1] Timaeus Locrus (100E) separates this pair from the others, remarking that, 'though they are distinguished by touch, reason defines them by the inclination towards, or away from, the centre', or 'by their tendency to move towards their own region' (ῥοπᾷ ποτὶ τὰν χώραν).

[2] Theophr., *de sens.* 61–62, 68.

[3] Taking ἤ with μᾶλλον. No part has the property of 'being above (or below) the centre', or has any better right to that description than a point on the opposite side. This is the counterpart of the statement above, that

62D.   When a thing [1] is uniform in every direction, what pair of contrary terms can be applied to it and in what sense could they be properly used ? If we further suppose that there is a solid body poised at the centre of it all, this body

63.   will not move towards any of the points on the extremity, because in every direction they are all alike ; rather, if a man were actually to walk round and round that body, he would repeatedly stand at his own antipodes and call the same point on its surface ' above ' and ' below '.[2] For the whole being spherical, as we said just now, there is no sense in speaking of one region as above, another below.

As to the source of these terms and the things to which they really apply and which have occasioned our habit of using the words to describe a division of the universe as a

B.   whole, we may arrive at an agreement, if we make the following supposition. Imagine a man in that region of the universe which is specially allotted to fire, taking his stand on the main mass towards which fire moves, and suppose it possible for him to detach portions of fire and weigh them in the scales of a balance. When he lifts the beam and

C.   forcibly drags the fire into the alien air, clearly he will get the smaller portion to yield to force more readily than the greater ; for when two masses at once are raised aloft by the same power, the lesser must follow the constraint more readily than the greater, which will make more resistance ; and so the large mass will be said to be ' heavy ' and to tend ' downwards ', the small to be ' light ' and to tend ' upwards '. Now this is just what we ought to detect ourselves doing

the centre cannot be called ' above ' the ' lower ' hemisphere or ' below ' the ' upper ' hemisphere. The next sentence asks to what these contrasted terms, above and below, can be applied, if neither the centre nor any part of the circumference exhibits any corresponding difference of nature.

[1] This paragraph is in general terms, referring to *any* spherical figure, at the centre of which is a solid body. It *applies* to the actual universe, because this has a solid body at its centre, viz. the Earth.

[2] The connection between the two parts of this sentence becomes clear if we take the first part to mean that there is no reason why the central body should *fall down* in any direction, because there is no ' down ' for it to fall towards. On the contrary, the supposed traveller will be using ' above ' and ' below ' with reference to every direction in succession, since at any moment he will think he is ' on the top ' of the body which is ' beneath him '. Neither word, accordingly, stands for any inherent difference between the parts of the central body or of the universe as a whole. Paraphrasing this passage, Aristotle (*de caelo*, 308a, 20) assumes as a matter of course that the body round which the traveller walks is the earth, which actually occupies the centre of Plato's world.

63c.   here in our own region. Standing on the Earth, when we are trying to distinguish [1] between earthy substances or sometimes pure earth, we are dragging the two things [2] into the alien air by violence and against their nature ; both

D.   cling to their own kind, but the smaller yields more readily to our constraint than the larger and follows it more quickly into the alien element. Accordingly we have come to call it ' light ' and the region into which we force it ' above ' ; when the thing behaves in the opposite way, we speak of ' heavy ' and ' below '. Consequently, the relation of these things to one another must vary, because the main masses of the kinds occupy regions opposite to one another : what is ' light ' or ' heavy ' or ' above ' or ' below ' in one region will all be found to become, or be, contrary to what is ' light '

E.   or ' heavy ' or ' above ' or ' below ' in the opposite region, or to be inclined at an angle, with every possible difference of direction.[3] The one thing to be observed in all cases, however, is that it is the travelling of each kind towards its kindred that makes the moving thing ' heavy ' and the region to which it moves ' below ', while the contrary names are given to their opposites. So much for the explanation of these affections.

    As for the qualities ' smooth ' and ' rough ', anyone, I suppose, could see how they are to be explained. Roughness is due to a combination of hardness and unevenness, smooth-

64.   ness to evenness combined with closeness of texture.

Plato's account of ' heaviness ' and ' lightness ' is based on the axiomatic principle, already many times invoked, that bodies of like nature have a natural tendency to come together, fire to fire, earth to earth, and so on. He also assumes that the strength of this tendency varies with the size of the mass of fire (etc.) in question ; the smaller of two similar masses makes the less resistance to the violence which would force it away from its kind. Given

---

[1] διστάμενοι. Fraccaroli and Tr. translate ' weigh ', but produce no proof that the middle ἵστασθαι ever has this sense. At *Rep.* 36E, the word means ' set in contrast ': διαστησώμεθα τόν τε δικαιότατον καὶ τὸν ἀδικώτατον . . . τίς οὖν ἡ διάστασις. Here I take διστάμενοι to mean ' trying to distinguish ' which of two lumps of an earthy substance is the heavier by comparing or contrasting their behaviour when weighed.

[2] ἀμφότερα. The two portions weighed against one another in the hands or in the two scales of a balance.

[3] Thus, Earth being at the centre and fire all round the circumference, so that ' the main masses occupy opposite regions ', the line along which a stone falls to earth or fire rises will be in a different direction for every point on the Earth's surface.

these objective properties of all bodies, he can explain why one mass *feels* heavier (or lighter) than another, and how it will depend on the situation of the observer whether it feels lighter or heavier.

He begins by dismissing the popular notion that the universe is divided into two regions, ' above ' and ' below ' or ' up ' and ' down ', and that all bodies move of themselves downwards, and can only be made to move upwards by force. For this he substitutes his own picture of the finite spherical universe (so ordered by Reason), where the opposition is between the centre and the circumference. All points on the circumference are equally ' opposite to ' the centre, and neither the centre nor any part of the circumference can properly be described as ' above ' or ' below '. How then, he asks, have we come to use these terms as we do ?

His answer involves the further doctrine, previously stated, that the main masses of the four primary bodies are situated in regions proper to them. We have learnt that this is a consequence of the natural tendency of like towards like, operating within the finite spherical shape imposed by Reason on the chaos of motions in the Recipient (58A–C). It has all through been implied, as an obvious fact, that in the ordered world the four bodies are arranged, not merely like with like, but in a definite order : fire round the circumference (where it is the chief constituent of the stars' bodies), next the spheres of air and water, and earth at the centre. It seems to me impossible to doubt this ; and it follows that the Earth in Plato's system must be (as Aristotle says) ' situated at the centre '. I cannot conceive that the proper region of the main masses of earth and water could be the point occupied at any given moment by a planetary Earth revolving at some distance from the centre ; while the rest of the central region would be occupied presumably by air ; for, as we have seen (pp. 124 ff.), there is no hint of any Central Fire. If Plato had meant this, he could not have failed to state clearly a view so foreign to his readers' natural assumptions. If we ask why the four bodies are arranged in this order rather than in any other, the answer may perhaps be found in the statement (56B) that, of the three bodies formed of the same elementary triangle, fire is the most mobile and ' the lightest, as being composed of the smallest number of similar parts ', air stands next, and then water. Obviously, the mere principle that like things come together will not by itself account for the arrangement actually observed. Plato may intend to account for it by supposing that, since the chaotic motions of the primary bodies are confined within the circular revolution due to the World-Soul, the more nimble and mobile bodies would tend to be thrust outwards, nearer to the surface, i.e. the circumference. If so, the actual arrangement could be connected with their intrinsic properties and structure. But this point is left in some obscurity.

This is not the same thing as to say that fire is 'absolutely light' or earth 'absolutely heavy'. Moreover, if the transition from chaos to cosmos never actually occurred and the four main masses have always occupied their present concentric spheres, the behaviour of smaller masses can be accounted for simply by the overpowering attraction exercised by the main mass in the region it actually occupies. The smaller mass will move towards the larger, not vice versa, and any part of the main mass will resist an attempt to tear it away into an alien region. We are here concerned with *sensible* qualities. The reason why a stone *feels* heavy lies in this resistance. Fire would feel heavy to a man standing on the inner surface of the main mass of fire and trying to lift a portion of fire into the air. In this way we may think of 'heaviness' as analogous to colour. A body has strictly no colour save when some eye is seeing it ; there is in the body itself only the 'power' to give rise to a perception of colour in co-operation with a sentient organ. Similarly, a body has intrinsically only the tendency to move towards its like ; by calling it more or less 'heavy' we may mean only the consequent resistance that we experience when we contribute, on our side, the effort that is resisted. In this sense, 'heaviness' is the name of an 'affection' that we feel, rather than of any property independently existing in the bodies outside.

64A–65B. *Pleasure and Pain*

So far, the only sensible qualities considered are those which are perceived by the sense of touch, diffused all over the fleshy parts of the body. The next paragraph deals with the pleasurable or painful character of the affections produced in the subject. We are still concerned with 'common affections of the body as a whole'. There are, in the first place, the motions set up in the particles composing various organs of the body. When these motions penetrate to the consciousness, sensation follows in the soul ; but they may die away and be lost before the consciousness is reached. Finally, sensation may or may not be attended by pleasure or pain.

64A. Concerning the affections common to the body as a whole the most important point that remains to be considered is the explanation of the element of pleasantness or painfulness in those which we have just discussed ; and further all those affections which, having attained to sensation through the

64A.   organs of the body, may be also accompanied by inherent
pains or pleasures.[1]

Now in seeking the explanation of any affection, whether
perceptible or imperceptible, we must begin [2] by recalling
the distinction drawn earlier between what is mobile in

B.   structure and what is immobile; all the explanations we
are bent upon discovering are to be sought along this line.
When something that is naturally mobile is invaded by even
a slight affection, it spreads it all round, one particle passing
on the same effect to another, until they reach the consciousness and report the quality of the agent. The immobile,
on the other hand, being too stable to spread the motion
round, merely suffers the affection without setting any of

C.   its neighbours in motion; accordingly, since the particles
do not pass it on one to another, the original affection remains
in them incapable of transmission to the living creature as
a whole and leaves the subject without sensation. This is
the case with bone and hair and all the other parts in our
bodies that are composed chiefly of earth; whereas the
previously mentioned conditions apply to sight and hearing
above all, because in them fire and air play the largest part.

The nature of pleasure and pain, then, must be conceived

D.   as follows. An affection which violently disturbs the normal
state, if it happens all of a sudden, is painful, while the
sudden restoration of the normal state is pleasant; these
are perceptible, whereas a gentle and gradual change of
either sort is imperceptible.

Any process, however, that takes place with great facility
yields perceptions [3] in the highest degree, but is not attended
by pain or pleasure. Such are the affections that occur in
the visual ray itself, which was, in fact, described earlier as
a body formed in the daylight in intimate connection with
our own.[4] No pain is set up by cuts or burns in this ray

---

[1] In this sentence the first part refers to the 'affections' above discussed,
viz. qualities of objects as perceived, and what is meant by calling these
pleasant or painful (capable of causing pleasure or pain to a sentient being).
The second half refers to 'affections' occurring within the body and transmitted through the organs to the soul, where they 'acquire' sensation with
(or without) pleasure or pain.

[2] ὧδε is explained by ἀναμιμνῃσκόμενοι: 'in the following way, namely
by recalling . . .' Cf. 61D, ὧδε σκοποῦντες . . . ἐννοηθέντες.

[3] Literally 'is perceptible', but the perception in the following instance
of vision is perception of colour, not of the disturbance, which yields no
sensation at all, either pleasant or painful.

[4] συμφυὲς ἡμῶν. At 45D, συμφυὲς τῷ ἀέρι meant 'coalescing with the air'.
The genitive ἡμῶν is supported by the analogy of συγγενής and σύμφυτος (A.-H.).

64D.   or by anything else that is done to it, nor yet pleasure when
   E.   it returns to its former condition, although there are intense
         and very distinct perceptions, according as it is acted upon
         and itself meets and touches any object ; for no violence
         whatsoever is involved when the ray is severed and comes
         together again.[1] On the other hand, organs consisting of
         larger particles, which yield to the agent reluctantly and
         pass on the motions to the whole, have pleasures and pains—
         pains while they are being ousted from their normal state,
   65.   pleasures while this is being restored. Those in which the
         departure from the normal state [2] or depletion is gradual,
         while the replenishment is sudden and on a large scale, are
         sensible of the replenishment, but not of the depletion, and
         so afford to the mortal part of the soul [3] intense pleasures,
         but no pain. This is plain in the case of sweet smells.
         Where the disturbance of the normal state is sudden, and
         the restoration gradual and difficult, the opposite results are
   B.   produced ; as may be observed in the case of cuts or burns
         in the body.

Plato here connects his own doctrine of bodily pleasures and
pains, most fully set forth in the *Philebus*, with his theory of the
particles, whose shapes make them comparatively easy or difficult
to dislodge. Sensation of any kind occurs only in the soul, as a
result of changes or movements transmitted through the bodily
organs from the objects outside. In perception, the active quality
($\delta\acute{v}\nu\alpha\mu\iota\varsigma$) of the object is thus finally ' reported ' to the consciousness : we see a colour, hear a sound, and so on. The first point
is that the organs and external media in the case of sight and
hearing consist of specially mobile particles (fire and air), and
consequently the qualities are reported with exceptional intensity

[1] I understand (with Tr.) $\delta\iota\acute{\alpha}\kappa\rho\iota\sigma\iota\varsigma$ and $\sigma\acute{v}\gamma\kappa\rho\iota\sigma\iota\varsigma$ to mean the dislocation
of particles by cuts, burns, etc., and their return to their normal condition.
This is not felt by us because it is so easily effected that no ' violence ' is
required on the part of the agent.

[2] $\dot{\alpha}\pi o\chi\omega\rho\acute{\eta}\sigma\epsilon\iota\varsigma$ $\dot{\epsilon}\alpha\upsilon\tau\tilde{\omega}\nu$ ' departures from themselves '. This phrase is simply
a variant for the $\dot{\alpha}\pi\alpha\lambda\lambda o\tau\rho\iota o\tilde{v}\sigma\theta\alpha\iota$ of the next sentence. $\dot{\epsilon}\alpha\upsilon\tau\tilde{\omega}\nu$ would be
superfluous if $\dot{\alpha}\pi o\chi\acute{\omega}\rho\eta\sigma\iota\varsigma$ meant ' wasting '. $\kappa\acute{\epsilon}\nu\omega\sigma\iota\varsigma$ is such wasting as occurs,
for example, in hunger. Neither word means here the evacuation of un-
assimilated food, which follows on eating and is not associated with any
possible pain of want.

[3] The addition of the lower faculties (by implication ' mortal ') to the
' immortal principle ' has been mentioned at 42A. The ' mortal part of the
soul ' is mentioned where that passage is recapitulated at 69c, and indeed
the expression has already been used at 61c. Tr.'s note here is therefore
irrelevant.

and clearness, little being lost by friction on the way. The most earthy parts of the body, such as bones and hair, absorb the shock, and the motion dies away in them before it reaches the soul. Hence no sensation or perception results.

Pleasure or pain may or may not attend on sensation or perception, when it does occur. Pain is due to a sudden and violent disturbance of the normal state. The nature of the disturbance is not specified, but it seems to be implied that it is a dislocation, and possibly a transformation, of the particles composing the organ. Pleasure is due to the sudden restoration. If either process is sufficiently gentle and gradual, no sensation occurs and consequently neither pleasure nor pain. In the *Philebus* the theory provides the basis for the distinction between the 'pure' pleasures and the mixed, namely those which are preceded or accompanied by pains of want. The pleasures of smell, for example, are pure. As Archer-Hind remarks, Plato 'seems to regard sweet odours as the natural nutriment of the nostrils, which suffer waste when those are absent; but the depletion is so imperceptible that it is only by a sudden restoration of the natural state that we become conscious that there has been any lack'.

An apparent exception to the rule that violent disturbances cause pain is offered by the visual ray, regarded as an extension of the organ of sight. When we look at a candle-flame or pass a knife before our eyes, why do we not feel pain from the burn or the cut inflicted on the ray? This has to be explained by the extreme fineness and mobility of the fire particles composing the ray. These, it seems, yield so readily that no 'violence' is called for on the part of the disturbing agent. So the ray yields no pleasant or painful sensation, although the perceptions of its proper objects are exceptionally intense and distinct.

65B–66C. *Tastes*

From the general account of tactual sensibility and of pleasure and pain we pass to sensations transmitted through special sense-organs: tastes, smells, sounds, colours. In each of these classes we distinguish a number of main groups by names such as (in the case of tastes) 'bitter', 'pungent', 'sour', 'sweet'. These names roughly indicate the quality of the sensations we actually experience. The theory now attempts to connect the felt quality of a given class of sensations with the physical process supposed to occur in the sense-organ, which is itself to be explained by the inherent qualities of the external objects, connected with their structure as described earlier.

Knowing nothing of the nerves, Plato supposes that the tongue possesses diminutive passages (' veins '). These are said to extend to the heart; but since nothing is said about their containing blood, they may be not blood-vessels, but very fine tubes conveying the liquid or liquefied substances we taste into the veins proper. The bulk of the nourishment we take travels down the gullet into the belly and is there digested and passed on into the blood-stream, which then feeds all parts of the body. But very small samples of it make their way directly into the blood-stream through these fine tubes, in which they are ' tested ' by the tongue and the sense of taste, so that we may be warned against swallowing unwholesome substances. Plato describes only the behaviour of various substances in the tubes, which gives rise to sensations of sourness, pungency, etc., as soon as it is reported to the central seat of sensation. Presumably the disturbances are transmitted through the flesh of the tongue in the same way as in the case of touch. It is not implied that the message has to pass through the heart to reach the brain.

65B.   Some account has now been given of the common affections of the body as a whole and of the names bestowed on the agents that produce them; we have next to explain, if we can, the affections that occur in special organs of our bodies and, on the other side, how they are caused by the agents concerned.

C.   First, then, we must make clear to the best of our power what we omitted earlier in speaking of flavours,[1] namely the affections peculiar to the tongue. These, like most of the others indeed, appear to be due to contractions and dilations of some sort; and further they have more to do than any of the rest with degrees of roughness and smoothness. When earth particles,[2] making their way in at the small veins which serve the tongue as a sort of testing-instrument and extend to the heart, come into contact with the moist and

D.   soft flesh, as they are melted down they contract and dry up

---

[1] χυμός means (1) juice, (2) flavour (residing in a juice), (3) taste (as a sensation). Some references were made to the characteristic flavours of the juices (60A, B) and of the varieties of earth compounds (60E); but nothing was said about the corresponding processes set up in the tongue.

[2] Taking γήϊνα μέρη as subject (with A.-H., Tr., Fr.), not with κατατηκόμενα (Rivaud, ' et y dissolvent les parties terreuses ') which is passive. The reference seems to be to compounds of earth, loosely enough compacted to be soluble by water (60E). The moisture from the flesh melts them down into a state liquid enough for flavour to be perceptible; for it is probable that (as A.-H. says) Plato holds with Aristotle (de anim. 422a, 17) that all taste is produced by substances in a liquid state, whether liquefied before or after entering the mouth.

65D. the veins. If comparatively rough, they are felt as 'astringent'; if their roughening effect is slighter, as 'harsh'.

Substances which rinse the small veins and cleanse the whole region of the tongue are called 'acrid', if they produce this effect in excess and attack the substance of the tongue to the point of dissolving some part of it; such is the

E. property of soda. Those which are less powerful than soda and rinse the tongue to a moderate degree are saline without acrid roughness and rather produce an agreeable sensation.

Others, which absorb the warmth of the mouth and are softened by it, becoming fiery [1] and in their turn scorching that which heated them, mount upwards by virtue of their lightness to the senses in the head, cleaving whatever they

66. encounter. On account of these properties all such substances are called 'pungent'.

Again, there are the particles [2] of substances reduced to a fine texture by decomposition before they make their way into the narrow veins—particles that are duly proportioned both to the earthy and to the airy particles which the veins contain, with the result that they set these in motion and cause them to be churned round one another, and, as they are being churned, to form an enclosure and, as particles of one sort find their way inside particles of a different sort, to produce hollow films stretched round those that pass into the inside. [3] Thus, when a hollow film of moisture, earthy

---

[1] There is no inconsistency, if the substances in question (which are not named) contain water or consist mainly of water, like the juices at 59E. The water particles can be transformed into fire. It has not (as Tr. alleges, p. 466) 'been assumed all along that things get their flavours from the earthy particles they contain (γήϊνα μέρη, 65D, 2)'. A.-H. instances the effect of mustard.

[2] Reading τὰ δὲ αὖ τῶν : τῶν δ' αὐτῶν libri. (Cf. ὅσα μὲν . . . τὰ δὲ . . . τὰ δὲ, beginning the three previous sentences.) But the neuter plural required by ἔχοντα might be found in τὸ τῶν προλελ = τὰ προλελ. (So Tr.) Unless the grammar is extremely irregular, καὶ before τοῖς ἐνοῦσιν must mean 'both'. The requirement seems to be that the entering particles of moisture shall have been reduced by decomposition to a grade fine enough to permit them to fill the interstices between both the earth cubes and the air octahedra in the passages. The nominative is left in suspense.

[3] The construction and meaning are here uncertain. This part of the sentence seems to describe the formation of a bubble by a stirring movement, producing a globular film of moisture enclosing air. περιπίπτειν can mean 'surround', or 'encounter' ('jostle against one another', A.-H.), but hardly 'change their positions'. The notion of jostling seems superfluous. The preceding περὶ ἄλληλα and the following compounds of περὶ suggest rather that περιπίπτειν means that some particles (water) take up a position round others (air), form an enclosure (περίβολος : περιπίπτειν being a possible

66B.   or pure as the case may be, is stretched round air, they
       form, as moist vessels of air, hollow globes of water. Some,
       composed of pure moisture making a transparent enclosure,
       are called ' bubbles ' ; while, if the moisture is earthy and
       stirs and rises all together, we speak of frothing and fermenta-
       tion. What is responsible for these effects is called ' acid '.
         An affection opposite to all those which have just been
  c.   described is produced by an opposite cause. When the
       structure of the entering particles in liquids, being conform-
       able to the normal condition of the tongue, mollifies and
       smoothes the roughened parts and relaxes or contracts those
       which are unnaturally shrunken or dilated, and so thoroughly
       establishes the normal state, any such remedy for violent
       affections is always pleasant and agreeable, and has received
       the name ' sweet '.

66D–67A. Odours

Individual odours are as easy to recognise as tastes ; but Greek,
like English, has fewer adjectives (analogous to ' bitter ', ' sour ',
' acrid ', etc.) denoting the general character of a group of smells.
The epithets we apply, if not drawn from the names of substances
like spice and balm, mostly express only our likes and dislikes,
especially the latter. Plato connects this difficulty of classification
with the peculiar character of the particles composing odours.

66D.   So much for that matter. In the case of the faculty residing
       in the nostrils no definite types [1] are to be discerned. A

passive of περιβάλλειν). Cf. περιστῆναι , B, 4, and ἀνεμωθέντος καὶ συμπεριληφθέντος
ὑπὸ ὑγρότητος of the bubbles described at 83D. κυκᾶσθαι is important (Theophr.,
de sens. 84, summarises the entire sentence in τὰ δὲ κυκῶντα ὀξέα), but ambiguous : ' mingle them together ' (A.-H.), ' mescolare ' (Fracc.), ' s' émul-
sionner ' (Rivaud). I have supposed that περὶ ἄλληλα implies a circular
stirring, which would account for the globular form of the bubble. The
effect of this stirring is that the moisture (which may or may not be con-
taminated with the earthy particles it finds in the veins) falls into position
round the circumference (περιπίπτειν) and, as the airy particles (ἕτερα) find
their way inside the different (watery) particles, the action of the moisture
produces out of itself and the air the final result—hollow films (κοῖλα) of
moisture stretched round the air particles which get inside. I doubt whether
εἰς ἕτερα ἐνδυόμενα (cf. εἰς τὰς φλέβας ἐνδυομένων, 66A, 3) ἕτερα κοῖλα ἀπεργάζεσθαι
can mean ' taking up other positions to form new hollows ' (A.-H., Tr.).
The form of the phrase rather resembles 64B, τὸ μὲν γὰρ κατὰ φύσιν
εὐκίνητον . . . διαδίδωσιν (sc. τὸ πάθος) κύκλῳ μόρια ἕτερα ἑτέροις ταὐτὸν ἀπεργα-
ζόμενα, if μόρια there be taken (as by A.-H.) in epexegetic apposition to τὸ
εὐκίνητον.

  [1] εἴδη, definite varieties of smell, which could be classified by names corresponding to ' sour ', ' pungent ', ' bitter ', etc., in tastes. εἴδει in the next
line plainly means type of regular figure (pyramid, octahedron, etc.).

66D.   smell is always a half-formed thing, and no type of figure has the proportions necessary for having an odour. The veins of smell have a structure too narrow for earth and water and too wide for fire and air; hence no one has ever perceived any odour in any of these bodies; odours arise from substances in process of being liquefied or decomposed or dissolved or evaporated. They occur in the intermediate

E.   stage when water is changing into air or air into water. All odours are vapour or mist, mist being that which is on the way from air to water, vapour what is on the way from water to air; consequently, all odours are finer than water, grosser than air. Their nature is plainly seen when a man forcibly inhales the air through something that obstructs the passage of the breath: then no odour filters through with it; nothing comes but the air robbed of all scent.

67.   Accordingly, the diversities of odour fall into two sets. They lack names because they do not consist of a definite number of simple types.[1] The only clear distinction to be drawn here is twofold: the pleasant and the unpleasant. The unpleasant roughens and does violence to the whole cavity lying between the crown of the head and the navel; the pleasant soothes this region and restores it with contentment to its natural state.

This account of the particles which enter the passages of respiration and give rise to sensations of smell adds some unexpected features to the earlier classification of sensible bodies. Hitherto we have understood that these fall into two classes: the four simple bodies and bodies compounded of two or more of these. The simple bodies consist each of regular polyhedra of uniform pattern in different grades of size. We are now told that no grade of any simple body is so proportioned to the size of the passages as to be capable of causing sensations of smell—a statement which seems very improbable. Thus all grades of pure fire, air, water, and earth are ruled out. We might expect that smell would be caused by some compounds of the simple bodies, analogous to the complex juices which we can taste. But compounds are also excluded, for no clear reason. Smell is occasioned only by vapour and mist, and these are apparently composed of particles of a third class, formed in the transitions between air and water, finer than the icosahedra of water and grosser than the octahedra of air.

That particles of some shape suitably proportioned to the

---

[1] I understand πολλά to mean a definite number of species in which smells might be classified. So Tr., p. 473.

passages are intended, is clear from the whole context. ' It would seem then ', as Archer-Hind says,[1] ' as if Plato conceived matter in its passage from air to water, or from water to air, to be made up of irregular figures intermediate in size between the particles of air and those of water.' In the earlier description of these transformations (56c ff.), Plato spoke as if the triangles into which the figures of the simple bodies are broken down could wander about by themselves until they could reform into some regular solid. But at that stage the statement of the theory was by no means complete; we had not even learnt that there are many grades of size for each simple body. When this was added (57c), it became clear that the loose triangles might recombine in figures of the same pattern, but of larger or smaller grades, and we were told that this actually occurs in the case of water (58E). It may be that Plato is now adding a further possibility, which would remove the absurdity of supposing that triangular surfaces which are not the surfaces of any solid can stray about by themselves. Are we to understand that the triangles, on the way from one regular form to another, at every moment compose a series of irregular solids of intermediate size [2]—a third class of particles which are neither simple bodies nor compounds of simple bodies ? Between the fire-pyramid and the air-octahedron there is one such figure, the double pyramid with 6 faces. Between the octahedron and the icosahedron of water (the intermediate stages with which we are concerned) there is a whole series. If on any face of the octahedron we plant a tetrahedron, we obtain 3 new faces instead of 1 old one ; that is, we increase the number of faces by 2. We can therefore obtain irregular figures of 10, 12, 14, 16, 18 faces between the octahedron and the icosahedron. In vapour and mist we shall then have perceptible bodies composed of such irregular particles. The fact that they are irregular and are rapidly shifting from one shape to another would explain the indefinite character of odours, which prevents us from giving their varieties distinctive names.

It seems not unlikely that Plato is here disclosing a further feature of that ' longer account ' which he held in reserve (54B). Even so, the explanation of smell remains unconvincing. It is difficult to imagine that all the irregular intermediate particles can

[1] On 66E, inconsistently with A.-H.'s previous suggestion that ' the agent which excites smell is actually unformed matter '. Is there such a thing as unformed matter in Plato's system ?

[2] Cf. Tr., p. 471 : ' It is just then, when they are neither icosahedra nor octahedra but passing from one shape to the other by a series of intermediaries which are not " regular " polyhedra that they neither slip through the φλέβια without contact nor are too big to get into them at all.'

be so different in size from any of the grades of water and of air that they can fit passages which no grade of water or of air can affect.[1]

67A–C. *Sounds*

The section on sound is short and simple. The only distinctions mentioned—high and low, smooth and harsh, loud and soft—are connected with the motions of particles. Nothing is said about the differences of quality or timbre which we detect in our sensations.

67A.   Third among the organs of sensation we are considering is hearing ; and the affections occurring in this field must now

B.   be explained. Sound we may define in general terms as the stroke inflicted by air on the brain and blood through the ears and passed on to the soul ; while the motion it causes, starting in the head and ending in the region of the liver, is hearing. A rapid motion produces a high-pitched sound ; the slower the motion, the lower the pitch.[2] If the motion is regular, the sound is uniform and smooth ; if irregular, the sound is harsh. According as the movement is on a

C.   large or a small scale, the sound is loud or soft. Consonance of sounds must be reserved for a later part of our discourse.[3]

The stroke transmitted through the air is said to be inflicted, not on the ear-drum, whose very existence is ignored, but upon the brain and blood, through the ears. Professor Onians[4] points out that in Homer and the earliest writers after Homer there is evidence for the belief that the breath of which sound consists ' passes through the ears not to the brain but to the lungs. This, though it may seem foolish to us, is in fact a natural interpretation of the anatomy of the head, which shows an air passage direct from the outer air through the ear to the pharynx and so to the lungs. Aristotle remarks that the ear " has not a passage ($\pi \acute{o} \varrho o \varsigma$) to the brain, but has to the roof of the mouth ". The passage is divided by the tympanum, its lower portion being known as the Eustachian tube. The sense of smell working by the indrawing

---

[1] Galen (*Hippoc. et Plat.*, pp. 625 ff., Müller) suggests that the four other senses correspond to the four simple bodies : sight to fire, hearing to air, taste to water, touch to earth. Since there is no fifth simple body, smell is provided with something between air and water.

[2] Cf. Archytas, *Vors.* 35B, 1, τὰ μὲν οὖν ποτιπίπτοντα ποτὶ τὰν αἴσθησιν ἃ μὲν ἀπὸ τᾶν πλαγᾶν ταχὺ παραγίνεται καὶ < ἰσχυρῶς > ὀξέα φαίνεται, τὰ δὲ βραδέως καὶ ἀσθενῶς βαρέα δοκοῦντι ἦμεν, κτλ.

[3] 80A. The processes by which Plato imagined sound to be conveyed will be discussed on that passage.

[4] *Origins of Gk. and Rom. Thought*, pp. 65 ff., 81.

of the breath would form an obvious basis for comparison and analogy '. On our passage Professor Onians remarks that the idea that the movement produced inside us extends from the head to the *liver* has not been satisfactorily explained. ' It may well be a relic of the beliefs traced above that sound was breathed in through the ears to the θυμός in the chest and that breath reached the liver, and this would be helped by the consideration, e.g. in the passages from Aeschylus just discussed, that painful news reaches the liver '.

Diogenes of Apollonia and Anaxagoras regarded the ear as a mere channel for sound. Diogenes said that the air inside the ears is set in motion by the air outside and transmits this motion to the brain ; Anaxagoras, that the sound penetrates to the brain, striking on the hollow skull surrounding it.[1] The author of the Hippocratic treatise *On Flesh*, xv, attacks this view on the ground that the brain, being soft and moist, cannot be resonant.

67c–68d. *Colours*

The earlier account of vision (45B) dealt only with the visual ray whose fire coalesces with the daylight. We have now to consider, from the side of the objects seen, the variety of colours.

The difficulty of the following section arises partly from the fact that Greek adjectives denoting colours do not coincide with our own terms and are not easily identified, partly from the procedure, which begins by describing several varieties of fire-particles whose action on the visual ray produces different colour sensations, and then goes on to speak of compound colours as if they were pigments such as a painter makes by mixing other pigments.

67c. There remains yet a fourth kind of sensation which demands classification, since it embraces a great number of diversities. They are known by the general name of colour, a flame which streams off from bodies of every sort and has its particles so proportioned to the visual ray as to yield sensation.

d. Earlier we have explained merely how the visual ray arises ; so it is natural and fitting to add here a reasonable account of the colours, as follows.

The particles that come from other bodies and enter the visual ray when they encounter it, are sometimes smaller, sometimes larger than those of the visual ray itself ; or they may be of the same size. Those of the same size are imperceptible—' transparent ', as we call them. The larger, which contract the ray, and the smaller which dilate it, are analogous to what is cold or hot to the flesh, and again to what is

[1] Theophr., *de sens.* 40, 28.

67D. astringent or burning ('pungent' as we call it) to the tongue.

E. These are black and white, affections which are due to those particles and are similar in character, though occurring in a different field and for that reason presenting themselves in a different guise. The names should be assigned accordingly: 'white' to what dilates the visual ray, 'black' to what contracts it.

When the more piercing motion belonging to a different variety of fire falls upon the ray and dilates it right up to the eyes and forcibly thrusts apart and dissolves the very passages in the eyeball, it causes the discharge of a mass of

68. fire and water which we call a tear. Itself consisting of fire, it meets fire from the opposite quarter leaping out like a flash of lightning, while the in-going fire is quenched in the moisture; and in this confusion all manner of colours arise. The effect we call 'dazzling'; the agent which produces it 'bright' and 'flashing'.[1]

B. Then there is the variety of fire intermediate between these two, which reaches the moisture of the eyeball and is mixed with it, but is not flashing. The radiance of the fire through the moisture with which it is mingled yields blood-colour, which we call 'red'.[2]

Bright mixed with red and white produces orange. In what proportions they are mixed it would be foolish to state, even if one could know; the matter is one in which no one could be even moderately sure of giving either a proof or a plausible estimate.

Up to this point colours have been described in terms of the larger or smaller fire-particles which stream off the coloured object. White, black, and red seem to be regarded as primary or simple colours,[3] familiar to Greek eyes from vase-paintings. The addition of 'bright' or 'flashing' is puzzling. We are at first told that the dazzling effect gives rise to 'colours of all sorts'; but in the

[1] 'Bright' and 'flashing' are ranked as colours. This supports the belief that Greek terms for colour have more to do with differences of tone and brilliance than with differences of shade.

[2] So Aristotle, *Meteor*. 374a, 3, says that white light seen through a dark medium looks red (and he regards water as dark); e.g. the sun appears red through smoke or mist.

[3] According to Aristotle, *Meteor*. iii, 2, 372a, 1 ff., the rainbow contains three colours, red (φοινικοῦν), green (πράσινον), purple (ἀλουργόν), and sometimes orange (ξανθόν) between the red and the green. Red, green, and purple are, he adds, the only colours which painters cannot make by mixing. Democritus (Theophr., *de sens*. 73 ff.) recognised four colours as simple: white, black, red, pale yellow-green (χλωρόν).

last sentence ' bright ' is treated as if it were a simple colour entering with others, like white and red, into compounds. The first of the compounds, orange, is still treated as a natural colour ; the proportions of the ingredients (which we should still naturally take to be various grades of fire) cannot even be plausibly guessed. Here the method changes. We hear no more of different varieties of fire-particles. Prescriptions are given for making compound pigments out of the simple colours already named and orange. To the process of mixing pigments the statement that no one could make even a probable estimate of the quantities required seems hardly to apply.

68c. Red blended with black and white is purple, or dark violet, when these ingredients are burnt to a further point and more black is added to the mixture.

Tawny is formed by blending orange and grey, grey being a mixture of white and black ; while yellow is a combination of white with orange.

White combined with bright and plunged in intense black results in a dark blue colour ; dark blue mixed with white, in pale blue-green ; tawny and black, in green (?).[1]

From these instances of the blending of pigments Plato now reverts to the colours (considered as mixtures of varieties of fire particles) which they ' represent ' or, as it were, embody. His concluding words seem to warn us that no practical experiments in mixing measured quantities of pigments can yield any certain inferences as to the exact quantities of fire-particles of various grades composing a colour. The proportions involved are, as he said just above, inaccessible even to conjecture.

68d. From these examples it will be sufficiently clear by what combinations the remaining colours should be represented so as to preserve the probability of the account. But any attempt to put these matters to a practical test would argue ignorance of the difference between human nature and divine, namely that divinity has knowledge and power sufficient to blend the many into one and to resolve the one into many, but no man is now, or ever will be, equal to either task.

---

[1] πράσιος is commonly taken to mean green like the leek (πράσον), though Aristotle uses the form πράσινος and the substantive πράσιον means ' horehound ', of which two varieties are described by Theophrastus, H.P. 6, 2, 5. If green is meant, the statement is not much more surprising than that the addition of black to red should produce a ' bilious ' colour (83b). Democritus compounded πράσινον of πορφυροῦν (crimson) and ἴσατις (woad-blue), Theophr., de sens. 77.

68E–69A. *Conclusion*

The second part here ends with a reminder that it has been concerned throughout mainly with ' what comes about of Necessity '. We must study necessary causes, though such study be only a sober amusement, because this is the only way of approaching the manifestations of rational purpose in Nature. Happiness will consist in apprehending these and conforming our own nature to the harmony which we find in the universe. Cf. 47B, c and 90B.

68E. All these things, then, being so constituted of necessity, were taken over by the maker of the fairest and best of all things that become, when he gave birth to the self-sufficing and most perfect god ; he made use of causes of this order as subservient, while he himself contrived the good in all things that come to be. We must accordingly distinguish two kinds of cause, the necessary and the divine. The divine we should search out in all things for the sake of a
69. life of such happiness as our nature admits ; the necessary for the sake of the divine, reflecting that apart from the necessary those other objects of our serious study cannot by themselves be perceived or communicated, nor can we in any other way have part or lot in them.

III. THE CO-OPERATION OF REASON AND NECESSITY

69A–D. *Recapitulation. Addition of the mortal parts of soul*

THE third part now opens with a brief recapitulation of the steps by which the account of the works of Reason in the first part led us to the same point that we have now reached once more, from the opposite quarter, in the analysis of what happens of Necessity : namely the point of contact between the individual soul and the external world in sensation and sense perception. In the first part the rational soul was framed by the Demiurge himself. The second part has analysed the bodily down to its foundation in Space, the Receptacle of all becoming, and then built it up again by introducing the element of regular geometrical shape, imposed upon the chaotic motions and powers. The interaction of the simple bodies so formed has been described mainly in terms of necessary causation with little reference to rational design. The third part is now to exhibit the co-operation of Reason and Necessity in the work of the created gods. Their task is to frame the mortal parts of the soul and the bodily organs to house them. Henceforward the interest of intelligent purpose again predominates. The distinction between the created gods and the Demiurge is not maintained. Throughout this last part of the dialogue, the work is done sometimes by 'the gods', sometimes by 'the god'; at one place (71A) plural and singular are used in the same sentence. Plato does not seriously mean that the divine souls of the stars take an active part in the making of other living creatures. Their creative function is as mythical as that of the Demiurge, from which it is no longer kept distinct.

69A. Now that the materials for our building lie ready sorted [1] to our hand, namely the kinds of cause we have distinguished, which are to be combined in the fabric of our remaining discourse, let us in brief return to our starting-point and rapidly trace the steps that led us to the point from which we have now reached the same position once more [2]; and

B. then attempt to crown our story with a completion fitting all that has gone before.

As was said at the outset, these things were in disorder and the god introduced into them all every kind of measure in every respect in which it was possible for each one to be in harmonious proportion both with itself and with all the rest. For at first they were without any such proportion, save by mere chance,[3] nor was there anything deserving to be called by the names we now use—fire, water, and the rest; but all these he first set in order, and then framed

C. out of them this universe, a single living creature containing within itself all living creatures, mortal and immortal. Of the divine he himself undertook to be the maker [4]; the task of making the generation of mortals, he laid upon his own offspring. They, imitating him, when they had taken over an immortal principle of soul, went on to fashion for

---

[1] L. and S. (1927) cite, for the metaphorical use of διυλίζω, Archyt., ap. Stob. 3, 1, 108, διυλισμένα ἀρετὰ ἀπὸ παντὸς τῶ θνατῶ πάθεος.

[2] The 'same position' is sensation and sense-perception, which we reached at the end of the first part (45B–47E), and have now reached again in the concluding paragraphs of the second part. The expression is condensed; but ταὐτόν can hardly bear any other meaning.

[3] The reference is to those transient semblances of order which might occur without design in the chaos described at 53A by the mere attraction of like to like, or in the Atomists' casual vortices, or in Empedocles' system by the elements rushing through one another (cf. Ar., *Phys.* B4, 196A, 20 ff.).

[4] There is no suggestion in the Greek αὐτός of the 'lowly peasant' (αὐτουργός) whom Tr. (p. 495) connects with 'the thought of God humbling Himself in the service of His creatures'.

69C. it a mortal body englobing it round about.[1] For a vehicle
they gave it the body as a whole, and therein they built on
another form of soul, the mortal, having in itself dread and
D. necessary affections : first pleasure, the strongest lure of
evil ; next, pains that take flight from good ; temerity moreover and fear, a pair of unwise counsellors ; passion hard to
entreat, and hope too easily led astray. These they combined with irrational sense and desire that shrinks from no
venture,[2] and so of necessity[3] compounded the mortal
element.

69D–72D. *The bodily seats of the two mortal parts of the soul*

The summary at the end of this section (72D) explains that it
is concerned with the bodily habitations of the mortal parts of the
soul and the reasons why they are situated in certain organs,
separately from the divine part in the head. In the earlier passage
above referred to (44D–45B), the skull was described as the ' spherical
body ' in which the revolutions of the immortal soul were confined.
The head, containing the brain and the divine part of the soul, is
the human counterpart of the spherical body of the universe containing the revolutions of the World-Soul. The rest of the human
body, as we have just been reminded, was treated as a ' vehicle '
(ὄχημα, 44E), added because the head, unlike the body of the
universe, requires to be carried about from place to place. So the
trunk and limbs were there regarded as a machine for locomotion ;
and the sense-organs situated in the fore part of the head, as
instruments enabling the soul to find its way about. Only the
eyes were dealt with in detail. The whole account was concerned
with soul and body from the point of view of movement.

But we learnt earlier, from the address of the Demiurge (42A),
that the implanting of the immortal soul in a body subject to
perpetual waste and repair would entail certain necessary conse-

---

[1] The head, the ' spherical body ' in which the revolutions of the immortal
soul were confined (44D). The trunk and limbs were then added as a ' vehicle '
to carry the head about. Cf. 73C, the god moulds the brain containing ' the
divine seed ' into a spherical ball (περιφερῆ πανταχῇ), and then περὶ τὸν
ἐγκέφαλον αὐτοῦ σφαῖραν περιετόρνευσεν ὀστεΐνην (E).

[2] ἐπιχειρητῇ παντὸς ἔρωτι. The recollection of Eros, the son of Poros,
ἀνδρεῖος ὢν καὶ ἴτης καὶ σύντονος (*Symp.* 203D) makes Tr.'s ' dare-devil lust '
seem further from Plato's meaning than A.-H.'s ' love that ventures all things '.

[3] Note ἀναγκαίως here and ἀναγκαῖα παθήματα above (c, 8). The words
echo the repeated references to necessity in the parallel passage (42A) here
specially referred to. The body and the concomitant desires and passions
of the mortal soul are a necessary (indispensable) adjunct to the immortal
part, if man is to exist on earth. Limited by this necessity, the gods have
now to establish the mortal soul, as best they can, in suitable organs.

quences: sensation and perception, due to 'violent affections' from without, pleasure and pain combined with desire, fear and anger and many other feelings and emotions, which would need control. These 'necessary affections' have now to be further considered. Sensation has already been exhaustively treated in the second part; pleasure and pain, as 'common affections of the whole body', were analysed before the special senses were taken in detail (64A–65B). It remains to specify the bodily seats of the emotions and of the appetites connected with nutrition. These are housed in the organs inside the trunk: heart, lungs, belly, liver, spleen, etc. The position, structure, and functions of these organs are described, not from a physiological standpoint, but in relation to the feelings and appetites of the two inferior parts of the soul. The emphasis falls on the purposes they serve as the seats of feelings and desires that contribute to moral conduct; little is said about their behaviour as indispensable means to the preservation of physical life.

Two groups of organs corresponding to the mortal parts of the soul. —The organs, accordingly, are taken in two groups, separated by the diaphragm, corresponding to the higher and lower parts of the mortal soul already distinguished in the *Republic*.

69D. Now fearing, no doubt, to pollute the divine part on their account, save in so far as was altogether necessary, they

E. housed the mortal apart from it in a different dwelling-place in the body, building between head and breast, as an isthmus and boundary, the neck, which they placed between to keep the two apart. In the breast, then, and the trunk (as it is called) they confined the mortal kind of soul. And since part of it has a nobler nature, part a baser, they built another partition across the hollow of the trunk, as if marking off the men's apartment from the women's, and set the midriff as a fence between them.

The Spirited part situated in the heart. The lungs.—Above the diaphragm are the heart and lungs. The heart is the seat of the Spirited element (τὸ θυμοειδές, θυμός), which answered to the lower class of guardians in Plato's commonwealth, the garrison or standing army, subordinate to the philosophic rulers and embodying, with their characteristic virtue of manly courage (ἀνδρεία), the element of force in government. They were mentioned in Socrates' recapitulation at 17D.

70A. That part of the soul, then, which is of a manly spirit and ambitious of victory they housed nearer to the head, between

70A. the midriff and the neck, that it might be within hearing of the discourse of reason and join with it in restraining by force the desires, whenever these should not willingly consent to obey the word of command from the citadel.[1] The heart,

B. then, the knot of the veins and the fountain of the blood which moves impetuously round throughout all the members,[2] they established in the guardroom, in order that, when the spirit should boil with anger at a message from reason that some act of wrong is taking place in the members, whether coming from outside or, it may be, from the desires within, then every sentient part of the body should quickly, through all the narrow channels, be made aware of the commands and threats and hearken with entire obedience, and so suffer the noblest part to be leader among them all.

Aristotle (de anim. 403a, 19), illustrating how soul is related to body as form to matter, gives anger as an example : the dialectician will define anger as desire for retaliation ; the physicist will describe the material aspect, ' a boiling of the blood or heat [3] in the region of the heart '. The two aspects are combined in Plato's description. The rational part, as the headquarters of sense-perception, first becomes aware that an act of wrong is taking place in some region of the body. It sends down a message to the spirited element in the heart. Then the blood begins to boil and rush outwards through all the veins, so conveying to all the fleshy parts the impulse to quell the disturbance.

70C. Moreover, for the throbbing of the heart when danger is foreseen or anger aroused, foreseeing that all such swelling of passion would come to pass by mèans of fire, they devised a relief by implanting the structure of the lung, soft and bloodless and moreover perforated within by cavities like a

---

[1] The comparison of the intelligence in the brain to a sacred image set up in the *acropolis* of the body is attributed to Hippocrates in the *Anec. Med.* edited by Fuchs. Wellmann (*Fr. d. gr. Aerzte*, p. 19) thinks it may have occurred in a lost Hippocratic work.

[2] Cf. [Hippocr.] π. καρδίης, 7 : the great artery and the thick vein are the fountains of man's nature, and the rivers by which the body is watered. They carry the life of man, and if they are dried up, he dies. From the many points of contact between this treatise and Plato and Diocles, Wellmann concludes that it was written under the influence of the Sicilian medical school and in particular of Philistion (*ibid.*, 107).

[3] According to π. καρδίης, 6, the 'innate fire' (ἔμφυτον πῦρ) is seated in the left ventricle of the heart, together with the intelligence (γνώμη) which rules the rest of the soul (10). The blood is not naturally warm, as some suppose (12).

70C. sponge, in order that by receiving breath and drink,[1] it
70D. might cool the heart and so provide refreshment and ease
in the burning. For this purpose they cut the channels of
the windpipe to reach the lung, and set the lung itself around
the heart as a sort of buffer, so that, when the spirit therein
was at the height of passion, the heart might leap against a
yielding substance and be cooled down, and so being in less
distress be the better able to help the spirited element in
the service of reason.[2]

Plato's 'spirited element' is a group of emotions and sentiments
whose most characteristic expression is seen in anger, indignation,
ambition for success and victory, the love of power. In the conflict
of motives, as the *Republic* showed, these feelings can be enlisted
on the side of reason against the impulses of lawless appetite, that
mob of lower desires which must be kept in order by restraining
force if reason is to rule. When he uses the term θυμός and locates
the θυμός in the heart and lungs, Plato is following a very old
tradition. In a brilliant study of the Homeric conceptions of soul
and body, Professor Onians[3] has drawn a distinction between the
ψυχή and the θυμός which corresponds, to a remarkable extent,
with Plato's distinction between the immortal principle of reason
seated in the head and the spirited element in the chest. He
points out that in Homer the ψυχή and the θυμός are regarded as
separate entities.

'The θυμός is constantly spoken of as feeling and thinking
and active in the lungs[4] (φρένες) or chest (στῆθος) of the living

---

[1] Galen, *Hipp. et Plat.* 722 ff., remarks that to say that *all* drink goes to
the lungs would be ridiculously contrary to observed fact, and that Plato
elsewhere (70D, 72E, 78A, B) says that drink goes with food to the belly.
Physicians and philosophers dispute whether some *part* of our drink goes to
the lungs, a little moisture trickling like dew down the trachea. Following
the π. καρδίης, 2, Galen adds that experiment has proved that this is so. If
you give an animal water stained blue and then kill it, you will find that
the lungs are dyed blue. At 91A Plato speaks of the passage of drink through
the lungs by the kidneys to the bladder. The view that drink goes to the
lungs was current. A.-H. cites Plut., *Stoic. rep.* 29 : Plato's view is supported
by Hippocrates, Philistion, Dexippus (Dioxippus, libr.) the Hippocratean,
and among the poets by Euripides, Alcaeus, Eupolis, and Eratosthenes, who
all speak of drink passing through the lung. The περὶ καρδίης, 2, holds
that only a small part of the drink goes to the lungs, the rest to the belly.

[2] That the function of respiration was to cool the innate heat residing in
the heart was taught by Philistion and Diocles. The same doctrine appears
in π. καρδίης, 5 (ix, 84L), see Wellmann, *Frag. d. Gr. Aerzte* 81.

[3] *Origins of Greek and Roman Thought*, Cambridge, pp. 85 ff.

[4] That φρένες in Homer means the lungs, not the diaphragm, Professor
Onians has conclusively proved earlier.

person and as departing at death, but is not spoken of in connection with the succeeding state; the ψυχή, on the other hand, is " in " the person but is not spoken of as being in the lungs or chest or as thinking or feeling while a person lives, but seems to represent rather a " life-principle " or soul not concerned in ordinary consciousness and to be what persists, still without ordinary consciousness, in the house of Hades, there identified with the εἴδωλον, the visible but impalpable semblance of the once living . . . θυμός and ψυχή are both said to leave the body as if separately and simultaneously. And we may add, what does not appear to have been noticed, that while the ψυχή leaves the body ἠΰτ᾿ ὄνειρος (as a phantom of the person such as is encountered in a dream) and persists in the house of Hades as an εἴδωλον, i.e. preserves its form and does not disintegrate, the θυμός is " destroyed " and " shattered " by death. . . . If, then, the ψυχή is not the θυμός or " breath-soul " proper but represents something else in the living man, we are left with something gaseous and so liable to be " breathed forth ", possibly identified with the shadow, as which after death it is in fact described by Homer, σκιά (cf. umbra, etc.) and which is relatively " cold ". . . . In any case, the ψυχή was, I suggest, associated more particularly with the head, whence it would naturally be " breathed forth ". It has apparently attracted little notice hitherto that while the chest (στῆθος) and its organs, the φρένες (or πραπίδες), also κῆρ, κραδίη, ἦτορ and the θυμός are continually mentioned in the poems as the seat of consciousness and intelligence, feeling and thought, the head is also important in a different way, is in fact regarded as in a unique degree precious or holy, identified with the person and equated with that soul or principle of *life* which the ψυχή appears to be.'

Professor Onians identifies the Homeric θυμός with the breath, breath-soul, or spirit naturally located in the lungs (φρένες). The breath contains a good deal of water-vapour, which the ancients would derive from the blood, ' the hot liquid which is in fact concentrated in the heart and around it in the lungs (φρένες). The latter are filled with blood and breath that interact, giving and taking from each other. A tradition of this relation probably . . . lies behind the Platonic explanation, " θυμός from the seething (θύσεως) and boiling of the ψυχή ".[1] This θυμός is not the blood-soul as opposed to the breath-soul, nor indeed mere breath, but something vaporous within blending and interacting with the air without.' It is the stuff of all consciousness, including thought, which was not yet differentiated from feeling.

[1] *Crat.* 419D; cf. *Rep.* 440C; *Timaeus* 70B.

Plato has transferred the function of rational thought to the immortal ψυχή lodged in the head. His θυμός covers a much restricted field of consciousness; it is now a part of the soul, no longer a material substance; but it is still housed in the chest, its fury is the boiling of the blood, and it is mortal. The lungs are falsely described as bloodless.[1] They receive the breath (which they dispense to the body, 84D) and some of the liquid we drink; but their function emphasised in this context is merely to serve as a buffer for the throbbing heart and to cool it down. θυμός is now more closely associated with the blood than with the breath; its seat is the heart, rather than the lungs.

The Appetitive part situated in the belly. The liver and the spleen. —Below the diaphragm, the stomach is compared to a manger, to which the lower mortal part, the appetitive, is tethered like a stalled beast. Its region, extending as far down as the navel, is also tenanted by the liver and the spleen, for which relevant functions have to be provided. These have no connection with the physical function of the appetitive, namely nutrition.

70D. That part of the soul whose appetite is set on meat and drink and all that it has need of for the sake of the body's nature, they housed between the midriff and the boundary

E. towards the navel, constructing in all this region as it were a manger for the body's nourishment. There they tethered it like a beast untamed but necessary to be maintained along with the rest if a mortal race were ever to exist. Accordingly, they stationed it here with the intent that, always feeding at its stall and dwelling as far as possible from the seat of counsel, it might cause the least possible tumult and clamour and allow the highest part to take

71. thought in peace for the common profit of each and all.

And because they knew that it would not understand the discourse of reason and that, even if it should somehow become aware of any such discourse, it would not be in its nature to take any heed, whereas it would most readily fall under the spell of images and phantoms both by night and by day, the god, designing to gain this very influence,[2]

---

[1] Aristotle's remark that those who imagine the lung to be bloodless are deceived by the observation of lungs removed from animals under dissection, the blood having all escaped (H. A. 496*b*, 5), is directed against the Sicilian school, whom Plato is following.

[2] The interpretation of τούτῳ θεὸς ἐπιβουλεύσας αὐτῷ is doubtful: 'lay in wait for this same weakness' (Tr.). But the analogy of ἐπιβουλεύειν

71.
B.

formed the liver and set it in the creature's dwelling-place, and contrived that it should be a substance close in texture, smooth and bright, possessing both sweetness and bitterness. The purpose was that the influence proceeding from the reason should make impressions of its thoughts upon the liver, which would receive them like a mirror and give back visible images. This influence would strike terror into the appetitive part, at such times as, taking a part in keeping with the liver's bitterness, it threatens with stern approach [1]; swiftly suffusing this bitterness throughout the liver, it would cause bilious colours [2] to appear thereon; make it all rough and wrinkled by contraction; and as it shrinks and
C. bows down the lobe, obstructs the vessels, and closes the entrance, produce pain and nausea. Sometimes, again, when some inspiration of gentleness from the mind delineates semblances of the contrary sort, it gives rest from the bitterness, because it will not stir up or have dealings with a nature contrary to its own; rather, using towards it a sweetness of like nature to the sweetness in the liver itself,[3] and setting

τυραννίδι, etc., suggests that τούτῳ αὐτῷ may mean the exercising of ψυχαγωγία, the last thing mentioned. Note the vagueness of θεός following the plural εἰδότες at the beginning of the sentence.

[1] I suggest that μέρει χρωμένη may mean something like 'playing a rôle' and τῆς πικρότητος be governed by συγγενεῖ. This would account for συγγενεῖ agreeing with μέρει, not with πικρότητος, and for the position of the clause before χαλεπὴ προσενεχθεῖσα ἀπειλῇ which describes the rôle in question. A.-H.'s 'making use of *the bitter element* akin to its own dark nature' gives an unnatural sense to μέρει τῆς πικρότητος. Tr.'s 'availing itself in some measure of this congenital bitterness' is hardly a fair paraphrase of words which, taken literally in this way, mean 'availing itself of a congenital part of the bitterness', as if other parts were not congenital; and the word 'part' seems superfluous. I cannot find the phrase μέρει χρῆσθαι elsewhere in the sense of *partes agere*; but ἀγγέλου μέρος (*Agam.* 291), 'turn of duty as messenger' (L. and S.), τὸ ἐμὸν μέρος, etc., mean the part taken in some action, office, function (*partes = munus*), and χρῆσθαι τέχνῃ means to *exercise* a trade. μέρος seems appropriate here because to threaten is only one of two parts that the influence can play, and the word suggests 'exercising a part of its function which is in keeping . . .'

[2] Cf. 83B for the meaning of 'bilious colour'.

[3] I understand γλυκύτητι τῇ κατ' ἐκεῖ (γλυκύτητι) συμφύτῳ πρὸς αὐτὸ χρωμένη because this phrase seems parallel to μέρει τῆς πικρότητος χρωμένη συγγενεῖ above. γλυκύς from Homer onwards denotes a quality of persons. Or we may understand with A.-H. that the influence uses upon the liver the sweetness which permeates it and is akin to (the sweetness of) the influence itself. Since the bitterness of the liver has just been called 'a nature contrary to' that of the influence, I agree with A.-H. that there is nothing ridiculous in this interpretation. In any case it is expressed or implied that use is made of the bitterness or sweetness in the liver itself.

71D. it right till all is straight and smooth and free, it makes that part of the soul that dwells in the region of the liver [1] to thrive in well-being and gentleness of mood, and by night to pass its time in the sober exercise of divination by dreams, since it had no part in rational discourse and understanding. For our makers remembered their father's injunction to make the mortal race as perfect as possible, and they tried to set even the baser part of us on the right path in this way, by establishing the seat of divination in this part, that it

E. might have some apprehension of reality and truth.

That divination is the gift of heaven to human unwisdom we have good reason to believe, in that no man in his normal senses deals in true and inspired divination, but only when the power of understanding is fettered in sleep or he is distraught by some disorder or, it may be, by divine possession. It is for the man in his ordinary senses to recall and construe the utterances, in dream or in waking life, of

72. divination or possession, and by reflection to make out in what manner and to whom all the visions of the seer betoken some good or ill, past, present, or to come. When a man has fallen into frenzy and is still in that condition, it is not for him to determine the meaning of his own visions and utterances; rather the old saying is true, that only the sound in mind can attend to his own concerns and know himself. Hence it is the custom to set up spokesmen to

B. pronounce judgment on inspired divination. These are themselves given the name of diviners [2] by some who are quite unaware that they are expositors of riddling oracle or vision and best deserve to be called, not diviners, but spokesmen of those who practise divination.

This, then, is the reason why the liver has such a nature and situation as we have described: it is for the sake of divination. So long as any creature is yet alive the indica-

---

[1] This phrase might support Galen's often repeated assertion that Plato regards the liver as the seat of the appetitive part (e.g. *Hipp. et Plat.* 569; *U.P.* iv, 13; *in Tim.* 10 Dar. 11 Schröder).

[2] ' Prophet ' would be a more natural word; but Plato restricts προφήτης to its proper sense (' spokesman '). Apollo was both μάντις and Διὸς προφήτης; but Plato associates the word μάντις with μανία, the divine madness of *Phaedrus* 244B, contrasted with the uninspired augury which draws inferences from observed omens. This rational procedure he calls οἰωνιστική, not μαντική, and it is comparable with the business of the interpreter. Euripides (frag. 973) had written ' the best diviner (μάντις) is he who makes a good guess ', and Antiphon (*Vors.* 80A, 9) is said to have defined μαντική as ἀνθρώπου φρονίμου εἰκασμός. Plato may here be thinking of such misuses of the word μάντις. Cf. Plut., *def. orac.* 432C.

72B. tions given by such an organ are comparatively clear [1]; but deprived of life it becomes blind and its signs are too dim
C. to convey any certain meaning.

Commentators who do not believe in divination have exaggerated what Archer-Hind calls 'the keen irony pervading the whole' of the passage describing it as the gift of heaven to human unreason. It is true, no doubt, that Plato despised diviners like Euthyphro, and that he ranks the seer low in the hierarchy of incarnations at *Phaedrus* 248D. On the other hand, the seer is there placed fifth, above all poets, artists, and craftsmen; and Socrates' earlier rhapsody (244B) which classifies inspired prophecy with poetry, love, and philosophy itself as forms of 'divine madness' should not be forgotten. It is possible to combine a sincere respect for traditional religion with a low opinion of its average professors.

Except for the passing dismissal of omens from the entrails of sacrificial victims, the whole account is confined to divination by dreams and visions. An earlier doctrine, which there is some reason to call Orphic, had attributed revelation of the future in dreams to the divine and immortal soul, which is not the seat of normal waking consciousness (Pindar, frag. 131). Aristotle (π. φιλοσ. frag. 10) says that one source of our belief in the gods is the inspirations and divinations of the soul in sleep. The soul then 'comes to be by itself', recovers its proper nature, and divines and foretells the future. It is also in this condition when it is in the act of being separated from the body at death. Thus in Homer Patroclus foretells the death of Hector, and Hector the death of Achilles. Plato refers to this belief in the mantic power of the soul at the moment of dying (*Apol.* 39C and *Phaedo* 85B). In our passage, however, the view is different. The seer who has the visions is not the divine and immortal part, but the irrational and appetitive, which receives warnings and admonitions, in these symbolic images, from the reason, and requires the aid of the reason's waking reflection to interpret them.

The next paragraph describes the spleen as a useful adjunct to the liver, just as the lung was treated as a buffer for the heart.

72C. Again, the structure of the neighbouring organ and its position on the left are for the sake of the liver, to keep it always bright and clean, like a napkin provided to wipe a

---

[1] The words ἑκάστου τὸ τοιοῦτον seem intended to include the *corresponding* organ in *any* (non-human) creature; for the rest of the sentence dismisses divination from the appearance of the liver in sacrificed animals, although their dream images could not be due to any influence from reason, which they do not possess.

72C. mirror and always laid ready beside it. So, when any impurities arise in the region of the liver from bodily disorders, they are all purged away and absorbed by the spleen, whose texture is not close, since it has cavities not containing

D. blood. Hence, when it is filled with these offscourings, it waxes swollen and festered, and, when the body is purged, subsides again and is reduced to its former state.

72D–73A. *Summary and transition to the rest of the body*

The following paragraph notes that we have so far been concerned with the special habitations of the two mortal parts of the soul, as distinct from the divine part situated in the head. The heart and lungs in the upper region, the belly, liver, and spleen in the lower, have been treated as the seats of emotions and of the desire for food, with comparatively little reference to what we should call their physiological functions. We are next to consider ' the remainder of the body '.[1] This phrase covers, in the first place, the rest of the contents of the trunk (as opposed to the limbs), namely the viscera below the navel, which was the lower boundary of the appetitive. Their utility to the soul is stated, and the discourse then proceeds, without further preface, to ' the remainder of the body ' in a wider sense.

72D. Concerning the soul, then, we have stated what part of it is mortal and what divine, and where, in what company, and for what reasons the two are housed apart. We could confidently assert that our account is the truth only if it were first confirmed by heaven ; but that it is the probable account we may venture to say now, and still more on further consideration. Let that claim, then, be taken as made.

E. The next part of our task must be pursued on the same principles : this was the manner in which the remainder of the body came to be.[2] Now the design that would most

[1] Tr. (p. 517) objects to this translation, and renders : ' it remains to tell how the body was made '. ' We have ', he says, ' not yet heard about any part of the body ᾗ γέγονε, but only οὗ ἕνεκα γέγονε.' But the next sentences proceed to describe the *purpose* (not the manner) of the formation of the lower viscera (not the body as a whole), on precisely the same lines as the foregoing section ; and the subsequent paragraphs do not go back to heart, lungs, liver, etc., and tell us *how* they were made, but deal with other parts of the body, bones, flesh, etc., whose purpose has not yet been described.

[2] The reference of ἦν is to 61C, ' we have not yet described the formation of flesh and all that belongs to it, or the mortal part of the soul ' (so A.-H.), not, as Tr. suggests, to 69C, 6, which is part of a summary of statements already made, with no suggestion of any task still to be performed. The

72E.   fittingly account for its construction would be this.   The framers of mankind knew what would be our intemperance in meat and drink and that, out of gluttony, we should use far more than the moderate or necessary amount.   Accordingly, to make provision against the danger that disease should bring swift destruction and the mortal race should

73.   forthwith come to an end in immaturity, they appointed the lower belly (as it is called) as a receptacle to hold the superfluity of food and drink, and wound the bowels round in coils, in order that the nourishment should not pass so quickly through as to constrain the body to crave fresh nourishment too soon, and thus making it insatiable render all mankind incapable, through gluttony, of all cultivation and philosophy, deaf to the command of the divinest part of our nature.

Aristotle similarly explains the gluttonous appetite of fishes by the straightness of their intestine, which allows the food to pass through too rapidly for complete digestion.   Dr. Ogle observes that an abnormally short gut is, in fact, a sufficient cause for a ravenous appetite.[1]   Plato emphasises the coiling of the bowels as the one feature designed for the sake of the higher interests of the soul, passing lightly over their ' necessary ' functions.

73B–76E.   *The main structure of the human frame*

In the above account of the two mortal parts of the soul and their habitations, the most striking point is that the appetitive element appears to be restricted to desires connected with nutrition, to the exclusion of reproduction.[2]   In the machinery of the myth, the creation of the sexual parts and of the desire for intercourse is postponed until the whole account of the human body is complete and the moment comes for the less satisfactory men to be reincarnated as women (90E).   This is not to be taken as historical fact ; we are not to suppose that there ever existed a generation of men before there were any women or lower animals.   Indeed, we have more than once been told that ἔρως is a necessary con-

---

αὐτό in the next sentence obviously cannot mean the body as a whole (τὸ τοῦ σώματος), which was not designed for the purpose mentioned ; it must therefore refer to τὸ τοῦ σώματος ἐπίλοιπον.

[1] *De part. anim.* 675a, 20, trans. Ogle.   Tr. also quotes *de gen. anim.* 717a, 20.

[2] Unless this is included among the appetites for ' meat and drink and *all that the soul has need of for the sake of the body's nature* ', 70D.   Cf. *Laws* 782D : with mankind all things depend on three needs and desires : the desires for food and drink, which date from birth, and thirdly sexual desire, which emerges later.   All three are there called νοσήματα.

stituent of the mortal soul (42A, 69D). We are left to conjecture
the reasons for this curious plan. It is not enough to say that
differences of sex are postponed because the whole account of the
human soul and body applies equally to men and women, though
this may be true. If that were all, there would still be no reason
against recognising, as a part of the appetitive element, the desire
for intercourse and reproduction, which is after all common to
both sexes.

A clue may, perhaps, be found in the next paragraph, which tells
us that the seed, the physical vehicle for the transmission of life,
belongs to a different system of organs. The seed is a part of the
marrow, which extends from the head (where it forms the brain)
throughout the whole length of the spine; this is once more clearly
stated at 91A, B. The marrow is the fundamental substance; in
it are fastened the very bonds of life, the roots of every part of the
soul. Moreover, a portion of it, the brain, is the seat of the
immortal element in the soul. The seed is the means by which
the living creature attains to such immortality as the mortal can
have by perpetuating its race in generation. Sexual desire, as
Diotima explains in the *Symposium*, is only the lowest form of the
passion for immortality.[1] At a higher level the same energy finds
an object in fame after death, for which men will sacrifice life
itself; and higher still Eros becomes the passion for wisdom,
philosophy, whereby the soul may regain the pristine purity of its
divine nature. In contrast with modern doctrines of sublimation,
Plato regards the highest form of desire as primitive and essential;
the lower forms exist only at levels to which the soul is fated to
sink when incarnate in a mortal body. The whole doctrine is
briefly resumed at 90B, C, before any mention is made of the
organs of sex and of the channel provided for the seed. Regarded
in this light as the passion for immortality in all its forms, Eros
could not be treated as merely an element in the appetitive part.
Its physical medium, the seed, does not belong to the sexual organs,

---

[1] *Laws* 721B briefly repeats the doctrine put into Diotima's mouth at
Symposium 207 ff. Aristotle follows, *de anim.* 415a, 26: 'It is the most
natural function in all living things to reproduce their species; animal
producing animal and plant plant, in order that they may, so far as they
can, share in the eternal and the divine. For it is that which all things
yearn after, and that is the final cause of all their natural activity. . . .
Since, then, individual things are incapable of sharing continuously in the
eternal and the divine, because nothing in the world of perishables can abide
numerically one and the same, they partake in the eternal and divine, each
in the only way it can, some more, some less. That is to say, each persists,
though not in itself, yet in a representative which is specifically, not numerically, one with it' (trans. Hicks). Also *de gen. anim.* II, i.

which merely provide an outlet and a receptacle. As actually a part of the marrow, it is continuous with the brain, the seat of the immortal and divine part.

The marrow, seed, and brain.—Starting from the marrow, we are now to have a more systematic description of the human frame. The skeleton is a bony shield protecting the marrow, and itself protected by the flesh and skin. Thus the whole body is regarded as a vessel with successive layers, guarding at its core the substance in which the bonds of life are secured.

73B. With bone, flesh, and all substances of that sort the case stands thus. The starting-point[1] for all these was the formation of the marrow, for the bonds of life, so long as the soul is bound up with the body, were made fast in it as the roots of the mortal creature; while the marrow itself is formed of other things. The god set apart from their several kinds those triangles which, being unwarped and smooth, were originally able to produce fire, water, air, and earth of

C. the most exact form.[2] Mixing these in due proportion to one another, he made out of them the marrow, contriving thus a mixture of seeds of every sort for every mortal kind. Next he implanted and made fast therein the several kinds of souls; also from the first, in his original distribution, he divided the marrow into shapes corresponding in number and fashion to those which the several kinds were destined to wear. And he moulded into spherical shape the plough-land, as it were, that was to contain the divine seed; and

D. this part of the marrow he named 'brain',[3] signifying that, when each living creature was completed, the vessel containing this should be the head. That part, on the other hand, which was to retain[4] the remaining, mortal, kind of soul he divided into shapes at once round and elongated, naming

[1] ἀρχή does not mean that marrow is the fundamental stuff in the composition of all the other tissues, as Tr. supposes (pp. 518, 531). Bone is steeped in it (73E), but there is no marrow in flesh (74C).

[2] No physical bodies in the visible world of becoming can have the exact perfection of the surfaces and solids of mathematics. This is one of the limiting conditions which prevent the works of Reason from reaching ideal perfection. The triangles composing the surfaces of visible and tangible bodies are only copies of the triangles whose construction was described earlier.

[3] 'Brain' (ἐγκέφαλον) because 'in the head' (ἐν κεφαλῇ).

[4] καθέξειν (so Tr.). At 74E the bones which contain marrow are called ἔμψυχα. The marrow is the life-substance in which all parts of the soul are rooted; but it is the actual seat only of the immortal part. The mortal part is located elsewhere, in heart and belly, and only linked to the marrow by anchor-cables.

73D.   them all 'marrow'.[1]   From these, as if from anchors, he
put forth bonds to fasten all the soul ; and now began to
fashion our whole body round this thing,[2] first framing
round the whole of it a solid shield of bone.

The sentence describing the formation of the marrow is of doubtful
meaning.   Taylor translates :

'Thus he devised a universal seed for all mortality (πανσπερμίαν
παντὶ θνητῷ γένει), fashioning the marrow from these' (i.e.
selected triangles).   'Next he implanted the varieties of soul (τὰ
τῶν ψυχῶν γένη) in it and bound them fast there ; also in the first
original distribution he divided the marrow itself into shapes
answering in number and quality to the several varieties' ('sc.
the different "patterns" (εἴδη) or "parts" in the soul').

On this interpretation (which is that of most editors) the whole
sentence refers only to the human soul and marrow.   But certain
phrases are difficult to understand unless we adopt the view suggested in Rivaud's translation that the marrow contains *seeds of all
sorts for every mortal kind* (of animal), the roots of *the kinds of souls*
(plural) [3] of beasts as well as of man, and a 'preformist' determination of the various shapes (types of body) which the souls of all
those *species* (εἴδη) were destined to *wear*.[4]   This interpretation
accounts for the rather emphatic phrase 'from the first, in his
original distribution'.   Provision was thus made in this fundamental substance for what is mythically represented later (91) as
a degeneration of the male human type into woman and the lower
animals.   These last fall into three main classes—land animals,
birds, and fishes—all of which include vertebrates ; and it is there
explained how the vertebrate pattern of body is distorted and
modified to suit their degenerate souls.   Our passage seems
intended to forecast these modifications of the highest pattern,
with its distinction of the round brain in its spherical skull from
the elongated columns of marrow in the spine and other bones.
Rivaud refers to 76E, where provision is made in the male human

[1] 'Shapes', plural, because there are columns of marrow in other bones
than the spine.

[2] τοῦτο, not the soul (as in Tr.'s translation), but the brain and marrow
(as in his note, p. 523).

[3] This plural occurs, I think, nowhere else.

[4] It is difficult to understand that the *two* shapes (spherical and columnar)
which are described in the following sentence can correspond to the *three*
parts of the soul, or that the two mortal parts, seated in heart and belly,
can be said to *wear* (σχήσειν) the columnar shape of the marrow in the bones,
to which they are merely rooted or anchored.   Tr. ignores this σχήσειν in
his translation, though in his note he says the subject of ἔμελλε σχήσειν is
probably τὰ τῶν ψυχῶν γένη.

body 'at the very birth of mankind' of structures which will become useful when women and beasts are 'developed' from men. This development never actually happened; and this is one of the places where the mythical machinery becomes embarrassing and entails the use of rather vague language. Plato may wish to indicate that the marrow is the fundamental life-substance in all animals and the same substance in all.

The doctrine that the seed [1] comes from the brain and the marrow of the spine was held by the Sicilian school of medicine. Alcmaeon of Croton is said to have called the seed a part of the brain (*Vors.* 14A, 13) and Hippo of Rhegium to have taught that it flows from the marrow (*ibid.* 26A, 12). The two views are combined by Diocles and Plato. The Hippocratean school, on the contrary, believed that the seed came from all parts of the body. [2]

Bone, flesh, sinews.

73E.   And bone he constructed as follows. Having sifted out earth that was pure and smooth, he kneaded it and soaked it with marrow; then he plunged the stuff into fire, next dipped it in water, and again in fire and once more in water; by thus shifting it several times from one to the other he made it insoluble by either. Of this, then, he made use, first to turn a sphere of bone to surround the creature's brain, [3] and to this sphere he left a narrow outlet; and

74.   further, to surround the marrow along the neck and back, he moulded out of bone vertebrae, which he set to serve as pivots, starting from the head through the whole extent of the trunk. Thus, to protect all the seed, he fenced it in a stony enclosure, and in this he made joints, availing himself in their case of the property of the Different, inserted between them [4] for the sake of movement and bending.

---

[1] That 'the divine seed' here means the semen is explicitly stated at 91B, 1. It is 'divine' as being part of the marrow which contains the immortal part of the soul, and also as being the vehicle and means of the immortality of the species.

[2] The evidence is collected in Schröder's note in his edition of Galen's Commentary on the *Timaeus*, p. 53. Diocles, frag. 170, Wellmann.

[3] αὐτοῦ can hardly mean 'of bone' since ὀστεΐνην follows. I can only understand it as referring, not to any word in the immediate context, but to the creature which Plato imagines being constructed (ἑκάστου ζῴου, D, 1). Cf. 78c, τῷ πλασθέντι ζῴῳ. Tr.'s 'on the spot', 'round the actual brain', seems to me impossible. Why not περὶ μὲν αὐτὸν τὸν ἐγκέφαλον?

[4] The spine is unlike the skull in consisting of *many* separate parts and being capable of *variable* movements in any direction. This curious phrase indicates that Plato saw something symbolic in this contrast with the single and solid sphere of the skull (analogous to the spherical body of the world),

74.   Again, considering that the constitution of bone was
B. unduly brittle and inflexible, and moreover that, if it should
become fiery hot and then cold again, it would decay and
quickly cause the destruction of the seed within it, for these
reasons he devised the sinews and the flesh in such a way
that, by binding together all the limbs with sinew contracting
and relaxing about their sockets,[1] he might enable the body
to bend or stretch itself out ; while the flesh was to be a
defence against burning heat and a shelter from wintry cold,
and also a protection against falls, like our borrowed trappings
of felt [2] : it would yield to bodies softly and gently, and it

adapted only to the constant revolutions of the rational soul. The lower
parts of the soul, connected with the spinal marrow, exhibit the characteristics of the ' wandering cause '. Cf. the contrast (at 44D) between the head
and its vehicle (the rest of the body) with limbs capable of travelling ' through
all the regions ' (up and down, forward and backward, right and left), added
because the creature was to possess ' all the motions there are '. This is
substantially Fraccaroli's view. He translates : ' *adoperando l'azione del
variabile per ottenere tra di esse col mezzo suo e movimento e flessione.*'

Here (as Tr. notes, p. 528) there is a covert polemic against the Empedoclean
notion of evolution by the survival of useful characters produced by chance.
Aristotle (*de part. anim.* 640a, 19), attacking the same view, instances
Empedocles' theory that ' the backbone was divided as it is into vertebrae
because it happened to be broken owing to the contorted position of the
foetus in the womb '. At 73C, ' in his original distribution ' may hint that
the various species of animals were not developed casually or one species
from another. Again at 75D the mouth was designed ' as it is now arranged '.

[1] The marrow (like the flesh and sinews) extends to other bones than those
of the skull and spine, notably the thighs. The reference here is to the
socket-joints at the συμβολαὶ τῶν ὀστῶν (74E) of arms and legs (' all the limbs ',
cf. 75D), as well as to those of the vertebrae. περὶ τοὺς στρόφιγγας, standing
between ἐπιτεινομένῳ καὶ ἀνιεμένῳ and καμπτόμενον, goes with both, rather
than with either to the exclusion of the other.

[2] τὰ πλητὰ κτήματα, ' acquired things manufactured by felting ' (hair).
Had Plato written the more obvious phrase τὰ ἐπίκτητα πιλήματα, editors and
lexicographers would not have missed the sense or imagined that κτήματα
could mean (here only, and where Galen quotes the phrase) ' coverings ' or
' materials '. The Division to define weaving at *Polit.* 279C classifies ' all the
things we manufacture and *acquire* ' (δημιουργοῦμεν καὶ κτώμεθα). One main
branch is defences (προβλήματα) of all sorts, including curtains, roof-coverings,
mats, wrappings (clothes), which may be manufactured, either by *felting*
(πιλητική, 280B) or by other processes. The meaning of κτήματα (acquire-
ments) is plain from *Laws* 942D, which recommends exposure to heat and
cold and hard couches in order not to spoil the natural powers of head and
feet, τῇ τῶν ἀλλοτρίων σκεπασμάτων περικαλυφῇ, τὴν οἰκείων ἀπολλύντας πίλων
τε καὶ ὑποδημάτων γένεσιν καὶ φύσιν. Our phrase means acquired coverings
manufactured of felted hair, in contrast with the ' native growth ' of our
own flesh, defending the bones, which themselves shield the marrow. Cf.
Chrysost. 1, 2, p. 140, τρίχας βλαστάνειν παρεσκεύασεν, ὥστε ἀντὶ πιλημάτων
τινῶν εἶναι τῇ κεφαλῇ. At 76C, D, the hair of the scalp is ' felted ' by cold

74C. contained in itself a warm moisture, which in summer it might sweat forth and so spread a native coolness all over the body by moistening it outside, while in winter, on the other hand, we should have this fire [1] as a fair protection against the assaults of the beleaguering frost outside. With this intent, he who moulded us like wax composed flesh, soft and full of sap, by making a duly adjusted compound with water and fire and earth, which he suffused with a ferment

D. composed of acid and saline.[2] The sinews, again, he made by mixing bone with unfermented flesh into a substance with properties intermediate between those two constituents, adding a yellow colour ; hence the sinews acquired a quality more tense and consistent than flesh, but softer and more pliable than bone. With these the god enveloped the bones and marrow, binding the bones to one another with sinews,

E. and he then buried them all under a covering of flesh.[3]

The uneven distribution of flesh.—The skeleton is now complete and clothed with flesh. The chief organs occupying the hollow of the trunk have been enumerated earlier as seats of the mortal parts of the soul. Some remarks are now added on the reasons for the uneven distribution of flesh. The bones containing the life-substance, marrow, in the largest quantities, viz. the skull and the spine, are comparatively ill-protected by flesh ; whereas others, such as the thighs, which contain little marrow, are thickly covered. This seems paradoxical, at first sight. The explanation throws an interesting light on the relations of Reason and Necessity. In order to shield the marrow and the life it contains, Reason uses, as indispensable means, the solidity of bone and the softer covering of flesh. But the necessary constitution of these integuments, as such, tends to defeat another purpose : that sensations shall be

($\tau\hat{\eta}$ πλήσει τῆς ψύξεως . . . συνεπιλήθη), as if to make a natural hat. The phrase evidently struck Galen, for after quoting it from the *Timaeus* at U.P. i, 27, he repeats it four times elsewhere.

[1] The native warmth, ' with a tacit antithesis to the fire on the hearth ' (Tr.). This phrase, like οἰκεῖον above, further illustrates the meaning of κτήματα.

[2] Tr.'s note on ζύμωμα (p. 531) assumes that marrow is the main stuff of which flesh is made, and that the earth, water, and fire *are* the added ' leaven '. But nothing is said in the text about the presence of any marrow at all in flesh. See note on 73B. ὑπομείξας (without καὶ before it) is simply the fourth in a string of aorist participles with no conjunctions. The object to be supplied with συμμείξας is σάρκα (A.-H.).

[3] The poetic use of κατασκιάζειν for burial may explain the curious phrase ' overshadowing from above ' (ἄνωθεν, not ἔξωθεν) and recall Empedocles' σαρκῶν ἀλλογνῶτι περιστέλλουσα χιτῶνι and σῶμα σῆμα.

easily and quickly transmitted to the marrow of the brain. The necessary means to that is thinness both of bone and of flesh. So two sets of necessary means ' refuse to coincide '. Necessity cannot be wholly persuaded by Reason to serve both its purposes, and Reason has to sacrifice the less important.

It should be noted that Plato never uses the word ' muscle ' ($\mu\tilde{\upsilon}\varsigma$). He seems to have thought of flesh as simply a covering and attributed all muscular action to the ' sinews '.

74E.   Now those bones in which there is most life he fenced about with the smallest amount of flesh ; those having least life within them, with flesh in the greatest abundance and of the toughest kind ; moreover at the joints of the bones, wherever no cogent reason appeared to require it, he caused but little flesh to grow. The purpose was that flesh should not hamper the bending of the joints and so stiffen the body as to make it hard to move about ; and, secondly, that the solidity of many layers of thick flesh packed close on one another should not cause dulness of sensation and produce hardness of apprehension and unretentiveness in the quarters of the mind. For this reason the thighs and shins and parts

75.   about the hips, the bones of the upper arms and fore-arms, and all other parts where there are no joints, and also all the bones within the body that are devoid of intelligence because they have so little soul residing in marrow—all these have a full complement of flesh. Those parts, on the contrary, which are the seat of intelligence have less—save where he formed a mass of flesh to be in itself an organ of sensation, as for instance the structure of the tongue. With most parts, however, it is as aforesaid ; for the constitution of this frame which of necessity comes into being and is

B.   reared with us in no wise allows dense bone and much flesh to go together with keenly responsive sensation. For if these two characters had consented to coincide, the structure of the head would have possessed them above all, and the human race, bearing a head fortified with flesh and sinew, would have enjoyed a life twice or many times as long as now, healthier and more free from pain. But as it was, the artificers who brought us into being reckoned whether they

C.   should make a long-lived but inferior race or one with a shorter life but nobler, and agreed that every one must on all accounts prefer the shorter and better life to the longer and worse. Hence they covered in the head with thin bone, but not with flesh nor yet with sinews, since it has no flexions.

75C. Accordingly the head they attached to the body of every
man is all the more sensitive and intelligent, but much
weaker. The sinews, again, on the same principle and for
D. these reasons, were set by the god all round the neck so far
as to the base of the head and welded by means of uniformity,[1]
and he fastened to them the extremities of the jawbones just
under the face ; while the rest he distributed among all the
limbs, connecting the joints. The mouth was equipped by
our makers for its office with teeth, tongue, and lips arranged
as now, for the sake at once of what is necessary and what
is best. They devised it as the passage whereby necessary
E. things might enter and the best things pass out ; for all
that comes in to give sustenance to the body is necessary ;
but the outflowing stream of discourse, ministering to
intelligence, is of all streams the best and noblest.

Skin, hair, nails.—The supreme importance of the head is
emphasised in the following description of skin and hair. These
are treated as if they were formed only on the skull for the protection of the brain.

75E. The head, however, could not be left merely of bare bone
because of the extremes of heat and cold in the seasons ;
nor yet could they suffer it to be so muffled in masses of flesh
as to become insensitive and dull. So from the flesh, which
76. was not entirely dried up in the process, there was separated
a film which was superfluously large [2]—' skin ' as we now

---

[1] ὁμοιότητι appears to be an instrumental dative (A.-H., Tr.) : no genuine
parallel is adduced for ὁμοιότητι = ὁμοίως (*aequaliter*, St. ; *symétriquement*,
Rivaud ; *uniformemente*, Fracc. ; *equally*, L. and S.). But why should
uniformity (A.-H.) or ' symmetrical disposition ' (Tr.) *weld* the sinews
together ? A. B. Cook (*Metaph. Basis of Plato's Ethics* 139) saw a contrast
between the skull, which ' has no flexions ', and the vertical column, which
the god made τῇ θατέρου προσχρώμενος δυνάμει (74A). ' Plato means that
the backbone is flexible, while the head is not.' This seems to imply that
κεφαλήν, not νεῦρα, is to be understood as the object of ἐκόλλησεν. If that
is possible, we might understand that the skull is ' welded together ' by its
uniform and continuous spherical shape (the sphere is the most ' uniform ' of
all figures, 33B), and so has no joints and needs no sinews, as the spine does,
κάμψεως ἕνεκα (74A). But it is difficult to believe that νεῦρα is not the object
of ἐκόλλησε.

[2] περιγίγνεσθαι, ' to be superfluous ' (not ' to be formed round ') goes with
μεῖζον, which could hardly stand without it. More skin is formed on the
fleshy parts of the face than these require, and it grows over the cranium
forming the scalp. If the drying of the flesh had been mentioned before,
the οὐ before καταξηραινομένης would be above suspicion ; but, as it has not, it
is odd to speak of the flesh as ' not in process of being entirely dried up ',

76. call it. This, owing to the moisture in the brain, grew and
closed in on itself so as to clothe the head all round ; and
the moisture rising up under the sutures watered it and
closed it, like a knot drawn together, on the crown. The
sutures are of very various patterns due to the action of the
revolutions and of the nutriment, being more or fewer in
number according as the struggle between those powers is
more or less intense.[1]

B. Now this skin was pricked all round with fire by the divine
part [2] ; and when the moisture issued forth through the
holes pierced in it, all that was purely moist and hot passed
away, but the part that was compounded of the same
ingredients as the skin was lifted by the motion and stretched
into a long thread outside, of a fineness equal in size to the
puncture ; but its movement was so slow that it was thrust
back by the surrounding air without and coiling back inside

C. under the skin took root there. To these processes is due
all the hair that grows on the skin : it is a thread-shaped
substance of the same nature as the skin, but harder and
denser as a result of the felting effect of the cooling, whereby
each hair is felted together [3] as it is detached from the skin.
When our creator made our heads shaggy with it, he used
the means above stated, but his thought was that this was
the right thing to serve, instead of flesh, as a covering to

D. protect the brain, both light and sufficient to provide shade
in summer and shelter in winter, without being an obstacle
to hinder readiness of perception.

Aristotle's account of the growth of hair is similar. ' No animal
has so much hair on the head as man. This, in the first place, is
the necessary result of the fluid character of the brain, and of the
presence of so many sutures in his skull. For wherever there is
the most fluid and the most heat, there also must necessarily occur
the greatest outgrowth. But, secondly, the thickness of the hair

when you mean that it was in process of being dried, but not entirely. Odd,
but not perhaps impossible ; and the οὐ may be kept as negativing the κατα-,
which would be out of place if οὐ were omitted : we should expect ξηραινομένης
as in Aristotle's τὸ δὲ δέρμα ξηραινομένης τῆς σαρκὸς γίνεται.. See Tr.'s note.

[1] The conflict which goes on in infancy, 43A ff. In this paragraph and the
two following the operation of ' Necessity ' comes to the front and Plato
speaks as if skin, hair, and nails had been developed by the blind action of
the primary bodies, unconsciously subserving a useful purpose.

[2] The fire in the brain, forcing its way upwards to seek its like. Cf. τὸ θεῖον,
69D, and τὸ θεῖον σπέρμα, 73C.

[3] So forming a natural felt hat (πῖλος), as Tr. remarks. See note on 74B
(p. 296, above).

in this part has a final cause, being intended to protect the head, by preserving it from excess of either heat or cold. And as the brain of man is larger and more fluid than that of any other animal, it requires a proportionately greater amount of protection ' (*Part. Anim.* 658b, 2, trans. Ogle).

76D. Further, where the fabric of sinew, skin, and bone is finished off [1] in fingers and toes, a compound of the three, when it is dried off, forms a single hard skin containing them all. Such were the means used in its making, but the true reason and purpose of the work was for the sake of creatures that were to be hereafter. For our framers knew that some day

E. men would pass into women and also into beasts, and that many creatures [2] would need nails (claws and hoofs) for many purposes ; hence they designed the rudiments of this growth from the very birth of mankind.

Such, then, were their reasons and purposes in causing the growth of skin, hair, and nails at the extremities of the limbs.[3]

The main structure of the human frame is now complete, with its covering of flesh, skin, and hair. We have next to consider the working of necessary functions entailed by the physical environment. The principal ones are the digestion of food and respiration. Various strands of necessary causation combine to produce the result that man must live in an atmosphere containing fire and air. This is necessary, not desirable ; he would be better off if, like the universe as a whole, he were not preyed upon from outside by heat and cold and all those ' strong powers ' which cause wasting away, disease, age, and death (33A). The world's body had no surrounding air that it must breathe, and no need of organs for receiving nourishment and getting rid of waste products (33C). But man is exposed to the assault of the elements and so needs constant repair.

[1] Taking καταπλοκή to be the point where a plait is finished off by making the ends fast (cf. καταστροφή). Hence Herod. iv, 205, τὴν ζόην κατέπλεξε, viii, 83, καταπλέξας τὴν ῥῆσιν. Hesych. καταπλακεῖσι· συνδεθεῖσι, περιπεπλεγμένοις.

[2] Beasts (not women), as Galen rightly understood (*U.P.* 1, 12[1]). Plato is neither anticipating Darwin nor following Empedocles. Women and beasts have not actually developed from men ; nor had anyone ever believed that they did. But Plato, having included transmigration in his mythical machinery, with the unusual and fantastic addition that men are imagined as existing at first alone, has to take this way of conveying that claws and hoofs in animals are more obviously useful to them than nails are to human beings.

[3] A.-H. inserted τε after τρίχας. I cannot believe (with Tr.) that ἔφυσαν δέρμα τρίχας is Greek for ' they made the skin grow into hair '. But if Plato could write γῆς πυρὸς ὕδατός τε καὶ ἀέρος (82A), no change may be necessary here. He takes many abnormal liberties with these conjunctions.

76E–77C. Plants

The account of respiration and digestion is prefaced by the mention of the necessary sustenance provided by plants, which are treated here because we cannot imagine man living with nothing to eat, and what he eats must be described before the machinery for disposing of food. No mention can be made of feeding on animals, since the beasts are to be postponed to the end.

76E. Now that all the parts and limbs of the mortal creature
77.   were united in a living whole, which, as the result of necessity, must spend his life surrounded by fire and air and be consequently dissolved and depleted by them and so waste away, the gods devised succour for him. They gave birth to a substance of a kindred nature to man's, but combined with other shapes and senses,[1] so as to be a living creature of a different sort. These are trees, plants, and seeds, now tamed and schooled by husbandry into domestication with us, though formerly there were only the wild
B.   kinds, which are the older. Anything that has life has every right to be called a living creature in the proper sense ; and the kind of which we are now speaking has the third form of soul, which, we said, is seated between midriff and navel ; this has nothing to do with belief or with reasoning and understanding, but only with sensation, pleasant or painful, and appetites. For it is always suffering all affections, but its formation has not endowed it with any power to observe the nature of its own affections and to reflect thereon [2] by

---

[1] Plants are very different in shape and appearance (ἰδέαις) from human beings and, although sensitive, they have not our organs of sense ; but their substance must be akin (συγγενῆ) to ours ; otherwise we could not feed on them. Cf. ἀπὸ συγγενῶν, 80D, 8, and τὸ συγγενές, 81A, 3, and B, 2, where the meaning is unmistakable. All plants and animals are composed of the same four simple bodies (cf. 82A and *Philebus* 29B, *Soph.* 266B).

[2] Reading φύσιν with W. and understanding (with A.-H.) κατιδόντι φύσιν τῶν αὑτοῦ which A.-H. rendered by ' observing its own being ' ; i.e. a plant, though conscious, is not self-conscious. A.-H. admitted that the expression is a little strange and doubted whether φύσει should not be preferred, as having an ' overwhelming preponderance of MS. evidence ' in its favour. On the other hand, φύσει adds nothing to the implications of γένεσις and creates a hiatus with οὐ following. See note on 20A. τῶν αὑτοῦ is vague ; it might mean ' its own concerns ' (cf. γνῶναι τὰ αὑτοῦ, 72A), or any ' parts of itself '. I have supposed that it takes its meaning from πάσχον διατελεῖ πάντα: τῶν αὑτοῦ (παθημάτων), ' its own passive affections ', which make up all that it is conscious of. Martin read φύσιν and translated : ' il ne lui a pas été donné de raisonner sur ce qui le concerne (λογισάσθαι τι τῶν αὑτοῦ) d'après la connaissance de sa propre nature (κατιδόντι φύσιν) ' ; but, as Tr. observes, κατιδεῖν φύσιν cannot mean ' observe its *own* nature '.

77C. revolving within and about itself, rejecting motion from without and exercising motion of its own. Therefore it lives, indeed, and is no other than a living creature, but it stands still, fixed and rooted, because it is denied self-motion.[1]

Plants are regarded as sensitive to the group of qualities discussed at 61C ff. as ' common affections of the whole body ' and attended by pleasure and pain (64A). They feel heat and cold, and some at least shrink from contact with hard or heavy or rough objects. The pleasurable or painful character of such contacts is supposed to be accompanied with some faint degree of desire to seek or shun. Galen observes that plants have the power of distinguishing and drawing to themselves congenial substances on which they feed, while rejecting those which are harmful. But plants have no perceptions such as we receive through the special organs of sense enumerated, with the corresponding qualities, from 65B onwards. Nor have they anything corresponding to the rational revolutions of the immortal soul seated in the brain of man. It may be for this reason that they are excluded from Plato's scheme of transmigration, though they were admitted to that of Empedocles.

77C–E. *Irrigation system to convey nourishment. The two principal veins*

The coming sections are obscure, at first reading, because Plato seems to be describing simultaneously digestion, the circulation of the blood, respiration and transpiration (through the skin), and even the transmission of sense-impressions. Some of these processes are dealt with very cursorily, and the anatomical connections between the various organs are left extremely vague.

The treatment of all these topics becomes more intelligible when we realise that Plato was partly occupied with a problem of hydraulics (ὑδραγωγία), which is to be explained by the interconnection of these systems. The whole body is nourished by the blood; blood is formed out of food in the belly and this is near the lower end of the trunk; how does the blood rise to the head and get distributed all over the body? Plato, like his contemporaries, drew no distinction between arteries and veins; nor had he any conception that muscular action of the heart had anything to do with the movement of the blood. The heart, in fact, is not

---

[1] Not all self-motion, since it has soul which is by definition the self-moving thing. Only motion from place to place is meant. As Galen remarks (*Comment*. p. 12, Daremberg), plants can grow upwards and downwards and attract nourishment.

named in this section from beginning to end ; the lung only once, as the destination of the windpipe.[1] The body is like a house with a cistern on the ground floor. If all unconscious and reflex muscular action is completely ignored, how is the water to be driven up a pipe to the attics, so that it may descend again to all the rooms ? The reader will be well advised to forget all he knows about the anatomy and functions of heart and lungs, veins and arteries, and set out in this state of ignorance, hardly exceeding Plato's own, to follow his solution of the mechanical problem, step by step.

From this standpoint it becomes clear why the discussion falls into the following sections. Plato first describes the two main conduits of the irrigation system (77C–E). We are then told how the blood is driven through these channels. The necessary force is supplied by the respiratory apparatus, which is here treated as if the pumping of blood were its main function (77E–79A). Respiration itself is explained as a mechanical operation. The motion is kept up by the natural movement of the internal fire towards its own kind, and by the circular thrust so imparted to the air (79A–E). There follows a digression on other examples of the circular thrust operating mechanically in lifeless things (79E–80C). We then learn how the action of fire in the belly converts the food and drink into blood, which this machinery drives through the veins to repair the wastage in all parts of the body (80D–81B). A final paragraph explains why growth occurs in youth, and later decay and old age set in, ending in natural death (81B–E).

The gods, having provided in plants the substances required to nourish all our tissues, like by like, have now to fashion an irrigation system, as a gardener cuts channels to carry water from the source of supply to every quarter of his garden. Plato describes only the two main vertical conduits.

77C.　Now when the higher powers had planted all these kinds as sustenance for our nature, weaker than their own, they made throughout the body itself a system of conduits, cut like runnels in a garden, so that it might be, as it were, watered by an incoming stream. First they cut as covered

D.　conduits, under the juncture of skin and flesh, two veins along the back corresponding to the twofold form of the body, with a right side and a left. These they brought down alongside the spine, enclosing between them also the generative marrow, in order that this might be kept in full

---

[1] 78C. The lung's office, ' to dispense breath to the body ' (as distinct from its cooling function, 70C), is mentioned for the first time at 84D.

77D. vigour and also that, by running downhill, the current might flow easily thence to the other parts and make the irrigation uniform.

E. Next, they split up these veins in the region of the head and plaited the ends so as to pass across one another in opposite directions, slanting those from the right towards the left side of the body and those from the left towards the right side. This was partly to provide the head with a bond helping the skin to connect it with the body, since there were no sinews holding it all round at the crown,[1] and further in order that the body as a whole might be informed of the effect of sense-perceptions coming from the members on either side.[2]

Aristotle,[3] after remarking on the extreme difficulty of tracing the course of the veins, reviews the statements of earlier writers, including Diogenes of Apollonia (frag. 6) and Polybus (*de nat. hom.*). Polybus traces four pairs of veins from the head downwards. Diogenes speaks of the two principal veins ' extending through the belly along the backbone, one to right, one to left ; either one to the leg on its own side, and upwards to the head, past the collar-bones, through the throat '. Plato's two veins appear to be these, which have been identified with the Hepatitis (right) and the Splenitis (left). Our knowledge of Diocles' doctrine about the blood-vessels is imperfect.[4] He regarded the heart as the source of the blood, and the aorta and the ' hollow vein ' as the two principal channels. Here he agrees with the older Hippocratic treatise *On Flesh*, which declares that ' there are two hollow veins from the heart, one called *arteria*, the other the " hollow vein " attached to the heart. The heart, where the hollow vein is, has the greatest amount of heat, and it dispenses the *pneuma* ' (chap. 5). Diocles also described the course of some of the minor veins.

Plato does not attempt to fill in even the barest outline of the circulatory system ; otherwise he must have mentioned the heart, which he called earlier ' the knot of the veins and the fountain of the blood which moves impetuously round through all the members ' (70B). This phrase suggests that the two dorsal veins here men-

[1] The sinews stopped short at the base of the skull (75D).

[2] We learnt at 70B that the blood, rushing outwards from the heart when anger boils up, conveys to all sentient parts a message from the brain, which has been warned by perception of some injury needing retaliation.

[3] *Hist. Anim.* iii, 2. See D'Arcy Thompson's notes in the Oxf. Trans. and Tr.'s notes.

[4] See Wellmann, *Frag. d. gr. Aerzte* 89 ff.

tioned must meet in the heart, as Aristotle [1] also held, adding that the heart can be regarded as part of them, since they extend both above and below it. We may add that one vein at least must extend below the heart to the belly, whence the blood formed there has to be raised to the head.[2] Thence a system of smaller conduits carries the blood ' downhill ' to water all the rest of the body and to convey to all the members the sense-impressions from the head. The passage is most easily understood, not as a grossly inadequate account of the circulatory system, but rather as formulating the mechanical problem of hydraulics. The blood can easily flow downhill through branches in all directions. But some force is needed to raise the blood from the belly to the top of the hill.

77E–79A. *Respiration as the driving power of the irrigation system*

The gods, accordingly, now provide for the carrying or driving of the water (ὑδραγωγία) along the conduits. The power is provided by the respiratory system, next to be described. Respiration, as Plato has already said, has other purposes. The lung was designed to receive breath and drink in order that it might cool the heart and so provide refreshment (70C). This function is barely mentioned here in a single word.[3] The only purpose dwelt upon is the mechanical one of keeping in movement the internal fire which is to digest food in the belly and raise the blood so formed from the belly into the veins. In this process the lung itself appears to play as little a part as the heart plays in circulation. The mechanism is to consist of currents of air and fire. It is first remarked that we may take it as possible for fire and air to penetrate the skin and flesh, because their fine particles will make their way through the coarser materials. There can thus be transpiration through the skin, as well as respiration through mouth and nose. This is essential to the process called the ' circular thrust '.

That the body transpires through pores all over the surface of the skin was taught by Empedocles :

'Thus do all things draw breath and breathe it out again. All have bloodless pipes of flesh spread over the surface of the body, and at their mouths the outermost surface of the skin is perforated all over with pores closely packed together, so as to

---

[1] *Hist. Anim.* iii, 3, 513a, 15. See Thompson's note.

[2] Diocles spoke of ' veins which receive the nourishment from the belly ', Galen viii, 187.

[3] ἀναψυχομένῳ, 78E. Aristotle held that cooling of the internal heat is the proper function of respiration.

keep in the blood while an easy passage is cut for the air to pass through. Thus whenever the thin blood rushes back from these, the bubbling air rushes in with an impetuous surge ; and when the blood leaps back again, the air is breathed out once more ' (frag. 100).

Empedocles' doctrine was reproduced by Philistion,[1] who taught that the purpose of respiration is to cool the natural heat of the body and that health depends on the unimpeded passage of the breath, not only through mouth and nostrils, but all over the body. Diocles also held that the body has a natural heat residing in the blood, which conveys life and movement in the veins throughout the whole frame. His account of the cycle of respiration was the same as Plato's : inhalation (or exhalation) through mouth and nose coincides with exhalation (or inhalation) through the pores. In opposition to the Coan school, which held that the breath first reaches the brain and is then dispersed throughout the rest of the body,[2] the Sicilians taught that the heart is the central seat of the breath of life or breath-soul (ψυχικὸν πνεῦμα), which passes thence to the rest of the body through the veins and is the power that moves the limbs.[3] This breath also conveys sense-perception. It is in perpetual motion, circulating through the veins together with the blood. According to this doctrine, then, the breath and the blood travel together through the same channels : respiration and the circulation of the blood are a single process ; and since the blood actually consists of the digested food, the same system conveys to all parts of the body their proper nourishment.

77E.
78.
They then proceeded to provide for the water-carrying in a manner now to be described, which we shall the more easily grasp if we first agree upon the following principle. All bodies composed of smaller particles are impervious to larger particles, but those consisting of the larger are not impervious to the smaller ; and of all the kinds fire has the smallest particles and consequently passes through water, earth, and air and all bodies composed of these, and nothing is impervious to it. This principle must be applied to our

---

[1] See O. Gilbert, *Meteorologischen Theorien*, pp. 344 ff. Wellmann, *Frag. d. Gr. Aerzte 70.*

[2] [Hippocr.] π. ἱερ. νουσ. 16. This treatise is held to be earlier than Diocles. Wellmann, op. cit. 77 ff.

[3] At *Crat.* 399D Plato connects ψυχή with τὸ ἀναψῦχον, as having the function of breathing and cooling the body ; and again with φύσιν ὀχεῖν (φυσέχη), because it moves the body. Cf. Diocles, frag. 17, and Anon., Lond. xxxi, 54 : the soul, being pneuma, is light and the whole body is carried (βαστάζεται) by the soul.

78. belly[1] : when food and drink fall into it, it keeps them in ;
 B. but it cannot keep in the air we breathe and fire, since their
    particles are smaller than those of its own structure.

The Weel or Fish-trap.—The god now avails himself of this
penetrating capacity of fire and air, to provide a mechanism supply-
ing the power to drive the blood-food upwards into the veins from
the belly. This mechanism is the respiratory system. It consists
of currents of air and fire, and of nothing else. The currents pass
in and out of the body by certain routes which, in a modern book,
would be represented by lines in a diagram. In Plato's dialogues
diagrams are not used[2] ; accordingly, the pattern formed by the
courses of the currents is visualised as the outline of a well-known
object, the fisherman's weel. The god is said to construct a network of this shape and then fix it in and around the ' living creature
he had moulded '. The next sentences, describing the fashioning
of the weel, are to be understood as if Plato were drawing a picture,
the lines of which stand for the routes followed by currents of air
and fire.

78B. The god accordingly made use of these (air and fire) for the
    water-carrying from the belly to the veins, weaving out of
    air and fire a network, after the fashion of a fisherman's
    weel. This had a pair of funnels (ἐγκύρτια) at the entrance,
    one of which again he made fork into two ; and from these
    funnels he stretched, as it were, reeds[3] all round throughout
    the whole length to the extremities of the network. The
 C. whole interior of the basket he composed of fire, while the
    funnels and the main vessel were of air.

In order to reconstruct Plato's diagram of the currents, it is
necessary first to be clear about the construction of the fish-trap
or weel (κύρτος, Lat. *nassa*) and the meaning of ἐγκύρτιον, above
translated ' funnel '.

The weel, like our lobster-pot, is a basket (πλόκανον), with a wide
opening at the top. Stretching down inside and below the opening

---

[1] Here and below κοιλία probably means, not the whole hollow of the trunk,
but the belly, where fire reduces meat and drink to blood (80D). Plato is
specially thinking of this as the first operation, which must be performed
before the blood is driven upwards (see 78E–79A). It is of course also true
that fire and air can penetrate the skin anywhere.

[2] Thus, even in the *Meno*, where a difficult problem of geometrical construction is discussed, the lines and areas are described but not indicated by
letters, as they would be in Aristotle.

[3] The *nassa* was sometimes made of reeds (*iunci*, Plin., *N. H.* xxi, 114).
Hesych. κύρτος· ἀγγεῖον σχοινῶδες, ᾧ οἱ ἁλεῖς χρῶνται. Oppian, *Halieut.* iv,
53.

there is passage in the form of a truncated cone, narrowing down to a hole just large enough to admit the fish. The principle is the same as that of the cheap inkpots in which the glass forming the wall is bent over and downwards to form a funnel which prevents the ink from escaping if the pot is upset.[1] The construction of the *nassa* is precisely described by Silius Italicus, to illustrate how the Romans were lured into the narrowing path between the cliffs and the shore of the Trasimene lake:

‘ So by the crystal waves the cunning fisherman weaves osiers to make his light weel (*nassa*) with its wide mouth. The inner part (*interiora*) he ties with special care ; he brings the ends together, making them taper gradually along the middle of the belly (*aluum*) and fastens them together. By the trick of this contracted aperture he keeps the fish drawn from the sea from making its way back, though it found the way in easy.’ [2]

Such being the weel, what is the ἐγκύρτιον ? Some commentators have been misled by Galen into supposing that ἐγκύρτιον means a smaller weel inside a larger one. He wisely recommends those who live near the sea to go and look at the fisherman's weel. Midlanders will at any rate have seen the kind of basket called τάλαρος.[3] If they will imagine this basket without the perforations at the base and with its mouth at the top opened wide, they will have a sufficient picture. Evidently he himself had seen a weel, and it is, he adds, a *single* basket (πλέγμα ἁπλοῦν) ; but he now tells us to imagine that it contains another small basket of the same shape, standing on the bottom of the big one, but with its opening much lower down. We are then to imagine yet another small basket inside the big one. These small baskets are, he says, what Plato means by ἐγκύρτια.

Galen can be convicted, out of his own mouth, of not knowing the meaning of ἐγκύρτιον. Plato speaks of it as being a part of the whole weel, the other part being the main vessel or belly (κύτος, Silius' *aluus*). That is to say, it is the name of an actual part of an actual thing, the weel in common use. But Galen himself tells

---

[1] Apostolides, *La Pêche en Grèce* (Athens, 1907), p. 51 (cited by A. W. Mair, *Oppian*, Loeb Library, 1928, p. xlvi), writes : ‘ *La pêche au moyen de nasses est bien simple, mais toutes n'ont pas la même forme : elle change suivant les poissons qu'on cherche à capturer. Ce sont des paniers, avec un orifice précédé d'une entrée conique, par laquelle, une fois entrés, les poissons ne peuvent plus sortir.*’

[2] Cf. Lucian, *Merc. Cond.* 3, τῶν κύρτων τὸ ἀδιέξοδον ἔκτοσθεν ἐπὶ σχολῆς ἀλλὰ μὴ ἔνδοθεν ἐκ τοῦ μύχου.

[3] τάλαρος commonly means a basket shaped like a wastepaper basket with a mouth much wider than its base. This is evidently not the sort of basket Galen means, which Daremberg understood as the basket used for making cream-cheese out of curdled milk (Hom., *Od.* i, 247).

us that the weel is a 'single basket', which does not contain these imaginary smaller baskets.[1] There existed, therefore, nothing to bear the name ἐγκύρτιον in the sense of 'a small weel inside a large one'. If a Byzantine scholiast had glossed the word ἐντάφιον as 'a small tomb inside a larger one', we might be suspicious; if he added 'though no such tombs actually exist', we should know he was romancing, because there are no current names for non-existent parts of well-known objects. Plato uses ἐγκύρτιον as a current name, not as one which he had invented for an imaginary addition to the structure of the actual weel. The gloss [2] describing ἐγκύρτια as 'plaited structures inside a weel' is vague; but it at least recognises that the name belongs to some existent part of a weel; and it also states that Plato applied the word to the pharynx, not to the cavities of the lungs and the belly, to which Galen's imaginary baskets are supposed to correspond.

The conclusion is that ἐγκύρτιον means the essential feature differentiating the weel from other baskets, namely the cone-shaped funnel. The 'inner part' of this, as Silius says, is made by bending inwards and downwards the ends of the osiers or reeds forming the wall of the belly. Their points, set round the small opening at the bottom of the funnel, repulse the trapped fish if he tries to escape.[3] There are two ways in which the funnel can be formed. The simplest is illustrated in the common lobster-pot, which is roughly spherical, the upright osiers of the sides being bent round, so that the funnel is sunk within the main outline of the basket. In another type, illustrated here, the funnel projects above the belly, like the wicker funnel which is said to have been affixed to the voting-urn, so as to admit only the ballot.[4] In either case the funnel is, as

---

[1] It is for this reason that Galen says that, even if you have seen a weel, it is hard to understand Plato's meaning.

[2] *Lex. Plat.* ἐγκύρτια· τὰ ἐν τοῖς κύρτοις ἐνυφάσματα· χρῆται δὲ ἐν Τιμαίῳ ἐπὶ τῆς φάρυγγος τῇ λέξει (repeated by Suidas).

[3] Oppian, *Halieut.* iv, 47, describes how the fish, trapped in the 'backward-plaited' weel (ἐν κύρτοισι παλιμπλεκέεσσιν), tries to escape. 'He dreads the sharp rushes (σχοίνους) which bristle around the entrance and as he comes against them wound his eyes, even as if they were warders of the gate' (trans. Mair).

[4] Schol. Ar., *Eq.* 1150: κημός . . . πλέγμα τι ἐκ σχοινίων γινόμενον ὅμοιον ἠθμῷ, ᾧ τὰς πορφύρας λαμβάνουσιν. Soph., frag. 504, Pearson: κημοῖσι πλεκτοῖς πορφυρῶν (Herw., πορφύρας codd.) θηρᾷ (Tucker, φθείρει codd.) γένος. Opp., *Halieut.* 5, 598: πορφύραι are caught with κυρτίδες ἠβαιαὶ ταλάροις ὅμοιαι. Ar., *H.A.* 534a, 20: eels are caught in earthen pickle-pots into the mouth of which the so-called ἠθμός has been inserted. κημός meant also the funnel at the top of the κάδος or κάδισκος (balloting urn): Ar., *Vesp.* 754, κἀπισταίην ἐπὶ τοῖς κημοῖς ψηφιζομένων ὁ τελευταῖος Jebb in Pearson, Soph., *Frag.*, vol. ii, 154, where most of this evidence is collected). Hesych., κημός

Plato says, ' at the entrance ' ($\varkappa\alpha\tau\grave{\alpha}\ \tau\grave{\eta}\nu\ \epsilon\check{\iota}\sigma o\delta o\nu$), whereas Galen's imaginary small baskets are not at the entrance ; one of them is at the bottom of the whole vessel with its opening a good way below the mouth of the main basket. This alone shows that Galen

was wrong to imagine that the $\grave{\epsilon}\gamma\varkappa\acute{\upsilon}\varrho\tau\iota\alpha$ stood for the cavities of the belly and the lungs.[1] The two funnels, as originally constructed, answer to the two passages for food and breath, through mouth and nose, which terminate in the throat, one leading into the

. . . $\tau\grave{o}\ \grave{\epsilon}\pi\iota\tau\iota\theta\acute{\epsilon}\mu\epsilon\nu o\nu\ \tau\hat{\eta}\ \tau\hat{\omega}\nu\ \delta\iota\varkappa\alpha\sigma\tau\hat{\omega}\nu\ \grave{\upsilon}\delta\varrho\acute{\iota}\alpha\ \pi\epsilon\pi\lambda\epsilon\gamma\mu\acute{\epsilon}\nu o\nu\ \pi\hat{\omega}\mu\alpha\ \pi\alpha\varrho\acute{o}\mu o\iota o\nu\ \chi\acute{\omega}\nu\eta$ (funnel.) The accompanying sketch is copied from an eighteenth-century map of Cambridgeshire in my possession, showing a man drawing out of shallow water a round wicker weel, surmounted by a funnel. A boy has thrust his hand inside and pulled out a captured eel. In the Ethnological Museum at Cambridge there is a string of plaited weels of this pattern from Africa. The device is no doubt world-wide. The other illustration is ' composed from two Roman mosaics, in both of which it is represented as lying half-buried among sedges in a shallow piece of water ', Rich, *Dict. of Rom. and Gk. Antiq.*, s.v. *nassa* (cf. Daremberg and Saglio, s.v. *colum*). Naturally neither picture attempts to show the narrowing projection of the cone inside the vessel. This is invisible, if the texture of the outer wall is close. It can be seen in the African weels above mentioned.

[1] Stallbaum rightly understood that they correspond rather to the *passages into* the lungs and the belly, though other features of his interpretation seem unsatisfactory. So L. and S. ' $\grave{\epsilon}\gamma\varkappa\acute{\upsilon}\varrho\tau\iota\alpha$, passages into the $\varkappa\acute{\upsilon}\varrho\tau o\varsigma$ or creel or fish-trap, to which Pl. compares the throat '. The reference to Galen (added in the last edition) is, however, likely to mislead. Rivaud translates $\grave{\epsilon}\gamma\varkappa\acute{\upsilon}\varrho\tau\iota\alpha$, ' *tuyaux* '.

œsophagus, the other into the trachea. Like the fisherman in Silius, the god constructs the funnels first. He makes a pair of them, instead of the usual one. He further makes one of them fork into two. This probably means the division of the breath-funnel into two entrances, through mouth and nose; for we can breathe, as well as swallow our food, through the mouth. As all agree, Plato's comparison is a somewhat clumsy substitute for a diagram.

The god then proceeds to construct the main vessel (κύτος) of the basket by ' stretching from the funnels reeds all round through-out the whole length to the extremities of the network '.[1] In the Roman *nassa* illustrated above the reeds supporting the funnel extend all along the weel and are brought together at the base. The shape is now complete. We are finally reminded that the entire structure consists only of fire and air. ' The whole interior ' (τὰ ἔνδον ἅπαντα) is of fire, while ' the funnels and the main vessel are of air '. That is to say, all the lines in the diagram represent channels or currents of air. The space within the figure is for the present to be imagined as occupied by fire. As we hear later, ' in every living creature the inner parts about the blood and veins are the hottest, like a well-spring of fire which it has within itself. It was, indeed, this that we likened to the network of our weel, when we said that the whole extent of the central part was woven of fire, while the parts on the outside were of air ' (79D).

This structure is now applied to the living creature, the funnels being let into mouth and nose, while the outline of the main vessel encloses the trunk on the outside. The funnels require a certain amount of adjustment and adaptation. At their lower ends they are prolonged to form the trachea and the gullet. At the upper end the breath-funnel has already been divided into two outlets, mouth and nose, in order that we may be able to breathe while the mouth is occupied with eating.

78c. This structure he took and set it about the living creature
he had moulded, in this way. The part consisting of the
funnels he let into the mouth; and this part being twofold,

---

[1] This sentence is ambiguous. The reeds were understood by Galen as representing arteries and veins extending from the belly and lung (Galen's ἐγκύρτια) *outwards* in all directions to the surface of the body and containing the ' rays of fire ' mentioned at 78D. But there is no corresponding structure in the fish-trap. κύκλῳ διὰ παντός can mean *along the whole surface* '; cf. Albinus, *Didasc.* xiv : συνέβη αὐτὴν (τὴν ψυχὴν) τὸ σῶμα τοῦ κόσμου κύκλῳ διὰ παντὸς περιέχειν καὶ περικαλύψαι, = *Tim.* 36E, κύκλῳ τε αὐτὸν ἔξωθεν περικαλύψασα. At 74A, 3, διὰ παντὸς τοῦ κύτους means ' along the whole extent (length) of the trunk '.

78c.   he prolonged [1] one of the funnels downwards by way of the windpipes into the lung, and the other alongside the windpipes into the belly.   The first funnel he divided into two parts, to both of which he gave a common outlet by the channels of the nose,[2] so that when the other passage was not working by way of the mouth, all its currents also might be replenished from this one.

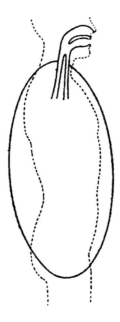

The funnels having been adjusted, the main vessel is now fitted round the trunk on the outside. Galen understands the vessel to represent the skin and the layer of air in contact with it. If we omit the skin (which does not sink into the body), we can accept his view, with the qualification that the outlines of the vessel really mark the limits of a coat of air enveloping the trunk.

78D.   The rest, the main vessel of the weel, he attached round all the hollow part of the body. And all this he caused at one moment to flow together inwards on to the funnels —softly, because they are made of air; while at another

---

[1] καθῆκεν, like καθιέναι πώγωνα, ' to let your beard grow long ', and καθῆκαν at 77D, 3.   It should be remembered that the lung contains no blood (70c).   Its passages are filled with air and fire.

[2] The breath-funnel was originally forked into two passages, the mouth and the nose (78B).   I understand the present operation to be the splitting of the nose-passage into the two nostrils (so A.-H.).   This completes the system of passages.

78D. moment the funnels flow back, and the network sinks in through the body—for this is porous—and then out again ; meanwhile the rays of fire stretched through inside follow the movement of the air in either direction. This process was
E. to continue without ceasing so long as the mortal creature holds together ; it is indeed the process which the name-giver entitled inhalation and exhalation.

This account is intelligible if we remember that the whole outline of the structure—funnels and main vessel—consists of air. When we breathe in through the mouth, the currents of air converging along all sides of the body flow together (συρρεῖν) inwards upon the 'funnels', i.e. the columns of air defined by the mouth and nose and the gullet and trachea. This happens 'softly', without shock, because it is only air that is pouring in upon air. When we breathe out, the columns of air flow back again towards, and out through, the mouth and nose. The external currents are thus reversed by the pressure of the breath, and the external air so displaced (represented by the outline of the main vessel) sinks into the body through its pores, to fill the space inside that would otherwise be left empty by the air we have breathed out. The process is then reversed again. The air which has entered through the pores passes out again *by the same route* and sets the external current moving again towards the mouth. Why this reversal takes place will be explained presently. Here it is only added that the 'rays of fire' stretched through the interior of the structure follow the movement of the air in either direction. The fire which occupies the interior is described as separated into rays because it passes along the same channels or pores as the air. In a diagram these might be roughly indicated by lines radiating outwards in all directions from the heart. The statement refers specially to the process which Galen describes as transpiration (διαπνοή) ; and the channels in question appear to be the veins (79D), or rather still finer passages, at the ends of the veins, which will keep in the blood, but just permit fire and air to get through.

The process of respiration is now to be connected with digestion and with the conveyance of nourishment in the blood through the irrigation system. Apart from the mention of the cooling effect which the lung has upon the heart (as we were told earlier, 70C) the emphasis falls on the mechanical use of respiration, as the force which maintains two processes : (1) by keeping the internal fire constantly in motion, it enables the sharp fire-particles to penetrate and cut up the food and drink which have reached the belly ; and then (2) to carry the blood so formed along with its own movement

314

through the veins. This theory is in substantial agreement with
Diocles.[1]

78E.   All this that our body does and has done to it results in its
being nourished and keeping alive as it is watered and
cooled ; for every time that, as the breath passes in and out,
the fire within connected with it follows its movement and
in its perpetual rise and fall passes in through the belly and
79.   takes hold upon the meat and drink, it dissolves them and,
dividing them up small, drives them through the outlets in
the direction of its advance, discharging them into the veins,
as water from a spring into runnels, and making the currents
of the veins flow through the body as through an aqueduct.

79A–E. *Respiration maintained by the circular thrust*

We have seen that the whole process of ' watering ' and feeding
the body is kept up by the rhythmical movement of respiration.
Plato next attempts to show that respiration itself is maintained
mechanically. He represents it as going on without any impulse
from muscular contraction or from the intervention of the will.
He invokes the principle of the ' circular thrust ' ($\pi\epsilon\varrho\iota\omega\sigma\iota\varsigma$), which
is now explained.

The principle, being purely mechanical, has wider applications
mentioned in the later paragraphs, e.g. to medical cupping-instru-
ments and projectiles. As Aristotle's discussion of the circular
thrust shows, certain problems had arisen from the denial of the
void. Parmenides had denied the existence of any void spaces
inside the universe, and he had rejected the possibility of any
motion. The two propositions were connected, since it was held
that nothing could move without some empty place to move into.
Accordingly, when the Atomists reasserted the void, they were
restoring the possibility of motion. Others, however, such as
Empedocles and Anaxagoras, believed motion to be possible while
still denying that there was any real vacancy.

The projectile presented two problems. How could it, or any
other body, move at all, if there was no vacant room to move
into ? And how could a missile continue in motion after it had
left the hand which hurled it ? Both were solved by the theory
of the circular thrust. The moving body, since it does not advance
into vacancy, displaces the air in front of it. At the same time
the air must be closing up the vacancy that would otherwise be
left behind the projectile. Thus a stream of air is formed, pouring
from the front of the moving body to its rear. The force of this

---

[1] Wellmann, op. cit. 89.

current is now invoked to account for the body continuing in motion after it has lost contact with the source of impulsion. The air pouring round to the rear is supposed to push the projectile farther on its way, until the force of the original impulse somehow dies out.

The application of this principle to the process of respiration is stated first. Our breath is a sort of projectile impelled out of our mouths. Since it does not issue into empty space, it must dislodge the air near the mouth. At the same time it must leave no empty space inside us. Apparently it was assumed that the contraction of the body in expiration did not suffice to close up the vacancy inside or to provide space outside for the dislodged air to move into. Hence the theory that the expelled breath is replaced by a current which enters the body through pores in the flesh.

79A. But let us once more consider the means whereby the effect of respiration has come to take place as it now does. It was in this way. Since there is no vacancy into which any

B. moving body could make its way, and the air we breathe does move out from us, the consequence is at once plain to anyone : it does not go out into vacancy, but thrusts the neighbouring air out of its place. What is so thrust keeps on displacing its neighbours successively, and in the course of this compulsion the air is all driven round and enters the place whence the breath came out, refilling it as it follows the breath. All this goes on simultaneously,[1] as when a

C. wheel is driven round, because there is no vacancy. Consequently, the region of the chest and lung, in the act of discharging the breath outwards, is filled again by the air surrounding the body, as it is driven round and makes its way inwards through the porous flesh. Again, when the air is turned back and is moving outwards through the body, it thrusts round the respiration inwards by way of the passage of mouth and nostrils.

The last sentence reminds us that our breath does not, after all, behave quite like a projectile, which continues on the same course so long as the circular thrust goes on and the impulse has not died away. We do not, in fact, go on breathing outwards indefinitely, although the circular thrust should make this possible. After quite a short time we begin to inhale again, and the current must flow the other way. There is nothing yet to explain this reversal. An explanation is now sought by considering the original source of the impulse which must be supposed to have started the whole

[1] Not leaving any interval of time during which there would be a space left unfilled.

process. This is found in the fire contained in all living things, as a sort of well-spring of movement. The fire has its own natural tendency to move towards its like. There is fire in the warm breath we exhale, and this carries the air with it outwards through the mouth and nose towards the main body of fire all round the universe. When the breath gets outside it encounters colder air, and the fire in it will presumably continue its journey and pass out of the expelled air. So the breath is cooled outside. Meanwhile the dislodged air is pouring round, by the circular thrust, into the body through the pores. It reaches the internal fire and is heated thereby.

We might now expect that this freshly heated air would travel out through mouth and nostrils and keep up a continual process of exhalation. The breath would thus behave like the projectile, which is precisely the result we are seeking to avoid. At this point the explanation becomes obscure, because it is tacitly assumed that the air which comes in through the pores must also go out through the pores and not join the current passing out through the mouth.[1] Perhaps the assumption is tacit because it seems so improbable. Once it is made, the reversal can be explained. The air which has come in through the pores, having now been heated by the fire inside, will pass out again through the pores to seek its like. This will reverse the circular thrust and drive the cooled air near the mouth back into the body in inhalation.

79C.   We must suppose that the starting of this process is to be explained as follows. In every living creature the inner

D.   parts about the blood and veins are the hottest, like a fountain of fire which it has within itself.[2] It was, indeed, this that we likened to the network of our weel, when we said that the whole extent of the central part was woven of fire, while all the parts on the outside were of air. Now we must agree that the hot naturally moves outwards towards its kindred in its own region ; and that, since there are two ways through, one leading out by way of the body, while the other is by way of mouth and nostrils, whenever the hot makes for

E.   the air in one quarter, it gives a thrust round to the air in the other quarter ; and the air so thrust round, falling into the fire, is heated, while the air which passes out is cooled.

[1] See Tr.'s note, p. 562.

[2] Presumably the principal seat of this fountain of fire in blood and veins is the heart, ' the knot of the veins and the fountain of blood ', whose throbbing when it swells with passion is ' caused by fire ' (70A–C). See Wellmann's reconstruction of Diocles' doctrine, which has many points of contact with Plato (*Fragm. d. Gr. Aerzte* 44 ff., and 219).

This last sentence applies equally to both routes. Whether the air enters by the mouth or through the pores, it alike reaches the central source of heat and is warmed there. And, whichever way it goes out, it is cooled when it gets outside. The obscurity lies in the next statement, because the assumption that air must go out by the same way that it came in is not openly made. This being granted, the warmth of either lot of air will be constantly changing. The air which comes in at either passage will be warmed when it reaches the internal fire, and cooled again when it passes out, by its own route, to the atmosphere outside. As soon as it has got warm at the fire it will seek to pass out again towards the main mass of fire and so reverse its direction and set up a thrust the other way. If we now assume that inhalation (or exhalation) by one route alternates with inhalation (or exhalation) by the other route, there will be a rhythmical reversal of the current, comparable to a wheel turning first one way, then the other.

79E. And as the warmth changes and the air which travels by way of the one outlet [1] gets warmer, this warmer air is the more inclined to take the reverse direction by that route, moving towards its like, and gives a circular thrust to the air which travels by the other passage. This again suffers the same effect and reacts every time in the same way. So it sets up, under the two impulses, a motion like that of a wheel which swings now this way, now that, and thus it gives rise to inhalation and exhalation.

Aristotle sums up the theory quite clearly. ' It is said (in the *Timaeus*) that when the hot air issues from the mouth it pushes the surrounding air, which being carried on enters the very place whence the internal warmth issued, through the interstices of the porous flesh ; and this reciprocal replacement is due to the fact that vacuum cannot exist. But when it has become hot the air passes out again *by the same route*, and pushes back inwards through the mouth the air that had been discharged in a warm condition. It is said that it is this action which goes on continuously when the breath is taken in and let out ' (*De Resp.* 472*b*, 12, Oxf. trans.).

Aristotle rightly saw that this paragraph is concerned with the question how the whole process of respiration can start and proceed mechanically without involving volition. The new-born child does not, in fact, begin to breathe deliberately. Plato probably did not know that the midwife slapped the baby on the back to make

---

[1] This must be the meaning of κατά, as at D, 7, κατὰ τὸ σῶμα and c, 6, not the air ' *near* ' or ' *about* ' the outlet, if this means *outside* it, because expelled air gets cooler, not hotter.

it gasp; and he supposed that the motion was mechanically started by the natural impulse of the internal fire seeking its like.[1] Moreover, breathing normally proceeds without any conscious effort. Not having our conception of reflex muscular action, he thought that the process is maintained throughout life by the blind action of inanimate particles. It is true, as Galen remarks, that we can voluntarily breathe faster or slower; but it is also true that the vast majority of the breaths we draw are not voluntary actions. It therefore seemed necessary to supply a mechanical cause. Galen comments on the fact that Plato makes the whole process involuntary and ignores muscular action (*Hipp. et Plat.* 715 ff.).

79E–80C. *Digression. Other phenomena explained by the circular thrust*

Plato now interrupts his account of the irrigation system by a short digression. The principle of the circular thrust will help to explain a number of other phenomena, which had been falsely supposed to involve the existence of void or of a power of 'attraction' which Plato will not recognise.

79E. To this principle, moreover, we may look for the explanation
80. of what happens in the cases of medical instruments for cupping, of the process of swallowing, and of projectiles, which keep moving after their discharge either through the air or along the ground.

Plutarch [2] explains how the circular thrust (or *antiperistasis*, as it was called in Aristotle and later), eliminates the need of a void in these various cases. (1) Some of the air inside the physician's metal cup is converted into fire and so becomes fine enough to escape through the pores of the metal and give a thrust to the air outside, which is passed on till it exerts pressure on the flesh. So the flesh rises inside the cup and fills the space vacated by the fire which has escaped. (2) Swallowing again involves reflex actions which Plato ignores. And how can our food get down, if there is no void to receive it? [3] According to Plutarch, when the tongue presses the food into the throat, the air underneath is squeezed out towards the roof of the mouth and then helps to thrust the food

---

[1] Aet. iv, 22, 1, ascribes to Empedocles an explanation of how the first animal drew its first breath, as well as an account of respiration similar to Plato's (*Vors.* Emped. A74).

[2] *Plat. Qu.* vii, 1004D ff.

[3] Cf. one of the arguments for a void in Ar., *Phys.* iv, 6 (213*b*, 19): 'Everyone supposes that growth is due to the void; for our food is a body, and two bodies cannot occupy the same place.'

down. (3) Projectiles are pushed forward in the same way. The circular thrust accounts for the missile continuing its flight, or a ball rolling along the ground, after it has left the hand.

The account of concord and dissonance in musical sounds, which was promised earlier (67c), can now be given, evidently because the circular thrust is somehow involved. Its obscurity is due to the fact that here, as so often before, Plato reduces his explanation of a highly complicated process to a single sentence. He does not describe how the whole process takes place, but concentrates upon the one most important feature. Unless we can reconstruct the process as a whole, we cannot even understand what feature he is explaining. He tells us that the agreeable concord of two sounds of higher and lower pitch depends on the ' likeness ' (ὁμοιότης) of certain motions in us. Assuming for the moment that ' likeness ' must mean *correspondence*, we may translate as follows :

80A. This principle will also explain why sounds, which present themselves as high or low in pitch according as they are swift or slow, are as they travel sometimes inharmonious because the motion they produce in us lacks correspondence, sometimes concordant because there is correspondence. The slower sounds, when they catch up with the motions of the quicker sounds which arrived earlier, find these motions drawing to an end and already having reached correspondence

B. with the motions imparted to them by the slower sounds on their later arrival. In so doing, the slower sounds cause no disturbance when they intrude a fresh motion ; rather by joining on the beginning of a slower motion in correspondence with the quicker which is now drawing to an end, they produce a single combined effect in which high and low are blended. Hence the pleasure they give to the unintelligent, and the delight they afford to the wise, by the representation of the divine harmony in mortal movements.

In reconstructing the acoustic theory here implied we must avoid attributing to Plato modern theories of the transmission of sound by vibrations or wave-motions.[1] Nothing whatever is said

---

[1] A.-H. (p. 300) rightly sees that, according to Plato, sound is transmitted by a travelling body of air ; but he speaks of this body as ' vibrating ' and communicating its vibrations to the *ear*. The theory of consonance he deduces is, he says, ' entirely unsatisfactory ', for obvious reasons which he states. Tr. (p. 576) speaks of *vibrations* propagated more or less rapidly through the air to our *tympanum*, and of consonance as resulting from a uniform *wave-movement*, compounded of two *waves*. He then declares that ' the whole theory is quite perverse ', again for obvious reasons.

about waves; and if we introduce vibrations, we must be quite clear what it is that vibrates.

'Sound', we were told (67A), 'is the stroke (or shock) inflicted by air on the brain and blood through the ears, and passed on to the soul; while the motion it causes, starting in the head and ending in the region of the liver, is hearing (ἀκοή).' The swifter the motion, the higher the pitch of the sound heard.[1] The present passage throws some more light on the two motions here distinguished. (1) One motion takes place in the air outside, between the source of sound and the stroke inflicted directly on the brain and blood (not the ear-drum, of which nothing is said: the ears are treated as open channels). (2) The other motion is set up by this stroke in the whole region of our bodies frcm brain to liver. This motion is called 'hearing', though there is no sensation unless, or until, the motion 'reaches the soul' (consciousness). These two motions must be separately considered.

(1) The first, external, motion Plato calls the 'sound' (φθόγγος)— a musical sound, not a mere noise. It starts with a blow inflicted on the air by (say) the string of a lyre when it is plucked. It ends with the blow inflicted by the disturbed air on the blood and brain. There are two ways in which a shock might be transmitted. Each particle of air in a chain from source to brain might pass on the shock to its neighbour; the shock would then travel, not the particles. This cannot be intended here, because there would then be no circular thrust involved. Apparently we must suppose that a portion, or portions, of air travel from source to brain.[2] Such a portion, as the context implies, is like other projectiles: its advance sets up a circular thrust which keeps it in motion till the impulse dies away.

It is clearly implied that the notes in question are not indefinitely prolonged, like those of a flute. We are to imagine the simpler case in which a string is plucked once for all. The question remains, however, whether only one projectile is despatched, or Plato takes account of the vibration of the string after the blow, and thinks

---

[1] It was added that a sound is 'similar' (ὁμοίαν) or 'regular' when it is uniform and smooth; irregular, when it is rough. The word ὅμοιος is there applied to a single sound. In the present passage it describes a relation between two sounds, on which their concordance depends. Hence 'similar' here must mean 'corresponding' in some way.

[2] So Beare, *Gk. Theories of Elem. Cognition*, 109: 'Not the rapidity of vibrations *in* air, but that of the mere onward movement of air or portions of air, seems to have been for Plato the producing cause of height in tones. Moreover Plato, like his predecessors, believed that a definite portion of air was projected forwards from the sonant body to the ear; not that a *mere movement* took place in the medium.'

of each vibration as despatching a fresh projectile after the first, until the string comes to rest. The sound yielded by a plucked string does, in fact, go on humming for a short time and then dies away. Plato recognises this where he speaks of the internal motion ('hearing') as 'drawing to an end' when it is overtaken by the later sound. This might be due only to the behaviour of the internal motion, continuing after receiving a single blow. But a sustained flute note could not be explained by a single shock. It is more likely that the vibrating string sends off a succession of projectiles, which would account for the continued humming, in contrast with the abrupt noise produced (say) when a book is dropped on the floor.

It is not rash to assume that Plato was acquainted with the acoustics of Archytas, who held a closely similar doctrine. There could, argued Archytas, be no noise of any kind if one body did not inflict a blow on another. All musical sounds are motions started by a blow, and these motions follow one another at shorter or longer intervals, making the sound heard correspondingly high or low. If a string is tuned too high, we lower the tension, so reducing the vibration and lowering the pitch. It follows that a tone consists of parts, which must be related in numerical proportion.[1]

We may, then, conjecture, with some probability, that Plato conceived a succession of projectiles travelling from source to brain, and inflicting a series of blows at lengthening intervals, for each projectile will start with a slightly weaker impulse than the one in front. Anyone curious about the propagation of musical sound could easily observe that, when he plucked a string and pulled it out of the straight, the string not merely returned to its original position but went on vibrating till it settled to rest. Each successive vibration can easily be imagined to discharge a fresh portion of air after the first, and it would be obvious that the rate of travelling would decrease as the string moved more slowly and gave a feebler blow. At the other end there would be a series of impacts at increasing intervals and each feebler than the last.[2]

---

[1] Archytas ap. Eucl. *Sect. canon.* Introd. (See Frank, *Platon u. d. soq. Pyth.* 174.) Diels-Kranz, *Vors.*[5] 47A, 19a and B, 1.

[2] Theon, p. 61, 11 (Hill.) = *Vors.*[5] 47A, 19a : Eudoxus and Archytas held that the proportion of the consonances is expressible in numbers ; for they too agreed that the proportions are in motions and that the quicker motion, since it delivers blows in unbroken succession (ἅτε πλήττουσαν συνεχές) and gives a sharper punch to the air, is acute (high) in pitch, while the slow motion, being more sluggish, is grave (low). So also Plut., *Pl. Qu.*, 1006B : The air receives a blow from the thing which sets it moving and delivers a blow, sharply if the shock has been violent, more softly if it was dull. Theon,

(2) We may now turn to the second or internal motion, which these ' travelling sounds ' start ' in us ' with the first impact on the brain and blood. It takes place in the region from brain to liver. There is no ground for supposing that it involves the travelling of any body from place to place ; the journey of the air or ' sound ' has ended at the point of impact. The internal motion may well be of a different kind.[1] Plato does not describe it here, beyond saying that it goes on for a time and then draws to an end, and implying that it has somehow a rate of speed which may or may not correspond with that of a second motion of the same kind. This is the motion earlier called ' hearing '. It is the physical side of hearing, a bodily motion which (if strong enough) is accompanied throughout by the sensation of hearing ; for we go on hearing the sound from the moment of the first impact until the humming dies away. The rate of this motion is determined by the speed of the external sound and the frequency and force of the blows it delivers. High and low pitch are strictly qualities of the sound as heard or of the sensation, and correspond to the rate of the internal motion. Hence Plato says that the swifter motions *present themselves to us* (φαίνονται) as higher in pitch.[2] The nature of this internal motion has been described at 64B as the transmission of shocks from one particle to another in those parts of the body which, unlike bone or hair, are soft and easily displaced. ' One particle passes on the same effect to another, until they reach the consciousness and report the quality of the agent.' This explanation, we were told, applies especially to sight and hearing.

We have now constructed, without recourse to unreasonable conjecture, a fairly complete picture of the external motion of the

Music. vi, p. 84, Dupuis : Adrastus ascribed to the Pythagoreans the theory that all sound is a movement in the air caused by a blow. The pitch depends on the speed of the movement ; the volume of sound, on its violence. The speeds of the movements and their intensities may, or may not, be in accordance with certain ratios. If they are not, the result is not properly a sound or note (φθόγγος), but a mere noise. Two sounds, struck on an instrument, are concordant when they ring or chime together (συνηχεῖ) according to a certain affinity and sympathy (κατά τινα οἰκειότητα καὶ συμπάθειαν). When they are struck simultaneously, a sweet and agreeable sound is heard from the mixture.

[1] Cf. the analysis of visual sensation at *Theaet.* 156c, where (1) the travelling motion (φορά), a quick motion passing between the eye and the object seen, is distinguished from (2) the ' slow ' change (ἀλλοίωσις) subsequently occurring in the organ and accompanied by sensation if it reaches the soul. There is a corresponding ἀλλοίωσις in the object, which is said to ' become coloured '.

[2] Cf. Ar., *de anim.* 420a, 30 : The high in pitch moves the sense much in little time, the low moves it little in much time. The quick is not the same thing as the high, nor the slow as the low ; rather the motion (of the sense) is ' high ' *because of* the swiftness, ' low ' *because of* the slowness.

travelling sound, terminating at the point of impact, and of the internal motion of blood and brain, starting from that point and lasting so long as the projectiles continue to arrive with sufficient force. Plato has taken nearly all this process for granted, though he has made clear that the two motions are distinct and that at least one of them (the external, in fact) involves the circular thrust characteristic of projectiles.

All that the present statement actually explains is the conditions that must be satisfied if a high and a low note are to be combined in a single harmonious affection of the hearing. Plato has already said that the concord depends on 'the correspondence of the internal motion'. This we shall now understand as meaning that the succession of shocks in brain and blood set up by a rapid sound outside must correspond with the shocks set up later by a slower sound. 'Correspond' evidently cannot mean that the rates are the same,[1] for that would result in unison, not concord. The correspondence meant can only be analogous to that of the vibrations of two strings, one of which vibrates (say) twice as rapidly as the other, so that each longer vibration coincides with every other shorter vibration and produces the concord of the octave.

The sentence describes how the blending takes place, on the explicit assumption that the higher of the two external sounds travels more rapidly than the lower and so reaches the point of impact in front of it. The only difficulty lies in the phrase which describes the earlier internal motion as '*having already reached correspondence*' with the later motion which starts when the slower sound arrives. Why does it need to *reach* correspondence, if correspondence existed from the outset between the two tuned strings? Something must have happened to disturb the correspondence of the external sounds on the course of their journey.

The data which are certainly assumed are as follows. (1) The two strings are plucked simultaneously; otherwise there would be no reason why the swifter sound should arrive first (προτέρων, A, 6). (2) The two strings must be in tune with one another, so that the proper concordant correspondence exists at the outset. (3) The higher sound, despatched by more vigorous and frequent impulses, travels the faster and arrives first. The lower follows more slowly with longer intervals between its projectiles. Why is not the original correspondence maintained throughout the whole process, as the projectiles in each stream arrive at lengthening intervals under the flagging impulse until they both cease? The only

---

[1] Martin (i, 393) and A.-H. understand ὁμοίαν to mean the *same rate*, and consequently dismiss the theory with a contempt which it would deserve if Plato could have overlooked so glaring an objection.

possible reason must lie in a further assumption, namely that the slower projectiles suffer on their journey a loss of speed more serious than the loss suffered by the quicker ones. Their original impulses were feebler and they are sooner tired, like a runner who at starting takes one stride to his competitor's two, but at the goal can only manage something less than one. In this way the original correspondence would be lost, and if the two notes are to sound concordant, it must be restored.

The explanation indicates how this restoration occurs. It opens with a description of the state of things existing at the moment when the slower projectiles begin to arrive. The quicker ones which arrived earlier have already set up the succession of internal shocks in blood and brain. But they are now arriving with diminished frequency, and the internal shocks are also slowing down. The correspondence will be now restored if, at the moment when the slower begin to arrive, the shocks have dropped to a rate which makes them chime in once more with the flagging steps of the new-comers. In our analogy of the race, the winner runs on beyond the goal (the point of impact) with slower and slower steps. Correspondence will be restored if, at the moment when the other reaches the goal, the winner has dropped to the original proportion of two steps to one.

The first sentence of the explanation is now clear :

'The slower sounds (projectiles), on catching up (at the point of impact), find the internal motions (shocks) set up by the quicker sounds which arrived earlier drawing to an end (getting continually slower) and by this time having reached a rate corresponding to the internal motions imparted to them [1] by the slower sounds on their later arrival.'

The next sentence describes the consequence—the way in which the two sets of internal shocks, now in restored correspondence, blend into a single harmonious affection. When the late-comers catch up the earlier at the point of impact, they set going a second internal motion. In thus 'intruding a fresh motion' they cause no disturbance (or jangling, as it were), because the first internal motion has already slowed down into correspondence. So the beginning of the slower internal motion fits on to the remainder of the first internal motion, overlapping it in due correspondence.

---

[1] The later internal motion is regarded as affecting the earlier one when it combines with it. The later sound brings in or interjects a fresh motion on the top of the other (ἄλλην ἐπεμβάλλοντες) ; but as the two motions correspond and fit together (προσάψαντες), the second does not 'disturb' or jangle the first.

The result is that the higher and lower sounds, both now audible together, 'combine in a single effect in which high and low are blended'.

The whole statement thus gives the conditions that must be satisfied, if we are to hear a concordant blend of high and low notes. All that is actually described is the state of things that must exist when we begin to hear the concord, and while it lasts. The rest of the process is taken for granted ; it is not even mentioned that the circular thrust is relevant because the external sounds are projectiles. The explanation above offered takes account of the distinction clearly drawn between the external and internal motions. The failure of other interpreters to produce any theory that is not 'entirely unsatisfactory' or 'wholly perverse' is partly the result of ignoring this distinction. The theory as above reconstructed is certainly open to objections ; but it is not obvious nonsense.

The digression now concludes with the mention of other phenomena involving the circular thrust and also the principle (explained at 57C) that the transformation of the primary bodies in their conflict results in their constantly interchanging the regions towards which they drift, like to like.

80B.   There are, moreover, the flowing of any stream of water,
C.   the falling of thunderbolts, and the 'attraction' of amber and of the loadstone at which men wonder. There is no real attraction in any of these cases. Proper investigation will make it plain that there is no void ; that the things in question thrust themselves round, one upon another ; that the several kinds of body, as they are disintegrated or put together, all interchange the regions towards which they move ; and that the results which seem magical are due to the complication of these effects.

Plato's chief concern seems to be to dispense with the alleged requirement of a void and to eliminate the hypothesis of 'attraction' [1] in all these cases by reducing all apparent pulling to pushing. At 58E the flowing of molten metal along the ground was due to the thrust of the neighbouring air. The flowing of water to its own region, between air and earth, is sufficiently explained by its natural tendency to seek its like. Here he may be thinking of water spreading (like the molten metal) along the level ground. It is not necessary to suppose that there must be empty space in

---

[1] Wellmann, *Frag. d. Gr. Aerzte* i, 37, gives evidence for the controversy about the occurrence of attraction in various phenomena. The author of *Ancient Medicine* (xxii) appears to attribute to the tapering shape of the cupping instrument its power to 'attract' blood from the flesh.

front for it to move into. Its advance will displace the air in front and this will, by the circular thrust, increase the pressure which keeps it moving. Plutarch suggests other ways in which the circular thrust might come in.

The downward fall of the thunderbolt calls for explanation as being contrary to the natural tendency of hot things to move upwards. Aristotle explains that clouds are densest on their upper side. · 'Just as the pips we squeeze between our fingers, though heavy, often jump upwards, so these things are necessarily squeezed out from the densest part of the cloud' (*Meteor.* ii, 9).

The so-called attraction of the loadstone is explained by Plutarch as the pushing of the iron towards the stone by a circular thrust set up by powerful effluences of air coming out of the stone. A theory of this type is attributed to Empedocles (*Vors.* A89).

80D–81E. *How blood is formed by digestion and conveyed through the veins. Growth and decay. Natural death*

After this digression, we now return to the point at which the account of respiration was complete. Plato has succeeded in reducing this to a purely mechanical process, which will go on throughout life, whether we are awake or sleeping, as an engine will run so long as the furnace burns. Like a piston in this engine, the fire inside us keeps up an oscillating motion as it accompanies the breath ; and its action is now invoked to effect the work of digestion—another process that goes on without the intervention of consciousness. The food in the belly is penetrated by the moving fire-particles and broken up into minute fragments. These actually form the blood, a stream of nourishment containing all the substances needed to replenish the waste in our tissues. The waste itself is due to the assaults of the elements outside the body, causing the escape of particles which fly off to seek their likes. The motion of the blood is still not connected with any action of the heart, but attributed to the oscillation of the respiratory system. As the stream moves through the veins, the various substances composing it are attracted, like to like, by the bodily organs, whose waste they thus repair.

80D. Now the effect of respiration, whence this discussion arose, takes place, as we said before, on this principle and by these means : the fire cuts up our food and oscillates [1] inside us as it accompanies the breath ; and by thus oscillating with it, fills the veins from the belly by discharging the cut-up

----

[1] αἰωρουμένου codd. Hermann's emendation αἰωρουμένῳ may be preferred, as by A.-H. and Fraccaroli : 'accompanying the oscillation of the breath '.

80D.   food from thence.   By this means, in any animal, the streams
of nourishment are kept flowing throughout the whole body.
The particles, being freshly divided and coming from kindred

E.   substances—from fruits or herbs which the god caused to
grow for this very purpose of feeding us—take on all manner
of colours owing to their being mixed together; but they
are chiefly pervaded by a red hue, a character inwrought on
moisture by fire that cuts and stains it.[1]   Hence the colour
of the stream throughout the body assumes the appearance
we have described; this we call blood, on which the flesh
and the whole body feed, so that every member draws water
therefrom to replenish the base of the depleted part.[2]   The
manner of this replenishment and wasting is like that movement of all things in the universe which carries each thing
towards its own kind.[3]   For the elements besetting us outside
are always dissolving and distributing our substance, sending
each kind of body on its way to join its fellows; while on
the other hand the substances in the blood, when they are
broken up small within us and find themselves comprehended

B.   by the individual living creature, framed like a heaven to
include them,[4] are constrained to reproduce the movement
of the universe.   Thus each substance within us that is
reduced to fragments replenishes at once the part that has
just been depleted, by moving towards its own kind.

Normal growth and decay.—Such being the process of nourishment,
it remains to explain why the young animal grows, whereas later
in life it dwindles and withers in old age.   The creature is originally
formed from the seed, which is a portion of the parent's ' generative
marrow ' (73B, 91B), and nourished on milk; so its consistency is
soft.   On the other hand, the triangles of the fire, air, water, and
earth which form the marrow are, as we were told (73B), exceptionally smooth and unwarped.   It is here added that their edges
fit closely together, so that the solid particles are firm and can more

---

[1] Cf. the account of red or blood-colour at 68B.

[2] τὴν τοῦ κενουμένου βάσιν, not ' the places that are left void ' (A.-H.).
The main food-conveying channels of the blood are sunk deep within the
flesh, which is watered and fed from beneath, as a plant from its roots.   Cf.
83E πυθμένων, the roots or foundations of the flesh.

[3] τὸ συγγενές, here and at B, 2, shows that συγγενῶν above (D, 8) means
' kindred ', not ' coaeval '.

[4] An allusion to 58A, which explained how all the four primary bodies are
comprehended by ' the circuit of the whole ', and how mutual attraction of
likes and the constant changes of direction of transformed bodies keep the
whole together and tend to allow no vacancy to be left unfilled.   The movement is reproduced here in the microcosm.

than hold their own against the incoming particles of which the food and drink consist. Accordingly, digestion is easy ; for the internal fire-particles, being close-knit and sharp, can penetrate and cut up the food-particles, whose triangles are older and weaker. Blood will thus be formed readily and in abundance, and the creature will grow. The onset of decay and the wasting of old age come when this situation is reversed and the incoming particles are too strong for the fiery ones inside to cut up. Rather these are shattered by the intruders. Digestion will be increasingly difficult and the blood will run thinner. The creature will now dwindle and waste away.

81B.   Now whenever there is more going out than flowing in, all things diminish ; when there is less, they grow. So when the frame of the whole creature is young and the triangles of its constituent bodies are still as it were fresh from the workshop, their joints are firmly locked together, although the consistency of the whole bulk is soft, having been but

c.   lately formed of marrow and nourished on milk. Accordingly, since any triangles composing the meat and drink, which come in from outside and are enveloped within the young creature, are older and weaker than its own, with its new-made triangles it gets the better of them and cuts them up, and so causes the animal to wax large, nourishing it with an abundance of substances like its own.[1] But when the root [2] of the triangles is loosened by reason of the many

D.   conflicts in which they have long been engaged with so many others, they can no longer cut up into their own likeness the triangles of the nourishment as they enter, but are themselves easily divided by the intruders from without. So every living creature is at this time overmastered and wastes away ; and this condition is called old age.

Death, when not due to violence or disease, is accounted for by the weakening and breaking down of the triangles of the marrow.

---

[1] The doctrine that each substance in our body is nourished by the accession of like substances already present in our food and drink was clearly asserted by Anaxagoras, and is alluded to at *Phaedo* 96D. Cf. Ar., *de gen. et corr.* 333*b*, 1 (Empedocles).

[2] ἡ ῥίζα τῶν τριγώνων χαλᾷ. This curious metaphor must describe the opposite of what was called above ' being strongly locked-together ' (σύγκλεισιν) so as to form a firm solid. This favours Tr.'s view (p. 586) that ' roots ' means the sides, which are the lines along which triangles are joined to compose a corpuscle. The metaphor seems to be taken from the loosening of a tree's roots. I cannot see that ῥίζα has anything to do with a ship, as Tr. would have it. The only phrase in the whole context that definitely suggests a ship is ἐκ δρυόχων ; but that was a current metaphor applied to other things.

This substance is the very seat of the life-principle; in it the 'bonds of life are fastened so long as the soul is bound together with the body' (73B). This whole passage is concerned only with the normal course of growth, decay, and death, which must occur in the healthiest creature. The entire process is explained in terms of the constituent triangles of the elementary bodies. Abnormal decay and death due to disease will be treated later.

81D.   And at last, when the conjoined bonds of the triangles in the marrow no longer hold out under the stress, but part asunder, they let go in their turn the bonds of the soul [1]; and she, when thus set free in the course of nature, finds

E.   pleasure in taking wing to fly away. For whereas all that is against nature is painful, what takes place in the natural way is pleasant. So death itself, on this principle, is painful and contrary to nature when it results from disease or wounds, but when it comes to close the natural course of old age, it is, of all deaths, the least distressing and is accompanied rather by pleasure than by pain.

Reviewing this whole account of digestion, nutrition, and respiration, we can see that Plato's main concern is with the hydraulics of his irrigation system. Completely ignoring the voluntary or reflex action of the muscles, he has discovered the motive power in the natural tendency of fire to make for its own region, assisted by the principle of the circular thrust. The problem presented by the upward movement of the blood from the belly to the head was analogous to that of the movement of water in the earth to the tops of mountains, whence it descends in streams. This troubled the mechanical genius of Leonardo da Vinci, who wrote: 'The water runs from the rivers to the sea and from the sea to the rivers, always making the same circuit. The water is thrust from the utmost depth of the sea to the high summits of the mountains, where, finding the veins cut, it precipitates itself and returns to the sea below, mounts once more by the branching veins and then falls back, thus going and coming between high and low, sometimes inside, sometimes outside. It acts like the blood of animals which is always moving, starting from the sea of the heart and mounting to the summit of the head; there, if the veins are burst, as one may see a vein burst in the nose, all the blood in the lower parts mounts to the height of the burst vein.' Like Plato, Leonardo regards the rising of the water in the mountains and of the blood

---

[1] At 89B we learn that each individual has his allotted span of life, 'since the triangles in every creature are from the outset put together with the power to hold out for a certain time, beyond which life cannot be prolonged'.

to the head as contrary to the natural movement of the fluids, and he accounts for it by invoking the vital heat : ' if the body of the earth did not resemble man, the water of the sea, being so much lower than the mountains, could not of its own nature mount to their summits. Hence we must believe that the reason which holds the blood at the summit of man's head is the same that holds the water at the summit of the mountains '. ' Water, the vital humour of the terrestrial machine, moves owing to natural heat.' ' Water is like the blood which natural heat holds in the veins at the top of the head.' [1]

Leonardo's comparison may have been prompted by Seneca (*Nat. Qu.* iii, 15 ff.), who holds that the Earth is analogous to the human body with its veins containing blood and its wind-pipes containing breath. In the Earth there are channels, some of water, some of breath, so similar to ours that the ancients spoke of ' veins of water '. The Earth contains as many different kinds of humours as our bodies ; both are subject to various kinds of corruption. Streams flow out of Earth like blood from our veins when these are opened. The Earth contains not only veins of water, but huge subterranean rivers, some of which emerge at the bottom of lakes. Seneca himself may well have been inspired by the myth in the *Phaedo*, which, as Stewart remarks, ' recommends itself to the " scientific mind " by explaining the origin of hot and cold springs, volcanic action, and, I think, the tides of the Atlantic Ocean '. We are there told that the Earth, at many places all over its spherical surface, is indented with hollows of every size and shape, into which water, mist, and air have flowed together ; they are the sediment of the pure ether which envelopes the true surface. All the hollows are connected by many channels, narrow or wide, bored through the Earth. Through these much water flows from one basin to another ; and there are many underground rivers of hot and cold water, or fire, or mud and lava, which fill each hollow in turn as the circulation (περιρροή) flows round to it. All these are kept in motion by a certain oscillation (αἰώρα) in the Earth produced in the following way. The largest chasm of all pierces right through the whole Earth. All the rivers flow together (συρρέουσι) into this chasm, and then out again, because the liquid has no base to rest upon [2] ; so it oscillates and surges up and down,

---

[1] These quotations from Leonardo's MSS. are taken from G. Séailles, *Léonard da Vinci*, Paris (1892), pp. 299 ff.

[2] ' There is no bottom at the centre of the earth. . . . We must keep in mind throughout this passage that everything falls to the earth's centre. The impetus (ὁρμή) of the water takes it past the centre every time, but it falls back again, and so on indefinitely ' (Burnet, *ad loc.*).

and the air or breath (πνεῦμα) connected with it does the same, accompanying it as it makes now for the other side of the Earth, now back again towards this side. ' Just as, when we breathe, the stream of the breath is perpetually exhaled and inhaled, so there the breath (πνεῦμα), oscillating with the liquid, produces terrible and irresistible winds as it comes in and goes out.' When the water retreats towards what we call the ' lower ' parts (i.e. the antipodes), the streams flow into the regions on that further side of the Earth, and fill them as irrigators fill their channels. Then, when the impulse is in the opposite direction, the streams fill the region on our side, and, flowing by the channels through the Earth, reach the hollows on the surface, forming seas, lakes, rivers, and springs. Thence they sink once more into the Earth, after making longer or shorter journeys round, and fall again to the main chasm at a higher or lower level, but always below the point at which they came out. They can always descend as far as the centre of the Earth, but no farther ; for the part of the chasm which is beyond the centre is uphill to a stream coming from either side.

The similarities between this passage and ours have often been remarked. The only cause suggested in the *Phaedo* for the oscillation of the waters is, as Olympiodorus [1] says, that ' the Earth is a living creature and as it were respires and causes certain refluxes by its inhalations and exhalations '. When the Earth inhales ' all the rivers flow together into the chasm, and then flow back again out of it ', as in the *Timaeus* the air surrounding the body ' at one moment flows together upon the funnels, while at another moment the funnels flow back '. There is, indeed, a not accidental resemblance between the hollows on the Earth's surface where the water collects to flow into the underground channels and the funnels of the pharynx where the air collects to pass into the irrigation system.

81E–86A. *Diseases of the body*

To the account of the normal course of decay in old age and natural death is now added a classification of diseases. The fundamental notion of nearly all Greek medicine was that health depends on a due balance or proportioned mixture of the ultimate constituents of the body. Where the schools differed was on the question, what these ultimate constituents are. According to Alcmaeon, the junior contemporary of Pythagoras, they are the

---

[1] *Comment. in Phaed.*, p. 204, Finckh. Wyttenbach on *Phaedo* 111E cites the Stoic Athenodorus' explanation of tides as due to inhalation and exhalation (Strabo iii, 7) and Pomponius Mela iii, 1, 15, *neque adhuc satis cognitum est anhelitune suo id mundus efficiat retractamque cum spiritu regerat undam undique, si, ut doctioribus placet, unum animal est.*

opposite powers (δυνάμεις), hot and cold, moist and dry, bitter and sweet, and the rest. Health is maintained by a sort of democratic equality (ἰσονομία); disease is caused by one opposite establishing its sole ascendancy (μοναρχία) over the other. Alcmaeon seems to have had an indefinite list of pairs of opposites; but the influence of Milesian cosmology gave special prominence to the two pairs— the hot and the cold, the moist and the dry—which were themselves associated with the seasons of the year and the cycle of life in all nature. The life of plants and animals does in fact depend upon a certain balance of the alternating powers of summer heat and winter cold, the rainy season and the times of drought.

The Coan school of medicine replaced the 'powers' by fluid substances, the humours (χυμοί).[1] The writer of the treatise on *Ancient Medicine*, attacking the intrusion of philosophic pre-conceptions into his art, points out that the hot, the cold, the moist, and the dry are not substances, but merely powers; the body consists of certain humours which *have* powers or properties. Health is the harmonious blending or balance of these substances, not of the powers. In the *Nature of Man*, attributed by Aristotle to Polybus, the humours are declared to be four in number: blood, phlegm, yellow bile, and black bile. These were, somewhat artificially, associated with the four principal powers.

The Italian and Sicilian school followed a different line. In Empedocles' system the four powers had been simply identified with the four elements, fire (hot), air (cold), water (moist), earth (dry). These elements are the components of the body, as of everything else. Philistion of Locri developed a medical theory on this basis, which is summarised as follows [2]:

Philistion holds that we consist of four 'forms' (ἰδεῶν), that is elements: fire, air, water, earth. Each of these has its own power: fire the hot, air the cold, water the moist, earth the dry. Diseases arise in various ways, which fall roughly under three heads. (1) Some are due to the elements, when the hot or the cold comes to be in excess, or the hot becomes too weak and feeble. (2) Some are due to external causes of three kinds: (*a*) wounds; (*b*) excess of heat, cold, etc.; (*c*) change of hot to cold or cold to hot, or of nourishment to something inappropriate and corrupt. (3) Others are due to the condition of the body: thus, he says, 'when the whole body is breathing well and the breath is passing through without hindrance, there is health; for respiration takes place not only through mouth and nostrils, but all over the body. . . .'

[1] See Loeb Trans. of Hippocrates by W. H. S. Jones, vol. i, pp. xlvi ff.
[2] Wellmann, *Frag. d. Gr. Aerzte*, 110, Philistion, frag. 4.

Philistion is believed to have practised at Syracuse, and Plato may have met him there.[1] It seems certain that he influenced Diocles of Karystus in Euboea, who was regarded by later physicians as a 'second Hippocrates'. Diocles practised chiefly at Athens, and wrote on all branches of medicine. His *floruit* is placed between 400 and 350 B.C. The many points of agreement between Diocles and Plato support the inference that both were influenced by Philistion. The common foundation is the four fundamental qualities of Empedocles, not the four humours, though Diocles was acquainted with these. In Plato's theory, accordingly, the humours play a secondary part. His first class of diseases is due to excess or defect or misplacement of one or another of the primary bodies. In the treatment of the remaining classes he brings into account his peculiar doctrine of the triangles composing the solid figures of the elements.[2]

(1) *Diseases due to excess or defect or misplacement of the primary bodies.*—This class corresponds to the first of Philistion's three classes. Diocles [3] also held that most diseases are due to 'anomaly of the elements in the body and of its condition' (Philistion's third class, due to disorders of respiration).

81E.   The origin of diseases is no doubt evident to all. Since
82.   there are four kinds which compose the body, earth, fire, water, and air, disorders and diseases arise from the unnatural prevalence or deficiency of these, or from their migration from their own proper place to an alien one; or again, since there are several varieties of fire and the rest, from any bodily part's taking in an unsuitable variety, and from all other causes of this kind. For when any one of the kinds is formed or shifts its place contrary to nature, parts that were
B.   formerly cold are heated, the dry become moist, and so also

---

[1] If Plato's *Second Letter* is genuine, it seems to imply that Philistion attended Dionysius II, and establishes Plato's acquaintance with him.

[2] Tr.'s notes should be consulted for evidence about fifth-century medical theories; but his belief that the contents of the *Timaeus* represent the thought of that century, not of Plato's own, has led him to take comparatively little notice of Diocles. He thinks 'the most likely view is that men like Philistion are responsible for all the main points in the medical part of the dialogue, and that they naturally enough followed the lead of Philolaus' (p. 599). Philolaus is called 'roughly contemporary' with Zeno, Empedocles, and Timaeus, who 'must be thought of as born at the very beginning of the fifth century' (p. 17). So the impression is conveyed that the court physician of Dionysius II was, so far as his doctrine is concerned, practically a contemporary of Empedocles. My own notes on the medical authorities should be taken as supplementary to Tr.'s. It is useless to repeat all the early evidence he has collected, especially about Pythagorean views.

[3] Aet v, 30, 2.

82B. with the light and the heavy, and they undergo changes of every kind. The only way, as we hold, in which any part can be left unchanged and sound and healthy is that the same thing should be coming to it and departing from it with constant observance of uniformity and due proportion ; any element that trespasses beyond these limits in its incoming or passing out will give rise to a great variety of alterations and to diseases and corruptions without number.

(2) *Diseases of the (secondary) tissues.*—We pass next from the simple bodies to the tissues composed of some or all of them. These secondary components of the body were known as ' homoeomerous ' substances, because they were believed to be indefinitely divisible into similar parts. Empedocles was the first to suggest that bone, sinew, flesh, and blood were composed of the four elements in definite proportions. Bone, for example, contained four parts of fire, two of earth, one of air, and one of water. Blood and flesh contained all four elements in equal or nearly equal quantities. In this tradition blood is not ranked with the other three ' humours '. In Empedocles it is the seat of the physical soul and consciousness.

Plato has already described (73B ff.) the composition of marrow., bone, sinew, and flesh ; blood is formed directly out of the digested food and contains portions suitable for the nourishment of all the tissues. He now tells us something of the manner in which the nourishing takes place. Sinew is produced from the fibrine, flesh from the coagulated blood when the fibrine has been removed. Sinew and flesh in their turn produce a viscous fluid, which glues flesh to bone and nourishes both the bones and the marrow they contain. Such is the natural order in which the several tissues are built up. Some points may have been suggested by Diocles' embryological observations :

' The first articulate formation of the embryo is observed at about the fortieth day. Up to nine days sanguineous threads can be traced ; at eighteen days flesh-like clots and some fibrinaceous formations can be seen, in which the pulsation of the heart is found. At thrice nine days, according to Diocles, a faint outline of the spine and head becomes visible in a mucus-like membrane. At four times nine days the whole body is distinct for the first time, or finally (adding another four days) about the fortieth day. Empedocles agrees with this account of the times at which the embryo becomes quite distinct.' [1]

Plato, however, is concerned with the normal process of nutrition, whereby the appropriate substances in the blood are built up into

[1] Oribasius iii, 78 = Diocles, frag. 175, Wellmann.

the several tissues whose waste they repair.[1]   Diseases of his second class are due to the unnatural reversal of this process, starting from the breaking down of flesh, which then discharges corrupt substances back into the blood.   These give rise to the noxious humours, the various kinds of bile and phlegm.   Finally, the trouble may go deeper and affect the bones and the very seat of life in the marrow.   I cannot find evidence that any medical writer had formulated this notion of a reversal of the normal course of nutrition as the cause of a special class of diseases, beyond the testimony, quoted by Taylor (p. 592), that Philolaus regarded bile, blood, and phlegm as the source of diseases, and bile (if present at all) as a 'useless' substance.   This is vague and inconclusive : blood seems to be ranked with the other humours, and all medical doctrine connected some diseases with misbehaviour of the humours. Plato's notion of a reversal appears to be associated with the opposition of growth (αὔξησις), resulting from normal nutrition, and decay (consumption, φθίσις) as the contrary process.   At 81B–E he traced the natural course of growth followed by the decay of old age, without the intervention of disease.   Decay was due to the inevitable wearing out and falling to pieces of the triangles. Here we are concerned with abnormal, morbid decay, in which the current that should foster growth is forced to flow backwards. The reverse of ' generation ' (γένεσις) is decay or corruption (φθορά.)

82B.   Again, seeing that secondary formations exist in nature, an
  c.   attentive consideration will discern a second class of diseases. Since marrow, bone, flesh, and sinew are composed of the bodies above named, and blood also of the same bodies, though in a different way, most of the diseases affecting them arise in the same manner as those just mentioned ; but the most serious afflictions take the form of a corruption of these structures, which occurs when the process of their formation is reversed.

In the natural course flesh and sinews arise from the blood —sinew from fibrine (for they are cognate), flesh from the
  D.   compacting of the blood from which the fibrine is being removed.[2]   From sinews and flesh, again, proceeds the

---

[1] I cannot see that this account of nutrition is inconsistent with the earlier description of the *composition* of the several tissues at 73B ff.   However they are composed, they must all be nourished ultimately by the blood.

[2] It is not explained what agency in the living body causes the blood without fibrine to be compacted into flesh. It may be the ' innate heat ', for when the blood is cold in the dead body blood without fibrine remains liquid (85D).   This is in accordance with the view which Diocles is presumed to have shared with Empedocles that the male embryo is more quickly

336

82D.   viscous and oily stuff which glues the flesh to the structure
of the bones and also feeds the growth of the bone itself
which encloses the marrow ; while at the same time the
purest part, consisting of triangles of the smoothest and most
slippery sort, filters through the close texture of the bones
and, as it is distilled from them in drops, waters the marrow.

E.   When the several structures are formed in this order, the
result, as a rule, is health.

Disease comes when the order is reversed. Thus, when
flesh is decomposed and discharges the results of its decomposition back into the veins, these are then filled with much
blood of every sort together with air ; this has a diversity
of colours and bitternesses, as well as acid and saline qualities,
and develops bile, serum, and phlegm of all sorts. All these
products of breaking down and corruption in the first place
destroy the blood itself, and providing the body with no

83.   further nourishment from themselves, they are carried everywhere through the veins, no longer observing the order of
natural circulation. They are at feud among themselves
because they can get no good of one another ; and they
make war upon whatsoever in the body keeps orderly array
and stays at its post ; so they spread corruption and
dissolution.

It will be seen that Plato regards bile and phlegm, two of the
fundamental substances in the humoral physiology, as morbid
products of the decomposition of flesh.[1] The next paragraphs
describe in some detail how the various recognised types of bile
and phlegm are so generated. There is some resemblance here to
the doctrine of Dexippus of Cos, the pupil of Hippocrates, who
regarded phlegm and bile as superfluous products of nutriment,
and held that the mixture of these two humours with the blood,
which changes colour accordingly, gives rise to the four species :
white phlegm, blood-coloured phlegm, yellow bile, and black bile.[2]
But Wellmann believes that Plato's theory of the humours is
probably due to Philistion and Diocles, who held that the humours
are produced by the action of the ' innate heat ' on the nutriment
in the veins. Blood is the mixture in normal proportions ; bile is

formed than the female, because the male is in the right (warmer) part of
the uterus. Diocles also held that overheating (ζέσις) *thickens* the blood
and so blocks the veins, causing indigestion. Wellmann, P.-W. Real Encycl.,
s.v. Diokles, 805-6.

[1] Cf. Tim. Locr. 102C, χολᾶς γὰρ ταὶ γενέσιες καὶ φλέγματος ἐνθένδε, χυμοὶ
νοσώδεες καὶ ὑγρῶν σάψιες . . .

[2] Wellmann, *Fr. d. Gr. Aerzte*, p. 75.

due to an excess of heat, phlegm to an excess of cold. Hence bile causes inflammations, and phlegm catarrhs (as in *Timaeus* 83c, 85b).[1]

83A. Now when the flesh which is decomposed has been formed a long time before, it resists concoction ; it turns black under long exposure to burning, and, being bitter because it is eaten through and through, it is dangerous in its assault
B. upon any part of the body that is as yet uncorrupted. Sometimes, when the bitter stuff has been fined down, the bitterness is replaced by acidity in the black colour. Sometimes, again, when the bitterness is steeped in blood it acquires a redder hue, and the mixture of the black with this redness give it the ' bilious ' colour.[2] Or again, a yellow colour may be combined with the bitterness when the flesh decomposed by the fire of the inflammation is of recent formation. To all these the common name ' bile ' has been given, either by
C. physicians, or perhaps by someone capable of surveying a number of unlike things and discerning in them all a single kind deserving a name ; while the several varieties of bile recognised [3] have been specially defined each according to its colour.

The serum of black and acid bile (in contrast to that of blood, which is a harmless lymph) is dangerous when combined with a saline quality by the action of heat ; this is called acid phlegm. There is also the product resulting from the decomposition of new and tender flesh, accompanied by air.[4] This is inflated by air and enveloped in moisture so

---

[1] Wellmann in P.-W. Real Encycl., *s.v.* Diokles, 803.

[2] The reading χολῶδες is supported by the use of τὰ χολῶδη for ξανθή as opposed to μέλαινα χολή in the passages quoted by Wellmann, *Fr. d. Gr. Aerzte*, p. 75, note 4, especially the testimony about Dexippus : [ὅταν δέ, φ(ησίν), ἡ χολὴ τ]ῷ αἵματι [ἐπιμειχθῇ, γί(νεται) τὰ] λεγόμενα χολῶδ [η . . . The alternative reading χλοῶδες can then be explained as due to someone who thought it absurd to call one class of bile ' bilious '. Plato has already mentioned the χολώδη χρώματα exhibited by the liver, in connection with the ' bitterness ' contained in that organ. Cherries might be similarly classified as white, black, and cherry-coloured.

[3] Namely the three species above described : black, ' bilious ', and yellow. The contrast is between the generic name ' bile ', and these three names for the species (εἴδη). τὰ ἄλλα means the species *as opposed to* the genus ; not that there are *other* species than those named. Diogenes of Apollonia and his contemporaries are said to have laid much stress on the colour of the complexion as a sign of temperamental humours and a symptom of corresponding diseases (*Vors.* 51A, 29a).

[4] As explained at 84E, air is produced *inside the body* by the decomposition of flesh. Cf. also 82E, 3.

83D.   as to form bubbles, individually too small to be seen. but becoming visible in the mass, as the froth so formed makes them appear white in colour. All this decomposition of tender flesh in combination with air we call white phlegm. Freshly forming phlegm, moreover, itself has a lymph, namely sweat, tears, and all other such flowing substances

E.   that are daily purged away.

     All these things become agents to produce diseases when the blood, instead of being replenished in the natural way from food and drink, takes its increase from the opposite quarter, contrary to the established use of nature.[1]

The next paragraph returns to the description of diseases due to this reversal of the normal process. The decomposition of flesh back into the blood, above mentioned, is less grave than the more deeply seated corruption of the fluid which binds flesh to bone and nourishes bone and marrow. Beyond that again lie the affections of the bones themselves, and finally of the marrow.

83E.   Now when the several sorts of flesh are broken down by diseases, so long as their roots hold firm, the mischief is but half done, for it still admits of easy recovery. But when

84.   that which binds flesh to bone falls sick and in its turn the stream that is separated off from flesh and sinews [2] no longer serves to nourish bone and bind the flesh thereto, but instead of being oily, smooth, and viscous, becomes rough and saline, parched by an unhealthy manner of living, then all the substance so affected crumbles back again up into [3] the flesh and sinews as it comes away from the bones; while

B.   the flesh, falling away with it from its roots, leaves the sinews bare and full of brine, and itself falls back again into the current of the blood, to aggravate the disorders before described.[4]

     Grievous as these affections of the body are, yet graver

---

[1] παρὰ τοὺς τῆς φύσεως νόμους : 'contrary to the laws of nature' is a mistranslation. All that is meant is the customary and normal process by which blood is healthily formed.

[2] Reading καὶ μηκέτι αὖ τὸ ἐξ ἐκείνων νᾶμα (αἷμα, codd : ἅμα St.) καὶ νεύρων ἀποχωριζόμενον. αἷμα is certainly corrupt, and νᾶμα has more point than Stallbaum's ἅμα. At 82D the fluid was said to drip from the bones and water the marrow.

[3] ὑπό, ' up into ', not merely ' under '. The natural movement—secretion of the fluid from sinews and flesh down on to the bones—is reversed.

[4] At 82E–83A. The more superficial diseases due to corruption of the blood by decomposing flesh are reinforced by these more deeply seated affections of the fluid.

84B. are those which go deeper and come when the density of the flesh does not allow the bone to receive enough ventilation. Through mouldiness the bone is overheated and decays ; no longer taking in its proper food, it goes rather the opposite

c. way and crumbles back into the nourishing fluid ; and that again falls into flesh, and the flesh into the blood, thus making the maladies of the parts previously mentioned [1] all more virulent.

Finally, the most desperate case of all is when the substance of the marrow becomes diseased by some deficiency or excess. This produces the most serious and deadly disorders, since the whole substance of the body is forced to flow in a backward course.

(3) *Diseases due to (a) breath, (b) phlegm, (c) bile. Fevers.*—The last section has described the origin of phlegm and bile in the reversal of the normal process, but no specific diseases have been named.   The third class now introduced contains groups of maladies due to the blocking of respiration or the morbid formation of air inside the body, and to the two noxious humours.

84C. A third class of diseases must be conceived as occurring in

D. three ways : (a) from breath, (b) from phlegm, or (c) from bile.

(a) When the lung, whose office is to dispense the breath to the body, is blocked by rheums and affords no clear passages, the breath fails to reach some parts and causes them to putrefy for lack of refreshment ; while too much of it passes into other quarters, where it forces its way through the veins and contorts them, dissolves the body, and is intercepted when it reaches the barrier at the centre.   Thus

E. are caused countless painful disorders, often accompanied by much sweat.

Often, too, when flesh is broken down,[2] air is formed inside the body and, not being able to make its way out, causes the same torments as those due to breath that has come from outside.   These are most severe when the air, gathering and swelling up round the sinews and the small veins

---

[1] τῶν πρόσθεν (governed by τὰ νοσήματα, not by τραχύτερα), the flesh and sinews and the viscous fluid.   The corruption of the bones, by travelling all the way back to the blood, aggravates *all* the less deeply seated diseases ; just as, in the preceding paragraph, the corruption of the viscous fluid aggravated τὰ πρόσθεν ῥηθέντα νοσήματα (B, 3).

[2] διακριθείσης, ' disintegrated ', cf. διακρινομένης τῆς σαρκός, 83E. So A.-H. Cf. 82E, 83C, D.

84E. there,[1] makes them stretch backwards the tendons of the back and the sinews attached to them. From the tension so produced the disorders of course derive their names, tetanus and opisthotonus. A cure is difficult; indeed, such cases are, for the most part, relieved by supervening fevers.

The first part of the above paragraph clearly refers specially to inflammation of the lungs.[2] The agreement of Plato's account of its causes with Diocles of Karystus once more points to the inference that both were dependent on Philistion of Locri, whose third class of diseases is due to the blocking of respiration (p. 333 above). The second group, including tetanus, is attributed to the formation of air inside the body as a product of decomposing flesh. Again Plato's doctrine has striking points of agreement with Diocles. The last remark, that such disorders are relieved by supervening fevers, is repeated in Hippocrates, *Aphorisms* iv, 57 (iv, 522L).

85A. (*b*) White phlegm, when intercepted, is dangerous, because of the air in the bubbles; but if it finds an escape to the surface of the body, is milder, though it disfigures the body by engendering white eruptions and kindred maladies. When it is mixed with black bile and is diffused over the divine circuits in the head so as to throw them into confusion, the visitation, if it comes in sleep, is comparatively mild, but an attack in waking hours is harder to throw off. As
B. an affliction of the sacred part, it deserves its name ' sacred disease '.

Acid and saline phlegm is the source of all disorders that occur by defluxion; they have received many different names according to the divers regions towards which the fluxion is directed.

Plato agrees with the author of the treatise *On the Sacred Disease* (epilepsy) that it is an affection of the brain and caused by phlegm, to which Plato (or his source) adds a mixture of black bile. He also defends the use of the name ' sacred disease ', but not on

---

[1] Does this refer to the network of small veins round the head (77E) and the sinews at the base of the skull (75D) ? ταύτῃ must presumably mean the neighbourhood of the shoulders, taking its meaning from ἐπίτονοι, which seems to mean whatever tendons or sinews were supposed to hold the back erect, like the back-stays of a mast. Wellmann (p. 11) quotes from the *Anec. Med.* 7, 544, the statement that all the ancient physicians held opisthotonus to be caused by τὰ ἀπὸ τοῦ ἐγκεφάλου πεφυκότα νεῦρα being filled with viscous humours, which the ψυχικὸν πνεῦμα knocks up against, so causing spasms. But the meaning of νεῦρα here seems to be doubtful.

[2] Martin ii, 355. Wellmann, *Fr. d. Gr. Aerzte* 9, 76.

the grounds attacked by the Hippocratean writer, that the disease was due to supernatural causes.[1]   Praxagoras and Diocles attributed epilepsy to the formation of phlegmatic humours in the ' thick artery '. These form bubbles which block the passage of the *pneuma* from the heart and so produce tremors and spasms of the body.[2]

85B.   (c) Inflammations of any part of the body, so called from its being burnt or ' inflamed ', are all due to bile.[3] If the bile finds a vent outwards, its seething sends up eruptions

c.   of various kinds; if shut up within, it engenders many inflammatory diseases. The worst is when the bile mingles with pure blood and breaks up the proper disposition of the fibrine. This substance is distributed throughout the blood to preserve in it a due proportion of thinness and thickness, in order that heat should not so liquefy it as to make it flow out through the porous texture of the body, nor yet should its excessive density make it too sluggish for ready

D.   circulation in the veins. The fibrine, by the composition of its substance, preserves the due mean; even after death when the blood is getting cold, if the fibrine is collected, all the rest of the blood is liquefied; whereas, if the fibrine is left, it quickly congeals the blood with the help of the surrounding coldness. Such being the action of fibrine in the blood, bile, which had its origin as old blood and is now dissolved back again into blood out of flesh, when it enters the blood in small quantities at first, hot and liquid, congeals under the action of the fibrine; and while this is happening

E.   and its natural heat is being quenched, it sets up internal chill and shivering. As the bile flows in with fuller tide, however, it overpowers the fibrine with its own hotness and by boiling up shakes it into disarray; and if it proves strong enough to obtain the mastery to the end, it penetrates to the substance of the marrow and in consuming it unlooses the soul from her moorings there and sets her free. When the flow is weaker and the body holds out against dissolution the bile is itself overpowered; then either it is expelled all over the surface of the body, or else, after being thrust through the veins into the lower or upper belly, banished

---

[1] Wellmann, op. cit. 26 ff. See also Tr.'s note, p. 602.

[2] Diocles, frag. 651, Wellmann.

[3] Not to phlegm (which is regarded by Plato as cold), in spite of the etymological kinship of φλέγμα and φλέγεσθαι. Polybus in the *de nat. hom.* connects phlegm with air, the coldest element, yellow bile with heat, black bile with dryness.

85E. from the body like an exile from a city at civil war, it causes
86. diarrhœa, dysentery, and all such disorders.

When the body has fallen sick chiefly through excess of
fire, it produces continuous heats and fevers; excess of air
causes quotidian fevers; excess of water, tertian, water
being more sluggish than air or fire. Excess of earth, which
ranks as the most sluggish of all the four, takes a fourfold
period for its purgation, and produces quartan fevers which
are hard to shake off.

The last paragraph on fevers has no connection with the previous
description of diseases due to bile. It belongs rather to the first
class of diseases, attributed above (82A) to excess or defect of one
of the four primary bodies. Wellmann [1] infers from a remark of
Galen's that Diocles recognised only the short periods (up to four
days) for intermittent fevers and regarded each kind of fever as
due to disorder of one of the four cardinal humours, and that this
doctrine was a development of Philistion's, followed here by Plato.
He deduces the following schéme for Diocles:

Fire	Water	Air	Earth
hot	moist	cold	dry
yellow bile	blood	phlegm	black bile
continuous fever	tertian	quotidian	quartan

It will be observed that Plato does not bring in the humours,
among which blood is not ranked with the others in his system.

86B–87B. *Disease in the Soul due to defective bodily constitution and
to bad nurture*

Next come those maladies of the soul whose origin can be traced
to a defective inherited constitution of the body or to a bad upbringing in youth. Plato here lays emphasis on the initial disadvantages
of a bad physique, producing a tendency to vice; and he connects
this topic with the Socratic doctrine, which he always maintained,
that no one is willingly (or wittingly) bad. Taylor finds that the
exposition of this maxim here ' explains away that very fact of
moral responsibility on which Socrates, Plato, Aristotle, and
Timaeus himself, when he is talking ethics and not medicine, are
all anxious to insist'. Timaeus, he thinks, not only contradicts
Plato and the rest, but states a view ' quite irreconcilable ' with
the earlier instructions to the created gods ' to steer the course of
the mortal creature nobly and well, except in so far as it shall be
the cause of evil to itself ' (42E). Taylor concludes that Plato,

[1] Op. cit. 92.

with a ' kindly irony ', is intentionally making Timaeus ' give himself away '. This seems an odd attitude to take towards an imaginary character whose creator has attributed to him a view ' glaringly inconsistent with itself' and irreconcilable with all that he says elsewhere. Archer-Hind, on the other hand, holds that Plato's view of vice as an involuntary affection of the soul ' well illustrates how admirably the various parts of his system fit together ', and that the interpreter's declaration in the *Republic* that ' responsibility lies with the chooser ; heaven is not responsible ', not only is not inconsistent with the maxim that no one is willingly bad, but is inevitably implied in it. In view of this divergence of opinion, it is important to consider carefully what Plato actually says here. The ' determinism ' which Taylor discovers in our passage was the last outcome of that materialistic view of the world which Plato regarded as the root of atheism and immorality. Even Epicurus shrank from such a conclusion and invented a physical basis for free will. That Plato should either accept such determinism himself or attribute it to a fifth-century Pythagorean is, in the last degree, improbable.

86B.   Such is the manner in which disorders of the body arise ; disorders of the soul are caused by the bodily condition in the following way. It will be granted that folly is disorder of the soul ; and of folly there are two kinds, madness and stupidity. Accordingly, any affection that brings on either of these must be called a disorder ; and among the gravest disorders for the soul we must rank excessive pleasures and pains. When a man is carried away by enjoyment or dis-

C.   tracted by pain, in his immoderate haste to grasp the one or to escape the other he can neither see nor hear aright ; he is in a frenzy and his capacity for reasoning is then at its lowest. Moreover, when the seed in a man's marrow becomes copious with overflowing moisture like the overabundance of fruitfulness in a tree, he is filled with strong pains of travail and with pleasures no less strong on each occasion (?) [1] in his desires and in their satisfaction ; for the most part of his life he is maddened by these intense pleasures and pains ;

D.   and when his soul is rendered sick and senseless by the body he is commonly held to be not sick but deliberately bad.

---

[1] καθ' ἕκαστον is difficult : ' des douleurs très grandes *chacune en particulier* ' (Martin) ; *in jeder Beziehung* (Müller) ; ' *from time to time* ' (A.-H.) ; *immer wieder* (Apelt) ; ' many a *specific* pang ' (Tr.) ; *a parte a parte* e nei desideri e negli effetti loro (Fraccaroli). In Plato the phrase normally means ' severally ' or ' individually ', as at 49B, 4, μᾶλλον ἢ καὶ ἅπαντα καθ' ἕκαστόν τε, 26C, 6, μὴ μόνον ἐν κεφαλαίοις ἀλλ' ὥσπερ ἤκουσα καθ' ἕκαστον (' in detail ').

86D. But the truth is that sexual intemperance is a disorder of the soul arising, to a great extent, from the condition of a single substance [1] which, owing to the porousness of the bones, floods the body with its moisture. We might almost say, indeed, of all that is called incontinence in pleasure that it it not justly made a reproach, as if men were willingly bad.

E. No one is willingly bad ; the bad man becomes so because of some faulty habit of body and unenlightened upbringing, and these are unwelcome afflictions that come to any man against his will.

Again, where pains are concerned, the soul likewise derives much badness from the body. When acid and salt phlegms or bitter bilious humours roam about the body and, finding no outlet, are pent up within and fall into confusion by blending the vapour that arises from them with the motion

87. of the soul, they induce all manner of disorders of the soul of greater or less intensity and extent.[2] Making their way to the three seats of the soul, according to the region they severally invade, they beget many divers types of ill-temper and despondency, of rashness and cowardice, of dulness and oblivion.[3]

Besides all this, when men of so bad a composition dwell

B. in cities with evil forms of government, where no less evil discourse is held both in public and private,[4] and where, moreover, no course of study that might counteract this poison is pursued from youth upward, that is how all of us

---

[1] The marrow, or that part of it which forms the seed, which the bones are not dense enough to retain and keep in its proper consistency. So A.-H. Since these words repeat τὸ σπέρμα ὅτῳ πολὺ καὶ ῥυῶδες περὶ τὸν μυελὸν γίγνεται (c. 4), I cannot understand why Tr. says that the substance meant is 'clearly' not the μυελός but the bones (p. 616). At 82D we learnt that the marrow is fed by the fluid which filters through the 'dense' substance of bone in drops. If the bones are too porous, the marrow will receive too much liquid, and also escape too freely by the channel which will be described later (91A).

[2] It is conjectured that this doctrine of vapours arising from the humours was held by Philistion and Diocles. See Wellmann, *Fr. d. Gr. Aerzte*, p. 78. Cf. the confusion caused in the soul's revolutions by the mixture of phlegm and black bile, causing epilepsy, 85A.

[3] It was a universal doctrine that lethargy was due to phlegm. Wellmann, op. cit. 80[1].

[4] Understanding ὅταν οὕτως κακῶς παγέντων (ὦσι) πολιτεῖαι κακαί. κατὰ πόλεις is usually either ignored by translators or rendered 'in the cities'. In this sense it seems to add nothing to ἰδίᾳ τε καὶ δημοσίᾳ. I suggest that λόγοι κατὰ πόλεις means 'discourses in conformity with (such) cities'. This provides λόγοι with the equivalent of ὁμοίως κακοί, which seems needed. The omission of τὰς before πόλεις is unobjectionable in the style of this dialogue, which treats the definite article with poetic freedom.

87B. who are bad become so, through two causes that are altogether against the will.[1] For these the blame must fall upon the parents rather than the offspring,[2] and upon those who give, rather than those who receive, nurture ; nevertheless, a man must use his utmost endeavour by means of education, pursuits, and study to escape from badness and lay hold upon its contrary.

The contents of the above section should be considered in the light of the two following, which recommend remedies to correct any disproportion of body and soul and the training of the divine part for its office as ruler. But it will be well to summarise here just what has been stated so far.

This section sets out to describe how ' disorders of the soul are caused by the bodily condition '. It is recognised, here and below, that when soul and body are united in the composite living creature, either can set up disorder in the other : intense intellectual activity may wreck the health, or a gross and too powerful frame may assert its interests to the point of causing dulness and stupidity in the mind (88A–B). After the earlier consideration of bodily diseases it is natural to pass on here to those disorders of the soul which have their origin in the condition of the body. It is not stated that *all* mental disorders are *solely* due to bodily states. Next it is added that ' folly ' (ἄνοια) must be recognised as disorder of the soul, and that there are two kinds of folly : ' madness ' and stupidity. ' Folly ' means any state in which the divine reason (νοῦς) is not exercising due control over the rest of the soul. The two main types are ' madness ' (μανία), which means frantic passionate excitement, not pathological insanity, and stupidity (ἀμαθία), that dull and lethargic ignorance which is incapable even of the desire for understanding. It is not said that these states of mind cover the whole field of what could be called ' disorder of the soul '.[3] They are the conditions which can arise from ' a bad habit of body ' and be encouraged by ' unenlightened upbringing ' in youth.

---

[1] The two causes are a defective constitution inherited from parents and bad upbringing, as is implied by the next sentence and by διὰ πονηρὰν ἕξιν τινά τοῦ σώματος καὶ ἀπαίδευτον τροφήν, 86E.

[2] *Laws* 755D : A man must be careful all through his life, and especially during the time when he is begetting, to commit no act involving either bodily ailment or violence and injustice ; for these he will inevitably stamp on the souls and bodies of his offspring.

[3] It should be remembered that νόσος is commonly used in a much wider sense than ' disease '. It is frequently applied, for instance, to passionate love and to political disorder. To have an unbridled tongue is a νόσος (Euripides). At *Laws* 782D the natural desires of food and sex are νοσήματα. In the same way ' badness ' (κακία) is a wider term than ' vice '.

346

In the moral, as in the physical, life of man there is, beside the operation of reason, much that 'comes about of necessity' and is repugnant to the inmost wish or will of the rational part. The *Timaeus* is primarily a physical rather than a moral treatise, and it is fitting that it should lay more stress than we find in the moral and political dialogues on the inevitable consequences of the immortal soul being housed in a body subject to the assaults of an environment composed of the same stuff. We have been told earlier that, when the infant soul is plunged into the stream of Becoming, its motions are thrown into such disorder that the rational revolution of the Same is completely arrested and robbed of all control, and even the inferior movement of the Different is so dislocated and distorted as to give rise to every sort of delusion and false judgment. 'Because of these affections, to-day as at the beginning, a soul comes to be without intelligence (foolish, ἄνους) at first, when it is bound in a mortal body' (43D–44A). Escape from this 'most grave disorder'[1] depends on a right upbringing at a later stage, when the revolutions have begun to settle down into their normal course. If this be neglected, a man lives maimed and imperfect, and returns to Hades 'in a state of folly' (ἀνόητος 44C).

No one holds the new-born infant morally responsible for starting life in folly and ignorance. The present passage adds that some individuals are further handicapped by inherited defects of body which make them peculiarly liable to excess of passion or to despondent lethargy. An abnormal condition of the bones and marrow may make sexual continence much more difficult for some, and their violent excitement will hinder reason from gaining control. Others may suffer from noxious humours inducing a melancholy and dispirited attitude and intellectual dulness. Such persons have not chosen their bodily habit and they are not to be blamed for it. The remedy lies in judicious training, both physical and mental, from the earliest years. If this is withheld and they are further surrounded by corrupting influences in an ill-governed state, again the blame should fall not on them, but on their elders. But they are not absolved from the duty, mentioned in the last sentence, of doing all they can by education and intellectual pursuits 'to escape from badness and lay hold upon its contrary'.[2] Here moral

---

[1] τὴν μεγίστην ἀποφυγὼν νόσον (44C) alludes to the mystic formula ἔφυγον κακόν, εὗρον ἄμεινον, to which Tr. recognises a reference in our passage: φυγεῖν μὲν κακίαν, τοὐναντίον δὲ ἑλεῖν (87B). Cf. also ὀρθὴ τροφὴ παιδεύσεως (44B) with ἀπαίδευτον τροφήν (86E).

[2] Tr. ignores this conclusion where he accuses Timaeus of 'the grievous blunder of drawing no distinction between the man who masters his temperament and the man who is mastered by it' (p. 616).

ρurpose will be exercised. But on this matter Plato has written at large elsewhere ; all that is relevant here is to give some account, in the following paragraphs, of the training, chiefly by diet and gymnastic exercise, needed to correct the prejudicial influence of physical defect.

In speaking, not of any and every form of vice, but of the inability to control excessive desire for bodily pleasure, Plato quotes the Socratic maxim, ' No one is willingly (or wittingly) bad '. The intemperance which has its origin in physical defect and grows for lack of remedial training is not to be attributed to the true will, whose inmost desire is always for the good. This desire,, which Plato and Aristotle after him call ' wish ' (βούλησις) and distinguish from the appetites deluded by an ' apparent good ', resides in the true self, the immortal part of the soul.[1] When we find men unable to control their desire for sensual pleasures, we should recognise that such desire has a physical source, and that in many individuals defects of inheritance and upbringing make it peculiarly difficult for reason to gain control. We are not to treat them as if their reason had from the outset deliberately chosen vice in preference to virtue. Such a choice is contrary to the nature of reason, and can only occur in the last stage of degradation when reason itself has become perverted and wholly enslaved to appetite. The condition is then past remedy.

The doctrine here is the same that is stated, for instance, at *Laws* 731B. The Athenian observes that every man has need to be both passionate and gentle. He needs passion if forced to defend himself against the wrong-doing of others when this is harsh and cruel, and to punish it when it is irremediable. ' But when men commit wrongs that are remediable, one should recognise that no wrong-doer does wrong willingly. For no one would ever willingly take to himself any of the worst evils, least of all in the most precious thing that belongs to him ; and to all men the most precious thing is the soul. So no one will voluntarily admit the worst evil into this most precious thing and live in the possession of it all his life long. In general the wrong-doer and he who has these evils is to be pitied, and it is permissible to show pity to the man whose evils are remediable, to restrain one's anger, and treat

---

[1] The distinction between ' wish ' (βούλησις) and ' doing what seems good to you ' is drawn in the *Gorgias*, 467. Aristotle retains the term at *E.N.* iii, 4, ' In the absolute sense the true object of wish (βούλησις) is that which *is* good ; but each man finds it in what *seems* good to him.' The sole judge is the virtuous man ' whose superiority lies precisely in his seeing the truth '. That the immortal part of the soul is the true self is stated at *Laws* 959B and repeated by Aristotle, *E.N.* x, 7, 1178a, 2, and ix, 8, 1168b, 35.

him gently, and not to keep on raging like a scolding wife ; although in dealing with one who is totally and obstinately perverse and wicked one must give free rein to anger.' This doctrine, which no one doubts to be Plato's own, is repeated at *Laws* ix, 860D, and there brought into relation with the more ordinary use of the terms ' voluntary ' and ' involuntary '. By calling all wrongdoing ' involuntary ', it is not meant that the law can disregard the distinction between doing an act on purpose and doing it by accident. The legal character of an act depends on its spirit and principle. The law must aim at curing evil intentions and inflict death only on the incurable. The doctrine of the *Laws* is in harmony with our passage. The evils here described are to be pitied because their origin lies in causes at work when a man cannot have begun to exercise rational control, and they are remediable if taken in hand before he becomes ' totally and obstinately wicked '. This is the answer to the criticism that Timaeus leaves out of account ' real wickedness ' and ' conceive of no wickedness that is more than weakness '.[1] The passage is not concerned with the ingrained and irremediable vice which calls for punishment or extermination. A physical treatise may confine itself to hygiene. All that is needed is the mild preventive remedies described in the next paragraphs.

87B–89D. *Disproportion between soul and body, to be remedied by regimen and exercise*

This is not the place to pursue further the topic touched upon in the last sentence—the corrupting influences of an ill-governed society and the reform in education needed to correct them. That belongs to a moral and political work like the *Republic* ; the *Timaeus* is a physical discourse, and Plato returns here to the living creature as a compound of soul and body, and in particular to the disorders due to a lack of proportion between the two components. These are to be corrected, not by the violent action of drugs, but by giving both soul and body the regimen and exercise they severally need.

87B. This subject, however, belongs to another kind of discourse ;
  c. here it is natural and fitting to set forth, on the opposite side, the countervailing treatment, the means whereby body and mind are kept in health ; for it is right to dwell upon good rather than upon evil.

    Now the good is always beautiful, and the beautiful never disproportionate ; accordingly a living creature that is to

---

[1] Tr., p. 615.

87C. possess these qualities must be well-proportioned. Proportions of a trivial kind we readily perceive and compute;

D. but the most important and decisive escape our reckoning. For health or sickness, goodness or badness, the proportion or disproportion between soul and body themselves is more important than any other; yet we pay no heed to this and do not observe that when a great and powerful soul has for its vehicle a frame too small and feeble, or again when the two are ill-matched in the contrary way, the creature as a whole is not beautiful, since it is deficient in the most important proportions; while the opposite condition is to him who can discern it the fairest and loveliest object of contemplation.[1] Just as a body that is out of proportion

E. because the legs or some other members are too big, is not only ugly, but in the working of one part with another brings countless troubles upon itself with much fatigue and frequent falls due to awkward convulsive movement, so is it, we must suppose, with the composite creature we call an animal. When the soul in it is too strong for the body and of ardent

88. temperament, she dislocates the whole frame and fills it with ailments from within; she wastes it away, when she throws herself into study and research; in teaching and controversy, public or private, she inflames and racks its fabric through the rivalries and contentions that arise, and bringing on rheums deludes most so-called physicians into laying the blame on the unoffending part.[2] On the other hand, when a large body, too big for the soul, is conjoined with a small and feeble mind, whereas the appetites natural to man are

B. of two kinds—desire of food for the body and desire of wisdom for the divinest part in us—the motions of the stronger part prevail and, by augmenting their own power while they make the powers of the soul dull and slow to learn and forgetful, they produce in her the worst of maladies, stupidity.

Now against both these dangers there is one safeguard: not to exercise the soul without the body, nor yet the body without the soul, in order that both may hold their own and

---

[1] Language and thought echo the passage describing the love of a beautiful person as the climax of musical education at *Rep.* 402D: 'when noble dispositions in the soul are combined in harmony with congruent features of outward form, this is the fairest object of contemplation for one who has eyes to see it . . . and the fairest is also the loveliest'.

[2] Note that the soul has its characteristic form of intemperance, which deranges the body, no less than the intemperance of the body, considered in the last section, disorders the soul.

88C. prove equally balanced and sound. So the mathematician or one who is intensely occupied with any other intellectual discipline must give his body its due meed of exercise by taking part in athletic training ; while he who is industrious in moulding his body must compensate his soul with her proper exercise in the cultivation of the mind and all higher education ; so one may deserve to be called in the true sense a man of noble breeding.[1] The several parts also should be cared for on the same principle, in imitation of

D. the universal frame. For as our body is heated and cooled within by the things that enter it, and again is dried and moistened by what is outside, and suffers affections consequent upon disturbances of both these kinds, if a man surrenders his body to these motions in a state of rest, it is overpowered and ruined. But if he will imitate what we have called the foster-mother and nurse of the universe [2] and never, if possible, allow the body to rest in torpor ; if he will keep it in motion and, by perpetually giving it a shake, constantly hold in check the internal and external

E. motions in a natural balance ; if by thus shaking it in moderation, he will bring into orderly arrangement, one with another, such as we described in speaking of the universe, those affections and particles that wander according to their affinities about the body ; then he will not be leaving foe ranged by foe to engender warfare and disease in his body, but will have friend ranged by the side of friend for the production of health.

89. Of motions, again, the best is that motion which is produced in oneself by oneself, since it is most akin to the movement of thought and of the universe ; motion produced by another is inferior ; and worst of all is that whereby, while the body lies inert, its several parts are moved by foreign agents. Accordingly, of all modes of purifying or bracing [3] the body, the best is gymnastic exercise ; next best the swaying motion of a boat or carriage which causes no fatigue ; while a third kind, though sometimes useful in extreme necessity, should in no other case be employed

B. by a man of sense ; I mean medical purgation by drugs. Disorders should not be irritated by drugs, save where

---

[1] ὀρθῶς, ' in the true sense ', not according to the vulgar use of καλὸς κἀγαθός for an upper-class person. But the words also bear their literal sense : the beauty and goodness characteristic of the well-proportioned body and mind (87C, D).

[2] Cf. 53A.   [3] συνιστάναι in this sense occurs in the medical writers.

89B. there is grave danger. For in general any disease has a settled constitution somewhat like that of living creatures. The composition of the living creature is so ordered as to have a regular period of life for the species in general [1]; and also each individual by itself is born with its allotted span, apart from inevitable accidents, since the triangles in every

c. creature are from the outset put together with the power to hold out for a certain time, beyond which life cannot be prolonged.[2] It is the same with the constitution of diseases : if this be deranged by drugs to the disregard of their destined period, it often results that slight maladies become grave or their number is increased. Hence, so far as leisure permits, one should manage and control all complaints by regimen,

D. instead of irritating a stubborn mischief by drugs.

The emphasis laid on exercise and regimen, as against drugs, is characteristic of the Sicilian school.[3] In this, as in other matters, they were followed by Diocles, who wrote a treatise on regimen. Some long extracts preserved by Oribasius [4] give much wise advice about diet and exercise, the preparation of food, and the care of the body generally, which is in full accordance with Plato's recommendations. The *Republic* had already dwelt upon the superiority of preventive training to drastic remedies applied when the harm was done, and also upon the need to bring the gentle and more spirited elements of the soul into harmony by cultivating both so as to correct the excesses of either.

89D–90D. *Care of the soul*

We now turn from the care of the whole living creature, and especially of its bodily part, to the care of the soul and its training for the rule it should bear. The main principle is one that was already announced in the *Republic*. Each of the three parts of the soul has its own legitimate sphere of interests and desires, and none of them should be thwarted or suppressed. If the energy of

---

[1] Cf. Ar., *de gen et corr.* 336b, 10 : ' The natural processes of passing-away and coming-to-be occupy equal periods of time. Hence, too, the times—i.e. the lives—of the several kinds of living things have a number by which they are distinguished. For there is an Order controlling all things, and every time (i.e. every life) is measured by a period ' (trans. Joachim). Fraccaroli and Tr. correctly explain that there is a fixed normal length of life for the individuals of each species, and also a peculiar expectation of life for each individual, according to his constitution.

[2] Cf. the account of natural death, 81D.

[3] See the passage from Aristoxenus in Iambl., *V.P.* 163–4, quoted by Tr., p. 629.

[4] Diocles, frag., 138 ff., Wellmann.

the soul is directed too much into one of the three channels, it can only be at the expense of the others. This doctrine had been so fully developed in the *Republic* that only a brief reference to it is needed here. The rest of the section is devoted to that innermost desire of the divine part, which (as Diotima explains in the *Symposium*) is the desire for the immortality or divinity that can be regained by the pursuit of wisdom.

89D. Let this suffice for the treatment of the living creature as a whole and of its bodily part, and the way in which a man may best lead a rational life, both governing and being governed by himself. Still more should precedence be given to the training of the part that is destined to govern, so that it may be as perfectly equipped as possible for its work of governance. To treat of this matter in detail would in itself

E. be a sufficient task ; but, as a side issue, it may not be out of place to determine the matter in conformity with what has gone before, with these observations. As we have said more than once, there dwell in us three distinct forms of soul, each having its own motions. Accordingly, we may say now as briefly as possible that whichever of these lives in idleness and inactivity with respect to its proper motions must needs become the weakest, while any that is in constant exercise will be strongest ; hence we must take

90. care that their motions be kept in due proportion one to another.

As concerning the most sovereign form of soul in us we must conceive that heaven has given it to each man as a guiding genius—that part which we say dwells in the summit of our body and lifts us from earth towards our celestial affinity, like a plant whose roots are not in earth, but in the heavens. And this is most true, for it is to the heavens, whence the soul first came to birth, that the divine part [1] attaches the head or root of us and keeps the whole body

B. upright. Now if a man is engrossed in appetites and ambitions and spends all his pains upon these, all his thoughts must needs be mortal and, so far as that is possible, he cannot fall short of becoming mortal altogether, since he has nourished the growth of his mortality. But if his heart has been set on the love of learning and true wisdom and he has exercised that part of himself above all, he is surely

C. bound to have thoughts immortal and divine, if he shall lay

---

[1] τὸ θεῖον the divine part of us, as at c, 4. At 76B, τὸ θεῖον meant the brain.

90C. hold upon truth, nor can he fail to possess immortality in the fullest measure that human nature admits[1] ; and because he is always devoutly cherishing the divine part and maintaining the guardian genius that dwells with him in good estate, he must needs be happy[2] above all. Now there is but one way of caring for anything, namely to give it the nourishment and motions proper to it. The motions akin to the divine part in us are the thoughts and revolutions

D. of the universe ; these, therefore, every man should follow, and correcting those circuits in the head that were deranged at birth, by learning to know the harmonies and revolutions of the world, he should bring the intelligent part, according to its pristine nature, into the likeness of that which intelligence discerns, and thereby win the fulfilment of the best life set by the gods before mankind both for this present time and for the time to come.

The passion for wisdom, the characteristic desire of the immortal soul, is symbolised in the *Phaedrus* by the wings which Psyche must receive from Eros. ' It is the function of wings to raise aloft that which is heavy to the region where the gods dwell. There is no bodily part that has more kinship with the divine ; and the divine is beauty, wisdom, goodness.' In our passage Plato connects this thought with his earlier account of the revolution and harmony of the heavens, after whose likeness we must re-establish the disordered movements of the incarnate soul. What lifts us towards our celestial affinity is the genius or *daemon* residing in the head ; and that Eros is a *daemon*, between mortal and immortal, we learnt in the *Symposium*. ' So in this tree of man, whose nervie root Springs in his top ', spiritual sustenance is drawn from contemplation of the heavens, as a plant draws its food from the earth. The life of reason can be fully enjoyed only after death when the spirit is released from the distractions of bodily needs[3] ; but our business here is to partake of immortality in the fullest measure that our mortal nature will admit. Our passage is echoed in Aristotle's final definition of human happiness (εὐδαιμονία) :

' If, then, among the forms of virtuous action, war and politics, although they stand out as pre-eminent in nobility and greatness, are yet unleisured and directed towards a further end instead of

---

[1] Reading ἀνθρωπίνη φύσις with APY. Cf. 69A, καθ' ὅσον ἡμῶν ἡ φύσις ἐνδέχεται. The reading of F ἀνθρωπίνη φύσει creates a hiatus with ἀθανασίας following and can be explained as intended to yield a commoner construction.

[2] The connection between εὐ-δαίμων (literally having a good δαίμων = luck) with δαίμων = guardian genius cannot be reproduced.

[3] *Phaedo* 66E. Cf. *Theaet.* 176A.

being desired for their own sakes, while the activity of reason, on the other hand, when it is speculative, appears to be superior in serious worth, to aim at no end beyond itself, and to contain a pleasure which is peculiar to it and so enhances the activity ; and if self-sufficiency, leisuredness, and such freedom from weariness as is possible to humanity, together with all the other attributes of felicity, are found to go with this activity ;—then, perfect happiness for man will lie in this, provided it be granted a complete span of life ; for nothing that belongs to happiness is incomplete.

Such a life as this, however, is higher than the measure of humanity ; not in virtue of his humanity will man lead this life, but in virtue of something within him that is divine ; and by as much as this something is superior to his composite nature, by so much is its activity superior to the rest of virtue. If, then, reason is divine in comparison with man, so is the life of reason divine in comparison with human life. We ought not to listen to those who exhort man to keep to man's thoughts, or a mortal to the thoughts of mortality, but, so far as may be, to achieve immortality and do what man may to live according to the highest thing that is in him ; for little though it be in bulk, in power and worth it is far above all the rest ' (*Nic. Eth.* x, 7, 7).

At this point, where the discourse of Timaeus has reached its climax, the thought recurs to his affirmation at the opening (29E) that the divine is not moved by any jealousy to withhold from the world or from man any perfection of which their nature is capable. Reason has endowed the world with harmony and beauty, and man with the capacity to reproduce them in himself. As the *Epinomis* (988A) urges, the study of the heavens, which the Athenians, under the influence of ' the Greeks' fear that it is wrong for mortal man to busy himself with things divine ', had proscribed as tending to atheism, ought rather to lead to the worship of the heavenly bodies themselves, a nobler religion than the established cult which had come from the barbarians. The divine power is not displeased by man's ability to learn, but feels ' a joy free from jealousy ' at his becoming good with heaven's aid.

90E–92C. *The differentiation of the sexes. The lower animals*

I have already (p. 292) suggested a possible reason why Plato relegates the differentiation of the sexes and the formation of the lower animals to this appendix. The highest form of Eros, the passion for divine wisdom and immortality, was dwelt upon in the last section. Its seat is the brain, at the head of the column of spinal marrow. The marrow is also the seed, the means by which the species maintains its immortality in time, as life is transmitted from one mortal individual to another. Provision has now to be made for this ' Eros of sexual intercourse ', by giving the seed an outlet and forming the male and female parts of generation. If my suggestion was right, Plato may wish to indicate that sexual passion is not the fundamental form of Eros, but an accidental appanage of existence in time. The individual human being requires all the faculties and functions that have hitherto been described ; but he could be imagined as complete without the organs of sex, which are added only for the sake of the species.[1] The organs are fantastically described as if they were additional ' living creatures ' appended to the already finished form of the human animal.

90E. And now, it would seem, we have fairly accomplished the task laid upon us at the outset : to tell the story of the universe so far as to the generation of man. For the manner in which the other living creatures have come into being, brief mention shall be enough, where there is no need to speak at length ; so shall we, in our own judgment, rather preserve due measure in our account of them.

Let this matter, then, be set forth as follows. Of those who were born as men, all that were cowardly and spent their life in wrongdoing were, according to the probable

91. account, transformed at the second birth into women ; for this reason it was at that time that the gods constructed the desire of sexual intercourse, fashioning one creature instinct with life in us, and another in women. The two were made by them in this way. From the conduit of our drink, where it receives liquid that has passed through the lungs by the kidneys into the bladder and ejects it with the air that presses upon it, they pierced an opening communicating with the compact [2] marrow which runs from the head down the neck and along the spine and has, indeed, in our

B. earlier discourse been called ' seed '.[3] This marrow, being instinct with life and finding an outlet, implanted in the part where this outlet was a lively appetite for egress and

---

[1] At *Laws* 783A the desires for food and drink are implanted from birth ; sexual *eros* is said to arise later. Plato apparently thought that this form of desire dates from puberty.

[2] συμπεπηγότα, forming one connected column, as contrasted with the marrow isolated in other bones.

[3] At 73C, 74B, 86C. At *de gen. anim.* 735a, 7, Aristotle says that the semen both has soul and is soul potentially.

91B. so brought it to completion as an Eros of begetting.[1] Hence it is that in men the privy member is disobedient and self-willed, like a creature that will not listen to reason, and because of frenzied appetite bent upon carrying all before it.

C. In women again, for the same reason, what is called the matrix or womb, a living creature within them with a desire for child-bearing, if it be left long unfruitful beyond the due season, is vexed and aggrieved, and wandering throughout the body and blocking the channels of the breath, by forbidding respiration brings the sufferer to extreme distress and causes all manner of disorders ; until at last the Eros of the one and the Desire of the other [2] bring the pair
D. together, pluck as it were the fruit from the tree [3] and sow the ploughland of the womb with living creatures still unformed and too small to be seen, and again differentiating their parts [4] nourish them till they grow large within, and thereafter by bringing them to the light of day accomplish the birth of the living creature. Such is the origin of women and of all that is female.

The lower animals.—The created gods now finish their appointed task by fashioning the remaining classes of living creatures : the birds of the air, land animals, and fishes. Timaeus' discourse has been concerned with the universe and with the nature of man, not with the whole field of natural history. So the lower animals can be briefly disposed of. Their forms are regarded as degraded types,

---

[1] τοῦθ' ἧπερ ἀνέπνευσεν ... τοῦ γεννᾶν ἔρωτα ἀπετέλεσεν.. The only satisfactory construction for τοῦτο is as object of ἀπετέλεσεν. Sexual desire is regarded concretely as a ' living creature ' with a life of its own (ἔμψυχον, A, 1–2). The gods open the communication from the channel of the drink to the living marrow or seed, which then itself produces the phallus, ' completing this part where it has found an outlet as an Eros of begetting '. The phallus is an embodiment (or ἄγαλμα) of this male Eros. I should not suggest this if the whole passage were not so fantastic, especially the latter part where the womb is called ' a living creature desirous of child-bearing '—the female counterpart of the male Eros—which is actually said to ' wander about the body '. In the *Symposium* Eros personified governs a genitive ; e.g. 200E, ἔστιν ὁ Ἔρως τινῶν, ' Love is (love of) some object '.

[2] ἡ ἐπιθυμία (feminine), the female ζῷον ἐπιθυμητικὸν τῆς παιδοποιίας, and ὁ ἔρως (masculine), the male ἔρως τοῦ γεννᾶν. The two co-operate, Eros sowing the seed, Ἐπιθυμία nursing and bringing to birth.

[3] Cf. 86c, where excess of seed was compared to overabundance of fruit on a tree. The condition is relieved by the plucking of the fruit. The marrow is, as it were, an inverted tree, with the brain for its root (90A) and the spinal column for its trunk. Democritus (Diels-Kranz, Vors.⁶ 68B, 5, p. 137, 13) spoke of plants and trees having their head rooted in earth.

[4] διακρίναντες, cf. Orib. iii, 78 = Diokles, frag. 175, Wellm. περὶ δὲ τὰς τέσσ' αρας ἐννεάδας ὁρᾶται πρῶτον διακεκριμένον ὅλον τὸ σῶμα (τῶν ἐμβρύων).

for the sake of the mythical doctrine of punishment by transmigration, announced to the souls before their first birth at 42c. The three classes correspond to the three parts of the soul, which the men condemned to such degradation have respectively misused.

91D.   Birds were made by transformation: growing feathers instead of hair, they came from harmless but light-witted men, who studied the heavens but imagined in their simplicity

E.   that the surest evidence in these matters comes through the eye.

Land animals came from men who had no use for philosophy and paid no heed to the heavens because they had lost the use of the circuits in the head and followed the guidance of those parts of the soul that are in the breast. By reason of these practices they let their fore limbs and heads be drawn down to earth by natural affinity and there supported, and their heads were lengthened out and took any sort of shape

92.   into which their circles were crushed together through inactivity. On this account their kind was born with four feet or with many, heaven giving to the more witless the greater number of points of support, that they might be all the more drawn earthwards. The most senseless, whose whole bodies were stretched at length upon the earth, since they had no further need of feet, the gods made footless, crawling over the ground.

B.   The fourth sort, that live in water, came from the most foolish and stupid of all. The gods who remoulded their form thought these unworthy any more to breathe the pure air, because their souls were polluted with every sort of transgression; and in place of breathing the fine and clean air, they thrust them down to inhale the muddy water of the depths. Hence came fishes and shell-fish and all that lives in the water; in penalty for the last extreme of folly they are assigned the last and lowest habitation. These are

c.   the principles on which, now as then, all living creatures change one into another, shifting their place [1] with the loss or gain of understanding or of folly.

92C. *Conclusion*

The closing sentence observes that, with the formation of the three lower kinds of animal, the World has now become what the

---

[1] μεταβαλλόμενα. Cf. *Laws* 903D, 904C, for μεταβολαί, meaning promotion or degradation to a higher or lower region, determined by the trend of our desires and consequent character.

Demiurge set out to make : the unique visible image of its model, namely, 'that (intelligible) Living Creature which embraces' and contains within itself all the intelligible living creatures, just as this (visible) world contains ourselves and all other creatures that have been formed as things visible' (30C).

92C. Here at last let us say that our discourse concerning the universe has come to its end. For having received in full its complement of living creatures, mortal and immortal, this world has thus become a visible living creature embracing all that are visible and an image of the intelligible,[1] a perceptible god, supreme in greatness and excellence, in beauty and perfection, this Heaven single in its kind and one.

---

[1] Understand (with Tr.) ζῷον ὁρατὸν τὰ ὁρατὰ (ζῷα) περιέχον, εἰκὼν τοῦ νοητοῦ (ζῴου), in accordance with 30C, D and 39E. Cf. Tim. Locr. 105A, κόσμω συμπεπληρωμένω ἐκ θεῶν τε καὶ ἀνθρώπων τῶν τε ἄλλων ζῴων, ὅσα δεδαμούργαται ποτ' εἰκόνα τὰν ἀρίσταν εἴδεος ἀγεννάτω καὶ αἰωνίω καὶ νοατῶ. For the reading νοητοῦ (not ποιητοῦ) see Tr.'s note.

359

EPILOGUE

THROUGHOUT the myth of creation here concluded we have watched the divine Reason bringing intelligible order into the world in so far as he could persuade Necessity to co-operate. I urged that, if Plato's words are not to be robbed of all meaning, Necessity must be recognised as standing for a factor in the existing world never completely subdued by Reason. Further, if this Reason can be identified with the reason in the World-Soul itself, that other factor can hardly be anything but an irrational element in the World-Soul, the source of wandering motions. There is at all times some chaos within the cosmos. Becoming was imaged as the child of a father and a mother, who correspond to Heaven and Earth, the first parents of more primitive myth. The father is from above, Olympian; the mother from beneath; and one of her names is Necessity. Already in Homer Zeus and the other Olympians are confronted by a power they cannot subordinate, called Destiny or Fate. Like Plato's Demiurge, the Homeric gods are not omnipotent; and it seems impossible to deduce from Homer any coherent account of the relation between their will and the thwarting opposition of Destiny. Here Homer left an unsolved problem to be grappled with by the only religious genius of classical Greece who can take rank with Plato. It is no accident that the greatest work of Aeschylus, the *Oresteia*, culminates in the reconciliation of Zeus and Destiny; and that the reconciliation is effected by divine Reason, in the person of Athena,[1] persuading the daughters of Necessity to co-operate in her beneficent purposes.

. In the introductory conversation Plato has provided a clue which may lead our thoughts back to the closing scene of the *Eumenides*. The legend of Atlantis, as Socrates remarks, is a theme well suited to the festival of Athena which is the occasion of the present meeting. The formal speeches delivered at the Panathenaea regularly recalled the leadership of Athens in the victory of Hellas over the barbarian invaders in the Persian wars. So, in Critias'

---

[1] The identification of Athena with wisdom (φρόνησις) goes back to the earliest allegorical interpretation of Homer by Theagenes of Rhegium, and was familiar to Plato, who says that, according to many interpreters of Homer, she is νοῦς τε καὶ διάνοια, θεοῦ νόησις, ἁ θεονόα (*Crat.* 407B).

legend, the idealised city of Athena, the only city ever ruled by reason incarnate in the lovers of wisdom, had led the Greek resistance against the hosts of Atlantis. In the next dialogue those inhabitants of the outer ocean are represented as filled with the insolence of riches and luxury. Their god is Poseidon, with whom their kings identify themse.ves by a sinister ritual, drinking the blood of a sacrificed bull. The contest of Athena and Poseidon [1] was figured on the western pediment of the Parthenon, which looks towards Salamis. The story of Atlantis, the central piece of Plato's triptych, is yet another symbol of the contest of reason with the ocean of lawless desires. The two forces met in unreconciled opposition, and both were overwhelmed together by flood and earthquake. The theme of civilised freedom triumphant over barbarism and tyranny was repeated in other sculptures of the Parthenon: the battles of Greeks and Amazons, of Lapiths and Centaurs, and, on the metopes of the eastern front, the battle of Gods and Giants. Here Athena stood in the centre beside her father Zeus, who blasted his enemies with the thunderbolt in a victory of superior force.

But, as Aeschylus knew, the triumphs of superior force are apt not to be final. In the dynastic succession of the gods themselves, Cronos had overthrown Ouranos, and himself been overthrown by Zeus; but 'where is there any joy of deities who have gained their awful throne by violence?' [2] One violent deed provokes another in revenge. This thought dominates the first chorus of the *Agamemnon*, which tells how the king at Aulis bowed his neck beneath 'the yoke of Necessity' and started the disastrous train— the sacrifice of his daughter, Clytemnestra's revenge, Orestes' divinely sanctioned murder of the murderess. The son, no less than the mother, could claim to be doing the work of Justice; but, if Justice means vengeance, where is this chain of dutiful crimes to end?

The answer is given in the *Eumenides*. Orestes, purified of guilt by Apollo himself, can yet find no peace in his soul. He is haunted and pursued by the Furies, hounded on by his mother's ghost, demanding blood for blood. The issue is brought to trial on the Hill of Ares, under the presidency of Athena, impersonating the wisdom of Zeus. Apollo comes to champion the cause of Orestes. He confronts the Furies with loathing and contempt.

---

[1] The *Critias* (109B) mentions the division of regions among the gods, but piously denies that it was 'by strife'. Shortly afterwards (c) comes the phrase οἷον οἴακι πειθοῖ ψυχῆς ἐφαπτόμενοι used of the gods' shepherding of mankind, in contrast with physical violence.

[2] Aesch., *Agam.* 192. The reading δαιμόνων δὲ ποῦ χάρις βιαίως σέλμα σεμνὸν ἡμένων; and the interpretation are suggested in Headlam's note.

Neither party can yield an inch of its claim. Nor can human justice reach a decision: the votes are equal. Both sides are in the right, though both may also be in the wrong. Athena now gives her casting vote for acquittal. Apollo vanishes; he has no more to say. The human protagonist, Orestes, is dismissed. The stage is left to the unappeased and furious spirits of vengeance, daughters of Night or of the Earth Mother, and, on the other side, Athena, the motherless child of the Father. Divine Reason is face to face with blind Necessity.

In wild confusion and desperate anger, the Furies threaten to blast the soil of Athens and poison the very springs of life. Athena turns to them, and her first words are: '*Be persuaded by me.*' She offers them a sanctuary and worship in a cave under the Hill of justice, where they may be transformed into powers of fertility and blessing. At first they cannot listen, but go on crying out for justice and revenge. Athena patiently repeats her offer. She reminds them that she alone knows the keys of that chamber where the thunderbolt is stored; but 'there is no need of that'. Violence will not remedy a situation that violence has created. Suddenly the Furies are converted, when Athena addresses their leader as follows:

> ' I will not weary of speaking good words. Never shall you say that you, the elder goddess, were cast out of this land by me, the younger, and by my mortal citizens, with dishonour.
> ' No; if you have any reverence for unstained Persuasion, the appeasement and soothing charm of my tongue—why then, stay here.'

To this persuasion the daughters of Necessity yield at last. The play ends with the song in which they promise fertility to the soil and citizens of Athena's land, and with the cry of triumph:

> ' So Zeus and Destiny are reconciled.'

Plato's trilogy, had it been finished, would have stood out as his masterpiece, throwing even the *Republic* into the shade. Aeschylus' masterpiece was finished; and the *Oresteia* still holds the supreme place in tragedy. The philosophic poet and the poet philosopher are both consciously concerned with the enthronement of wisdom and justice in human society. For each there lies, beyond and beneath this problem, the antithesis of cosmos and chaos, alike in the constitution of the world and within the confines of the individual soul. On all these planes they see a conflict of powers, whose unreconciled opposition entails disaster. Apollo and the Furies between them can only tear the soul of Orestes in pieces.

EPILOGUE

The city of uncompromised ideals, the prehistoric Athens of Critias' legend, in the death-grapple with the lawless violence of Atlantis, goes down in a general destruction of mankind. The unwritten *Hermocrates*, we conjectured, would have described the rebirth of civilised society and the institution of a State in which the ideal would condescend to compromise with the given facts of man's nature. So humanity might find peace at the last. And the way to peace, for Plato as for Aeschylus, lies through reconcilement of the rational and the irrational, of Zeus and Fate, of Reason and Necessity, not by force but by persuasion.

APPENDIX

22D, ἡμῖν δὲ ὁ Νεῖλος εἴς τε τἆλλα σωτὴρ καὶ τότε ἐκ ταύτης τῆς ἀπορίας σῴζει λυόμενος.

When the inhabitants of mountainous and dry regions are destroyed by scorching drought, the Egyptians are preserved by the Nile being 'set free' or 'unloosed'. Both ancient and modern commentators have been at a loss to understand *from what* the Nile is set free at such times. We may also ask *by what* it is set free.

(*a*) If, as is commonly assumed, the conflagration is the agent, there seems to be no sense in Porphyry's suggestion (Proclus i, 119[16]), followed by Archer-Hind, that the Nile is set free from the fountains at its source. As Taylor says (p. 53), there is no apparent reason why the Nile should be set free more copiously from such fountains in a time of drought and heat than at other seasons. On the supposition that heat is the cause, the only reasonable view is that which Porphyry rejected: 'the melting of the snows (ἡ χιὼν λυομένη) causes the abundance of water'. Porphyry, like Proclus, could not believe in snow so near the equator. Here they followed Herodotus (ii, 22), who knew no more than the Egyptians whom he questioned about the snows and the rainy season in Ethiopia. But the snow theory had been propounded by Anaxagoras (*Vors.*[4] 46A, 91), and Seneca remarks that it was adopted by Aeschylus (*Suppl.* 565, Egypt is λειμὼν χιονόβοσκος. φασὶ γὰρ λυομένης χιόνος παρὰ Ἰνδοῖς πληροῦσθαι αὐτόν, Schol. *ad loc.* Frag. 300), Sophocles (Frag. 797N = 882P. Why does Pearson say the theory cannot have originated with Anaxagoras?), and Euripides (*Hel.* 3; Frag. 228). Headlam observes that the belief was widely known and canvassed in antiquity and remained until our own day for the truth of it to be proved by Sir Henry Stanley. It might be argued that λυόμενος, which can mean 'being melted' as well as 'being set free', is a singularly appropriate word. One reason which led Proclus to reject the snow theory was the statement just below at E, 2, 'In this country the water does not fall from above upon the fields either then or at other times; its way is always to rise up over them from below.' This does not seem to me to mean that the waters of the Nile well up from subterranean sources, instead of being fed by melting snows; but only that there is no rain in Egypt, and the fields are watered by the inundation of the rising Nile. Hence when rains from heaven flood other parts of the earth, Egypt escapes destruction.

(*b*) Professor Stephen Glanville, when I consulted him, at once suggested that the river was ' unloosed ', not by the extreme heat, but by human hands opening the artificial dams and sluices which held up the water in normal times. Isocrates, *Busiris* 13, contrasts Egypt with other less fortunate regions, some of which are deluged by rains, others devastated by droughts. The Nile puts the Egyptian on a level with the gods with respect to the tilling of his soil ; for whereas to all other peoples rains and droughts are dispensed by Zeus, every Egyptian can control both these matters for himself. Plato may have had this passage of the *Busiris* in mind. It implies a universal system of irrigation. Is there possibly a reference to this in Chalcidius' paraphrase, aduersum huiusmodi pericula *meatu irriguo* perennique gurgite obiectus arcet exitium ? For irriguus meaning ' irrigating ' cf. Virg., *Georg.* iv, 32, irriguumque bibant violaria fontem ; Tib. ii, 1, 44, bibit irriguus fertilis hortus aquas ; Ov., *Am.* ii, 16, 2, irriguis ora salubris aquis. Dr. Heichelheim has kindly informed me that, according to F. Hartmann, *L'agriculture dans l'ancienne Egypte* (1923), 113 ff., the opening of the dams before the flooding of the Nile is mentioned as a good deed as early as the Book of the Dead. The irrigation system must therefore go back to the beginning of dynastic times. Mr. J. M. Edmonds writes that Tebtunis Papyri 49, 6, and 54, 16 (2nd cent. B.C.), are quoted by Preisighe for ἐκλύω in the sense of letting out the water by opening the sluices.

It is true that, if either of these interpretations is adopted, λυόμενος remains rather obscure. But λυόμενος is ' the only reading that has any real authority ' (Taylor), and no tolerable correction has been proposed. Cook-Wilson's αὐξόμενος must be rejected. There is no reason why such an obvious and intelligible word should be corrupted ; and the hiatus σῴζει αὐξόμενος in the clausula is without a parallel in this dialogue, where hiatus is very carefully avoided.

(2) 25D, διὸ καὶ νῦν ἄπορον καὶ ἀδιερεύνητον γέγονεν τοὐκεῖ πέλαγος, πηλοῦ κάρτα βραχέος ἐμποδὼν ὄντος, ὃν ἡ νῆσος ἱζομένη παρέσχετο.

κάρτα βραχέος is a modern reading attributed by Stallbaum to Edit. Bas. 2. The MSS. have (1) κάρτα βαθέος A in an erasure, or (2) κατὰ βραχέος Y, marg. A, καταβραχέος F, Proclus.

Archer-Hind defended κάρτα βραχέος as a possible, though unparalleled, expression for ' very shoaly mud ', i.e. mud covered only by shallow water. As the *Meteorologica* 354a, 22, says, outside the Pillars of Heracles there are ' shallows due to the mud ' (βραχέα διὰ τὸν πηλόν). He rightly remarked that βαθέος is pointless : ' Surely the question that would interest a sailor is how near the mud was to the surface ; its depth he would regard with profound indifference.' I cannot follow Taylor's reason for preferring βαθέος : ' the layer of mud is deep, and therefore abundant ; *this* is why the navigation presents difficulties '. ' Abundant ' here must mean ' extensive ' ; and why should Plato write ' very deep ' when he meant ' very extensive ' ? Deep mud need not be extensive. But

APPENDIX

I agree with Taylor against Archer-Hind that you cannot say
'shallow mud' when you mean 'mud covered by shallow water';
and it seems to me impossible to doubt that Plato did mean that.
Another objection, urged by Wilamowitz, is that κάρτα is a tragic
word. This is said to be its only occurrence in Plato.

The whole trouble is due to the assumption that βραχέος agrees
with πηλοῦ, and consequently that κατὰ βραχέος is meaningless. The
mere fact that this reading should have been preserved, though
unintelligible to most readers, is in its favour, provided that it can
bear the sense required. I suggest that κατὰ βραχέος can mean 'at a
little depth', 'a little way down', as κατὰ βραχύ means 'to a small
extent', διὰ βραχέος 'at a short interval', πρὸ βραχέος 'a short time
before', ἐν βραχεῖ 'in a short space'. κατά with the genitive has
lost its original sense of motion in κατὰ γῆς 'beneath the earth' or
τὸ κατ' ὕδατος 'the part (of the building) under water' (Hdt. ii, 149).
The nearest parallel, given me by Prof. Robertson, is Ar., Meteor.
339b, 12, κἂν εἴ τι κατὰ βάθους ἄδηλον ἡμῖν ἐστιν 'and any water there
may be hidden from sight at a (considerable) depth (in the earth)'.
The same phrase is used metaphorically at 2 Cor. viii, 2, ἡ κατὰ
βάθους πτωχεία..

The καταβραχέος of F is an attempt to give the true reading some
sense and construction on the false assumption that βραχέος agrees
with πηλοῦ. Wilamowitz (Platon ii, 387) supported καταβραχέος as a
new coinage of Plato's, analogous to κάτομβρος, κατάσκιος. Κάτομβρος
means 'rained down upon'; κατάσκιος 'overshadowed' or 'overshadowing'. 'Over-shallowed mud' is not a very convincing
expression.

(3) 41A, Θεοὶ θεῶν, ὧν ἐγὼ δημιουργὸς πατήρ τε ἔργων, δι' ἐμοῦ γενόμενα ἄλυτα
ἐμοῦ γε μὴ ἐθέλοντος.

The above is Burnet's text. In the first part of this famous
sentence, down to ἔργων, there is no sign of uncertainty about
the text in the MSS. or (so far as I know) in the ancient citations.
The variety of readings from that point onwards is probably
due to the difficulty of construing the sentence. After ἔργων
APYW read ἅ, but if this relative is retained there is no complete sentence. If ἅ is omitted with F (as by Burnet, Rivaud,
Taylor), the sentence can just be construed by understanding ἔργων
to stand for ἔργα attracted into the case of ὧν : 'works of which I
am maker and father, having come to be by my own agency, are
indissoluble save with my consent'. The attraction causes obscurity ; 'having come to be by my own agency' seems rather redundant
after 'works of which I am maker'; some reason has to be found
for the intrusion of ἅ; and we are still left with the main difficulty—
the meaning of θεοὶ θεῶν. We know from Proclus that ancient
critics were puzzled about the sense and construction of the whole
sentence, and that the phrase θεοὶ θεῶν in particular bore to the
Greeks themselves no obviously certain meaning.

APPENDIX

(a) Some held that ' gods of gods ' means that the cosmic gods (the heavenly bodies) are *likenesses of* the intelligible gods, just as the whole cosmos is ' an *agalma* of the everlasting gods '. This is obviously impossible, and the intelligible gods are a neo-Platonic invention.

(b) Others held that ' the most universal Henads ' are called gods of the cosmic gods, as it were ' lords of lords ', or ' kings of kings '. Linguistically this is (as Tr. remarks) the only defensible interpretation of the words θεοὶ θεῶν. Cf. *Critias* 121, θεὸς ὁ θεῶν Ζεύς. But Proclus raises the obvious objection that all the gods, visible and invisible, are included among those addressed. Archer-Hind's suggestion of rhetorical pomp—' Gods of gods ' signifying the transcendent dignity of the celestial gods as first-fruits of creation— is not supported by any satisfactory linguistic parallel.

(c) It is noteworthy that Proclus does not even mention the interpretation ' Gods, sons of gods ', which satisfied the Latin Cicero (*uos qui deorum satu orti estis*) and is favoured by some moderns. Archer-Hind rightly objects that the only father of the gods is the Demiurge himself ; ' the plural θεῶν is without propriety or meaning '. Tr. adds that there is nothing in the word θεοί to indicate that the genitive is one of origin : θεοὶ θεῶν is as impossible as ἵπποι ἵππων meaning ' horses sprung from horses '.

The upshot is that neither ancient nor modern critics have produced any satisfactory sense for θεοὶ θεῶν. Badham's emendation θεοὶ ὅσων . . . ἔργων, ἅτε δι' ἐμοῦ κτλ. creates an objectionable hiatus between the first two words and will not commend itself to anyone who observes the rhythm of the sentence. The whole address is composed with exceptional care in markedly poetical language. The dominant rhythm is Cretico-Paeonic. This is established in the opening phrase, which is in pure Cretic metre :

θεοὶ θεῶν | ὧν ἐγὼ | δημιουρ|γὸς πατήρ τ' | ἔργων.

Compare the opening of the *De Corona* : πρῶτον μέν, ὦ ἄνδρες Ἀθήναιοι, | τοῖς θεοῖς | εὔχομαι | πᾶσι καὶ | πάσαις,

which Dionysius illustrates by the grammarian's stock Cretic verse :

Κρησίοις | ἐν ῥύθμοις | παῖδα μέλ|ψωμεν.

Alcman has a longer phrase of the same pattern :

Ἀφροδί|τα μὲν οὐκ | ἔστι, μαρ|γὸς δ' Ἔρως | οἷα παῖς | παίσδει.

The rhythm is continued in the rest of the sentence (keeping ἅ) :

ἃ δι' ἐμοῦ | γεγόμεν' ἄλυτ' |
ἐμοῦ γε μὴ | 'θέλοντος.

The closing phrase has a parallel in the epodes of Pindar, *Ol*. ii :

ἐσ|λῶν γὰρ ὑπὸ | χαρμάτων | πῆμα θνάσ|κει
παλίγκοτον | δαμασθέν.[1]

---

[1] Cf. also Simonides, frag. 31, Bgk, 88 Edmonds, a poem in a mixture of metres : Κρῆτά μιν καλέουσι τρόπον, || τὸ δ' ὄργανον | Μολοσσόν.

APPENDIX

The whole sentence, in fact, is practically in Cretico-Paeonic verse ; and the rest of the speech could be reduced to a lyrical passage in a mixture of metres, not very unlike a strophe in the *Second Olympian*. In such a passage Plato might well adopt an order of words or a compressed construction which would not be quite natural in unrhythmical prose.

Since θεοὶ θεῶν has no acceptable meaning, it remains to try the expedient of detaching θεῶν from θεοί and placing the comma before θεῶν instead of after it. This was done by some ancient critics, who, according to Proclus (iii, 202²⁸), connected θεῶν with the following words, taking the whole as θεοί, ὧν θεῶν ἐγὼ δημιουργός. Proclus does not tell us what reading these critics adopted in the rest of the sentence ; but his own criticism shows that he understood them as making θεῶν simply a repetition of θεοί: ' Gods, of which gods I am maker ', i.e. ' Gods, of whom I am maker '. It is hard to believe that anyone could credit Plato with writing Θεοί, θεῶν ὧν when he meant no more than ' Gods, of whom '. But they may have been right to detach θεῶν from θεοί. θεοί by itself is no more abrupt than θεοὶ θεῶν or the γυναῖκες at the beginning of a tragic rhesis.

Suppose, then, that we punctuate : θεοί, θεῶν ὧν ἐγὼ δημιουργὸς πατήρ τ' ἔργων and understand this as a compressed form of θεοί, θεῶν ὧν ἐγὼ δημιουργὸς ἔργων τε (ὧν ἐγὼ) πατήρ. This would be quite intelligible if the words were in that order ; we have only to suppose that πατήρ τ' ἔργων is substituted for ἔργων τε πατήρ for the sake of the metre. Translate : ' Gods, of gods of whom I am maker and of works the father '. This leaves the genitives requiring some subject to govern them. After ἔργων appear the first signs of confusion in the MSS. and citations : ἃ APYW, om. F., τάδε margin of A. The simplest remedy is to read τὰ for ἃ and to take τὰ δι' ἐμοῦ γενόμενα as the subject governing θεῶν ἔργων τε : ' Gods, of gods of whom I am maker and of works the father, *those* which are my own handiwork are indissoluble, save with my consent.' ' Gods and works of which I am father and maker ' means the whole universe— the created gods and all the other works of the Demiurge who is ' maker and father of this universe ' (28c) and has just been called ὁ τόδε τὸ πᾶν γεννήσας (41A). Similarly at 69c the Demiurge is said to have framed the whole universe as a living creature containing all other living creatures mortal and immortal ; ' and of the divine he was himself the maker, while the task of making mortals he laid upon his own offspring '. So here, among all the creatures making up the world, some are made directly by the Demiurge himself—all those works, in fact, which have been created up to this point : the soul and body of the divine universe and the heavenly gods. These are τὰ δι' ἐμοῦ γενόμενα—' the works of my own hands '. And this sentence tells us that they are indissoluble save by his consent. This gives the words δι' ἐμοῦ γενόμενα a valid and appropriate sense. They cease to be a mere repetition of ὧν ἐγὼ δημιουργός, so that they might as well be omitted. Probably it was because ἃ δι' ἐμοῦ γενόμενα appeared to be a mere repetition that it was omitted in some ancient citations. This reading and interpretation have the advantage over some others of making the first sentence a general statement which does not anticipate the next, where it is applied to the created gods, who though not strictly indissoluble, will not in fact be dissolved.

Archer-Hind suggested reading τὰ for ἃ as a milder expedient than Badham's to produce a complete sentence with a subject τὰ δι' ἐμοῦ γενόμενα and a predicate ἄλυτα (ἐστί). But, not knowing that F omits ἃ, he hesitated to alter the text; and he did not see that the main difficulty, θεοὶ θεῶν, could be cured by making θεῶν a partitive genitive.

(4) 52C, τἀληθὲς λέγειν, ὡς εἰκόνι μέν, ἐπείπερ οὐδ' αὐτὸ τοῦτο ἐφ' ᾧ γέγονεν ἑαυτῆς ἐστιν, ἑτέρου δέ τινος ἀεὶ φέρεται φάντασμα, διὰ ταῦτα ἐν ἑτέρῳ προσήκει τινὶ γίγνεσθαι . . .

The difficulty here lies in the phrase αὐτὸ τοῦτο ἐφ' ᾧ γέγονεν.

(a) Archer-Hind boldly declared that the construction seemed to him ' a very simple and very Platonic σχῆμα πρὸς τὸ σημαινόμενον. What is meant by αὐτὸ τοῦτο ἐφ' ᾧ γέγονεν? Of course the παράδειγμα, and the whole phrase governs ἑαυτῆς just as if παράδειγμα had been written: ' since it is not the original-upon-which-it-is-modelled of itself '.

If the words would bear this construction, the sense would be reasonably good. But proof is needed that ἐφ' ᾧ γέγονεν can be equivalent to ᾧ εἴκασται or ἀφωμοίωται, and I know of no instance of ἐπί with the dative in this sense. Also why should ἑαυτῆς be there at all?

(b) Cook-Wilson, approved by Taylor, took the phrase to mean ' the very thing it was meant for ', ' what it was meant to be ', namely an image. So the whole phrase is equivalent to εἰκών and governs ἑαυτῆς: ' since it is not the very-thing-it-was-meant-for of itself '. Wilamowitz (Platon ii, 392) interpreted in a similar way, but with much hesitation. Ritter (Platon ii, 265): ein Bild, weil dieses nicht einmal in dem was es leisten soll selbständig ist. Susemihl (cit. Ritter, ibid.): nicht einmal seinen Zweck, um dessen willen es entstanden ist, in sich selber hat.

The words can certainly bear the meaning suggested; but it is hard to believe that Plato would write such a phrase when all he meant was εἰκών. Why not write simply ἐπείπερ οὐχ ἑαυτῆς ἐστι (sc. φάντασμα), ἑτέρου δέ τινος ἀεὶ φέρεται φάντασμα? But the real objection is that the resulting sense is wrong. If an image were an image of itself—a supposition which borders on nonsense—it would require a medium in which to exist just as much as it does being an image of something else.

(c) I suggest that αὐτὸ τοῦτο ἐφ' ᾧ γέγονεν means ' the very principle (or condition or terms) on which it comes to be '. This is a natural and common sense of ἐπί with the dative. ἑαυτῆς is a possessive genitive. An image comes to be on the same principle or condition as a reflection, which requires an object to cast it and a medium (ἀλλοτρία ἕδρα, *Rep.* 516B) to contain it. These conditions do not lie within or belong to the image itself. The genitive can be illustrated by *Phaedo* 92D, ' We agreed that the soul exists before entering the body as surely as the being we call " essence " *belongs to it* ' (οὕτως . . . ὥσπερ αὐτῆς ἐστιν ἡ οὐσία κτλ).

Exception might be taken to the reflexive ἑαυτῆς ; but it can be defended, particularly since εἰκών is the subject of γέγονεν immediately preceding ἑαυτῆς, though not of the main verb ἐστιν, and continues as the subject of φέρεται. Mr. J. E. Powell, whom I consulted, kindly sent me the following note on this point. ' The nearest passages I know to the text as it stands are those discussed in my Studies (C.Q. xxvii, 221, and xxviii, 173). These show— and a collection of reflexives from fourth-century authors, especially Aristotle, would swell the list enormously—that the reflexive *need not* always refer to the grammatical, nor even the logical, subject ; and in the passage of Aristotle quoted [*Rhet.* i, 5, 7, ὅρος . . . τοῦ οἰκεῖα εἶναι ἢ μή, ὅταν ἐφ᾽ αὑτῷ (vel ἑαυτῷ) ᾖ ἀλλοτριῶσαι, ' when it is in one's own power to alienate '] it would be impossible, as it is in *Timaeus* 52C, to construct a sentence in which both subject and reflexive would refer to the same thing. From a purely grammatical standpoint, therefore, I do not think the text can be proved false, though the value of MS. authority on the question of an Attic reflexive is always low.' Reflexives not referring to the subject are found at 73B, ταῦτα ὁ θεὸς ἀπὸ τῶν ἑαυτῶν ἕκαστα γενῶν χωρὶς ἀποκρίνων, 89A, τῶν κινήσεων ἡ ἐν ἑαυτῷ ὑφ᾽ αὑτοῦ ἀρίστη κίνησις, 85C, ὅταν (ἡ χολὴ) τὸ τῶν ἰνῶν γένος ἐκ τῆς ἑαυτῶν διαφορῇ τάξεως. The temptation to substitute ἐπ᾽ αὐτῇ (*penes ipsam*) for ἑαυτῆς should be resisted. The reference in Proclus, *in Parm.* 129A (p. 170, Cousin) : οὐδὲ γὰρ ἐν ἑαυτῇ ἐστιν ἡ εἰκών, καθόσον ἐστίν, ὥς φησιν (Τίμαιος), εἰκών, ἀλλ᾽ ὥσπερ ἄλλου ἐστίν, οὕτως ἄρα καὶ ἐν ἄλλῳ ἐστίν. is indecisive, and would not really warrant our writing γέγονεν ⟨ἐν⟩ ἑαυτῇ ἐστιν.

INDEX

Adamant, 251

Adrastus on geometrical proportion, 45

Aeschylus, *Eumenides*, 361 ff.

ἄγαλμα, meaning of, 99 ff.

ἀντιπερίστασις, 319

Armillary sphere made by Demiurge, 74

Astronomy in *Rep.* vii, 81

Atlantis, 18

Atomism, absence of moving cause in, 168

Attraction of amber and loadstone, denied, 326

Beasts as degraded types, 357

Becoming, cause of, 22, 26
  ambiguity of, 24

Being and Becoming contrasted, 22, 24 ff.

Bile, a morbid product, 337

Blood consists of digested food, 327

Bone, 295

Bowels, 291

Brain, 293

Causes :
  accessory—
    contrasted with purpose, 156 ff., 172
    not completely subdued to Reason, 209
  necessary and divine, 279

Central Fire :
  in Pythagorean system, 121 ff.
  not mentioned by Plato, 124, 265
  at centre of Earth, 126

Chaos :
  taken over by Demiurge, 35
  described, 197 ff.
  an abstraction, 203

Circular Thrust, 315 ff.

Circulation of blood, 327

Colours, 276

Concord of musical sounds 320 ff.

Critias, 1

Cube numbers, 46 ff.
  symbolising body in three dimensions, 68

Death, natural, 329

Demiurge, 34 ff.
  not omnipotent, 36
  required as moving cause, 197

Destructions of mankind, periodic, 15

Determinism, not complete in Atomism, 169

Difference, three kinds of, 60 ff.

Different :
  circle of, 72 ff.
    symbol of true belief, 95
    in human soul, 148
  motion of :
    a single motion, 82, 112
    identical with actual movement of Sun, 83

Digestion, 327

Diocles, 334

Diseases, 332 ff.
  settled course of, 352

Divination, 288

δύναμις :
  = square, or square root, 46
  = active property of body, 53

373

INDEX

INDEX

Printed in the United States
109041LV00005B/45/A